GOVERNMENT AND ENVIRONMENTAL POLITICS

GOVERNMENT AND ENVIRONMENTAL POLITICS

ESSAYS ON HISTORICAL DEVELOPMENTS SINCE WORLD WAR TWO

EDITED BY MICHAEL J. LACEY

The Woodrow Wilson Center Press
Washington, D.C.

The Johns Hopkins University Press
Baltimore and London

Editorial Offices
The Woodrow Wilson Center Press
370 L'Enfant Promenade, Ste. 704
Washington, D.C. 20024 U.S.A.

Order from:
The Johns Hopkins University Press
2715 North Charles Street
Baltimore, Maryland 21218-4319
Telephone 1-800-537-JHUP

Printed in the United States of America.
♾ Printed on acid-free paper.

97 96 95 94 93 6 5 4 3 2

Library of Congress Cataloging-in-Publication Data

Government and environmental politics : essays on historical
developments since World War Two / edited by Michael J. Lacey.
p. cm.
ISBN 0–943875–16–1 (alk. paper). — ISBN 0–943875–15–3 (pbk. :
alk. paper)
1. Environmental policy—United States—History—20th century.
2. Natural resources—United States—Management—History—20th
century. I. Lacey, Michael James.
HC110.E5G686 1989
363.7'08'09730904—dc20 89–22761
CIP

A catalog record for this book is available from the British Library.

The Center is the "living memorial" of the United States of America to the nation's twenty-eighth president, Woodrow Wilson. The U.S. Congress established The Woodrow Wilson Center in 1968 as an international institute for advanced study, "symbolizing and strengthening the fruitful relationship between the world of learning and the world of public affairs." The Center opened in 1970 under its own presidentially appointed board of directors.

In all its activities, The Woodrow Wilson Center is a nonprofit, nonpartisan organization, supported financially by annual appropriations from the U.S. Congress and contributions from foundations, corporations, and individuals. Conclusions or opinions expressed in Center publications are those of the authors and do not necessarily reflect the views of the Center staff, fellows, trustees, advisory groups, or any individuals or organizations that provide financial support to the Center.

Woodrow Wilson International Center for Scholars
Smithsonian Institution Building
1000 Jefferson Drive, S.W.
Washington, D.C. 20560
(202) 357-2429

For Wallace D. Bowman,
quiet mentor to many.

CONTENTS

ACKNOWLEDGMENTS

A great many individuals and institutions provided assistance in different phases of the project that resulted in this book. A special word of thanks is due to Wallace D. Bowman, who has given more useful advice, information, and encouragement to researchers on environmental issues in Washington, D.C., than anyone else in recent decades. He conceived the idea for the project and helped to get it started. Thomas Jorling and Malcolm Baldwin were also involved, and assisted in sorting out themes, topics, and appropriate participants. Authors of the essays presented here have worked hard to make the overall project a genuinely collaborative one, and have suffered patiently through a series of frustrating delays on the road to publication. Vital funding support was provided by the United States Department of Energy and by the Shell Oil Company Foundation. Others who played various but important roles in bringing the whole effort to fruition include Peter W. House, James Billington, Stuart Eizenstat, Michael Bean, Martin Melosi, Daniel Dreyfus, Ann Louise Strong, Thomas B. Stoel, Susan Abbasi, William Eichbaum, Steven Quarles, J. Clarence Davies, and Richard N. L. Andrews.

THE ENVIRONMENTAL REVOLUTION AND THE GROWTH OF THE STATE: OVERVIEW AND INTRODUCTION

Michael J. Lacey

This book began in a scholarly conference held at The Woodrow Wilson Center, a gathering designed as an experiment in the analysis of contemporary U.S. history. Its goal was not to ventilate the pros and cons of the policy changes achieved or attempted in the heated environmental controversies that embroiled government during the Reagan era, but rather to establish a sense of historical perspective on the development of environmental politics and institutions as a whole for the period from the end of World War II to the present.

The history of the present is for many reasons the hardest kind to write. While all alert contemporaries know something of their own time, establishing *historical* perspective on it does not come easily or automatically. Even in quiet periods it is an achievement that requires an effort to get beyond the limits of personal experience, to sort out the routine from the genuinely novel, to compare commonsense knowledge of how things seem to be working at the moment, so much of which is taken for granted, with how they worked in the past, in the area beyond the reach of private memory.

In stressful times this effort to step outside the present is all the more difficult, and the postwar era ranks among the most eventful and discordant in the nation's annals. The middle-aged reader of today came to maturity in a period of massive political mobilization that brought to the forefront of American life the issues of the civil rights and antiwar movements, the women's movement, the environmental movement, the consumer movement, and those of a host of public campaigns involving the problems of the poor, the elderly, the handicapped, and many other newly active groups agitating for change and innovation in public policy. The period was marked by its suspicions and a widespread loss of confidence in the efficacy of public institutions, even while countless new obligations were being set for them. Measured in terms of changes in government and legislation, it has been a time of extraordinary ferment, so durable and multifaceted as to make earlier times of great upheaval (the Progressive Era and the New Deal years, for example) seem simpler, if no less important, in comparison.

Environmental politics represent one aspect of the political awakening of postwar society, and the chapters that follow are intended to explore that aspect and place it in broad historical context. To do so, it was necessary at the outset to set some limits on topics and style of treatment.

While environmentalism represents just one department of American life in the postwar world, it is a very large department, indeed, and many have devoted their lives to working on the complexities of small parts of it. So vast and varied are its ideologically charged concerns, involving so many articulate groups contesting so many specific issues in so much technical argument, that no one person can hope to encompass the whole responsibly.

Thus selectivity was essential, as was the need for some common focus of interest. All those who wrote essays have been active in environmental affairs, some as government officials, some as professional experts in particular areas of policy controversy, some as academic scholars specializing in different segments of the new field. All were interested in the connections between society and politics that drive the formation of public policy. All were interested in marking the changes in those connections that had pressured government to create new roles and responsibilities for the evolving American state.

It was the process of change and accommodation to new values within the evolving state that we wished to examine. The "state," of course, is an abstract term that is comfortably at home only in formal social and political theory. It is at home there, however, because of its inclusiveness. In recent scholarship the tendency has grown to see the state not as something over and above the "private sector" that makes up civil society, but rather as something deeply enmeshed in all operations of social life, working increasingly as "the container of social processes," and representing a set of facilities "through which society can exercise some leverage upon itself and promote social welfare."[1]

Since as a practical matter government and the "state" are much the same in the history of American thought, it was decided that to get some insight into the attempts of postwar society to leverage itself, it would be well to concentrate attention on some of the changes in governance wrought by the emergence of environmental concerns at the level of the federal government, where a powerful trend of expanding jurisdiction has been underway. Because statutory law, together with the administrative procedures and regulations derived from it, represents the day-to-day "language" of the state, we gave special attention to the many-sided conversation there, to changes registered in the basic legal context and operating protocols of federal bureaucracy. For the topical areas covered, the authors do three things: they intercept the flow of events in the early postwar era, contrasting what was emerging with the received heritage; they describe the leading features of the developing postwar environmental policymaking apparatus; and, finally, to a greater or lesser degree, they speculate on likely future developments.

Within this general framework, it was necessary to choose between a comprehensive survey of the federal institutions concerned with environmental issues and areas of special concern that might be treated in some depth. The authors opted for the latter course, and chose to write on broad topical areas that lay within their special interests and competence. As a result, this book does not offer separate treatment of Congress or the courts, which played and continue to play such major parts in shaping environmental affairs, nor of the Environmental Protection Agency, which arose after 1970 as the principal focus of a new type of regulatory politics in the field of environmental policy. There is no chapter on the National Environmental Policy Act, nor are there separate analyses of air and water pollution controls and their evolution, topics that dominate much of the attention of those concerned with the new regulation. A great many studies have been published on these institutions and subjects, however, and while they are not accorded separate treatment in what follows, they nonetheless make their presence felt in many of the stories that actually are told.

The overall picture that emerges from these treatments is one of dramatic change from the status quo ante. Three factors are especially important in accounting for the change: the rapid diffusion of new "postindustrial" social values supportive of environmental regulation; the emergence of an ecological consciousness troubled by the uncertain impacts of many trends at work in the day-to-day functioning of industrial society (a consciousness that brought with it a powerful emphasis upon the always-contested political implications of new scientific research in many fields); and finally the appearance of new organizations (together with the rebuilding of old ones), which helped to give focus and voice to the whole movement.

The opening essay by Samuel P. Hays offers a brief *tour d'horizon* of postwar environmental politics. Basing his account on the careful examination of a prodigious accumulation of evidence from varied sources, Hays sets the scene by emphasizing those social and demographic factors that contributed to the political mobilization of the era. He points especially to the emergence then of a new mass-middle class, larger and far better educated, more affluent and active, than its predecessor. This class became the carrier of new environmental values and impulses, and Hays's basic historical insight is derived from this observation. He sees these new values and impulses not as directed *against* the dynamism of economic growth, the premise of most media reportage of dramatic tensions in environmental controversies, but rather as natural outgrowths of the dynamism itself, struggling to refine its force and to organize its energies in pursuit of a higher standard of living in which amenities and "quality of life" issues come to the forefront of concern. Thus the aims of the new environmentalists are understood to supplement rather than displace the inherited agenda of social and economic development.

Hays contrasts the prewar conservation epoch with the more diversified and complex environmental era that developed after the war. The inherited conservation tradition was itself a multifaceted one, but, on the whole, the concerns within it for nature preservation and recreational use were dominated powerfully by a professional and scientific leadership that stressed ideas common in the public policy debates of the time. Those ideas grew out of reactions to the first great period of corporate consolidation at the turn of the century, and were heavily influenced by the antimonopoly currents of thought then prevalent. Among them were ideas that emphasized the uses of publicly owned lands to counter the waste and inefficiencies of free market resources development. One argument fashioned by Theodore Roosevelt and Gifford Pinchot, for example, held that via governmentally supervised use for commodity production of those natural resources that were publicly owned, the people through leasing arrangements would rent resources to private corporations (rather than renting from them), and thus exercise a new kind of influence on the course of development.

The new environmental era, on the other hand, vastly complicated this tradition and worked toward its overhaul. It did so by emphasizing novel kinds of consumer values that could not be easily accommodated within the rough and ready "production" orientation of the earlier period, those that concerned not just quantities of resources, however efficiently regulated by public agencies, but rather amenities and "quality of life" concerns.

Hays points therefore to a fundamental transformation in the meaning for public policy of "natural resources" that was underway, as the phrase was pushed beyond the narrow confines of its original meaning as a factor of economic production and adapted to include new kinds of aesthetic and recreational *consumption* for which there was growing demand. The change was most dramatically evidenced by the drive for expansion of natural environments (wilderness, wetlands, wildlife refuges) in the midst of development, where their value shoots upwards, all aimed not at stopping change, he argues, but at establishing a new set of balances within it.

In Hays's account the new environmental interests shaped political issues around three major themes: values, science, and technology. Activists sought new objectives and refinements in the making of public policy that turned on pushing forward the frontiers of scientific inquiry and technological innovation. Their demands met resistance from both private and public institutions, many old-style conservation managers among them, who were comfortable with the natural resources rationale of the past, and dubious about the threatening elasticity of its successor.

The main achievement of the struggles of the period was the creation of an extensive machinery of public management, which entailed major revisions in the responsibilities and operating protocols of the old-line federal agencies, as well as the construction of new agencies set up to cope with the massive expansion of federal jurisdiction into problems of air and water pollution, toxic substances, and many other "commons" issues. Federal administrative agencies came to be the focal point of environmental politics, with officials and environmentalists often separated by barricades of mutual suspicion. Environmentalists used the legislature and the courts to enhance their objectives, while the agencies and the executive office of the president often sought to exercise a restraining influence.

While the changes wrought by the environmental revolution in American life have been substantial, Hays gives considerable emphasis to the strength and success of the opposition, its strategies of containment, and the widespread skepticism about further innovation at the upper levels of government and the corporate sector that generated an atmosphere of stalemate in the 1980s. The likely future course of environmental affairs is difficult to discern, but given the deep roots of the impulses at work as uncovered by Hays's social and demographic approach to the question, it is clear that they have become a permanent feature of public life in postindustrial society, certain to grow in importance, rather than to recede.

Throughout the postwar period much of the push and energy driving innovation came from new-style "public interest groups" that came into being and found a place in the maturing interest-group system that permeates American public life. Prominent among them were the modern environmental lobbies, some of them updated versions of groups with roots in the conservation epoch, some of them newcomers founded only in recent decades. Serving as switchboards, busy channels of communication, and pressure points for leveraging policy debate, together they worked to organize and focus the diffuse concerns of a rising popular constituency. Many hundreds of such groups came into existence at local and state levels in the postwar era, but national attention was generally captured by the campaigns that flowed from the work of a modest number of such voluntary organizations that took up headquarters in Washington.

Robert Cameron Mitchell in his essay explores in broad outline the world of Washington's environmentalist lobbies. He selects twelve groups active on the national scene, which among them by the mid-1980s represented several million members and supported the work of nearly one hundred lobbyists. Ranging from the long-established, like the Sierra Club, the National Audubon Society, the National Wildlife Federation, and the Wilderness Society, to such new types of organizations as the Natural Resources Defense Council, Environmental Action, Friends of the Earth, and the Environmental Defense Fund, these groups made up the core of the mainstream environmental movement's Washington presence throughout most of the postwar era. As shown by other essays in this volume, representatives of these groups repeatedly intervened at strategic points to force changes in the developing context of governance.

The groups reflected a variety of priorities and interests. All were suspicious of the developmental strategies of business and industry and wary of the uses of administrative discretion in the governmental bureaucracies. While in Mitchell's view the political outlook of the groups as a whole was "not too far left of the political center" by the standards of the time, they nonetheless represented a spectrum of ideologies that extended from the urgent activism of Friends of the Earth at one extreme, to the more moderate views of the National Wildlife Federation, with its massive membership, at the other.

To give the reader some insight into this aspect of the political mobilization that influenced the course of environmental affairs, Mitchell provides a brief historical account of these organizations, discusses the varied ways in which they sustained themselves financially, and comments on the division of labor that arose among them. He also describes the tactics of advocacy they adopted, giving special attention to the importance for them of changes in tax laws governing the activities of not-for-profit associations and changes in the legal rules of "standing" that occurred in the mid-1960s and opened up the courts to new kinds of lawsuits brought on behalf of environmental interests.

Mitchell notes the growing professionalization within the environmental lobbies, their acquisition of the legal and scientific experts so necessary for those who wish to participate in the complex "conversation" of government policymaking. In estimating the reasons for the success and durability of the environmental lobbies (which he suspects may be threatened in the future by looming problems of fundraising), however, Mitchell accords primacy not to tactics and specialized skills, but to the potency of environmental issues themselves and the roles they have come to play in the critique of modern society.

It never occurred to the statesmen of the early republic who wrestled with the problems of the public domain that one day vast holdings of federally owned lands would be prized for their qualities as natural environments, as assets for recreation in the mass culture of an industrialized society that came to take its "leisure time" seriously. Public affairs were one thing; what one did with one's own time was a private matter, and providing places for it was no concern of government. The early preservationists themselves had no idea of the scope and scale of demands that the future would bring, nor of just how powerfully the notion would grow that in fact the state did have responsibilities in this area. Joseph Sax tells the complicated story of how the current publicly owned national recreation system took its present configuration.

Sax devotes some attention to the experience of the National Park Service, which only reluctantly and with faltering vision made the transition from its role as prewar manager of a limited number of sites of spectacular natural beauty (in contrast to the states and localities that then struggled with the management of recreation in less attractive areas within easy visiting range of their populations), most of them the "crown jewels" of the nation's topography—the great national parks in the West—to its contemporary role as steward of an immense collection of areas of varied natural endowment widely distributed throughout the nation as a whole. Many of these new sites were established by Congress in light of the inability of state and local governments to keep up with postwar demand. Sax also discusses the situation of the Bureau of Land Management, which in view of its origins as an institution devoted principally to the needs of mineral production and the grazing of livestock, developed rather unexpectedly in these years to take an important place among the recreation management agencies of the present day.

The heart of Sax's account, however, deals with the experience of the United States Forest Service, oldest of the federal government's conservation agencies and the carrier of a proud tradition of scientific management in the public interest forged as a set of policy ideas in the Progressive Era and changed relatively little until recent decades. The story of the Forest Service since World War II is perhaps the best single indicator of the force of the changing values and arguments that came into play during the period.

A bureaucracy lives in the shadow of its "organic act," the fundamental statutory legislation that sets out its basic public responsibilities. Such foundings are part of the evolving language of the state, and are normally formulated in broad, general terms, leaving the vocabulary of nuance and detail to be worked up by administrators, often in an environment supercharged with watchfulness on the part of affected interest groups. Politically appointed administrators are in turn subjected to the rigors of legislative oversight, the possibilities of appropriations cutbacks, and the ever-present threat of mandated changes in their interpretations of just what it is that the people, through their elected representatives, have actually said, so far as the conduct of the public business is concerned. In these conditions the delicate plant of social consensus lives and has its being, and the "conservatism" of bureaucratic life takes its root.

Sax's discussion makes it clear that the Forest Service is a thrice-born servant of the people, the offspring of what amounts to a series of organic acts that were written in 1897, 1960, and 1976, each a response to the mix of ideas current in the public policy conversations of the time. The Forest Service Organic Act of 1897, for example, represented a break with the nineteenth-century tradition of disposal of public lands to private individuals and groups, and its basic achievement was to contribute instead to the growing belief in the need for their retention under permanent government ownership. The act gave the government the responsibility to regulate the occupancy and use of forest reserves. So far as the purposes served by retention were concerned, however, the act spoke mainly of the need "to furnish a continuous supply of timber for the use and necessities of the citizens of the United States."

The lengthy and complex provisions of the National Forest Management Act of 1976, on the other hand, make it clear that a host of new reasons for retention had been discovered in the interval, many of which go far beyond notions of the "use and necessities" of the citizenry as understood at the turn of century. Among other things, as Sax points out, the 1976 act requires that advisory boards "representative of a cross section of groups interested in the planning . . . and management . . . and the various types of use and enjoyment of the lands" be formed.

From the Progressive Era leadership of Gifford Pinchot onwards, the Forest Service had appeared from the outside to be one of the most stable and confident of the government's agencies. What accounted for its need to be reborn in the postwar era? Congress and the courts were caught up in the political mobilization that affected all institutions, and both played important roles in the process of change. Sax speaks of a congressional "style" that became characteristic of the statutes of the period, a style that was expressed repeatedly in provisions calling for more planning, more public participation, more restriction on traditions of administrative discretion, and more diversification within the official missions of the public-lands bureaucracies. So far as the courts were concerned, Sax points out that the mid-1960s' changes in standing to sue, which gave environmental groups new leverage on affairs, caused dramatic changes in conditions of life within the bureaucracy. Through changes in the rules of standing, the administrative agencies were gradually converted from experts "above the fray" to defendants forced to explain in detail to federal judges the rationale behind their decisions.

While changes in the organic acts of the land management agencies thus marked the advance of new environmental values, in many ways the most important statute of the postwar period was the Wilderness Act of 1964, to which Sax devotes considerable attention. The Wilderness Act was the result of congressional and interest group activism, and created a category of public land management new to the statute books, new even to the techniques of recreation management developed throughout the century by the National Park Service. Under the act Congress became the most important participant in decision-making about the values to be served in administration of the lands, while the Forest Service became the principal casualty of the wilderness movement. The act reserved to Congress the right to designate lands within the public domain that were to be devoted primarily to preservation, putting them beyond the reach of both commodity exploitation and recreational development as parks, and left the management of such areas to the agencies under whose jurisdiction they fell before the wilderness designations were made. Beginning with some nine million acres, in roadless areas of over five thousand acres "untrammelled by man," the wilderness preservation system has since grown steadily.

Many millions of those acres are under Forest Service jurisdiction, and Sax details the ironies behind their current disposition there. He points out that the service in effect invented the wilderness designation decades before wilderness politics rose to national prominence. It had done so independently of congressional pressure, as one among many administrative classifications in its evolving multiple-use principles of operation, and also to ward off the possibility that such lands might be transferred to the jurisdiction of the National Park Service. Multiple-use doctrine meant, however, that such primitive areas were only one element in the administrative calculus, and thus that they might one day be devoted to other uses in that calculus via administrative discretion.

This was the worry that mobilized wilderness advocates, led by the Wilderness Society. In the controversies that ensued, Sax argues, the Forest Service failed to understand the new social context within which it was operating, particularly the symbolic potency of the preservationist cause. Forest Service leadership generally considered the wilderness advocates as ideologues, and their ideology as unrepresentative of public sentiment in any case. "It did not occur to them," Sax argues, "that multiple-use management was an ideology too, a set of values just like wilderness preservation, and that they were engaged in a controversy about values, not about science."

Frank Gregg, a former director of the Bureau of Land Management (BLM), an agency which will soon be responsible for the management of 240 million acres, over half of the total of federally owned lands, is concerned in his essay primarily with the politics of commodity production in the Western region of the nation in the postwar era. To provide orientation for the reader unfamiliar with the scope and heritage of federal lands, he opens his discussion with a brief overview of the historical development of the public domain, culminating in a comparison of the legal, political, and administrative environments within which the two principal agencies responsible for commodities conservation in the "gospel of efficiency" tradition, the Forest Service and the Bureau of Land Management, now find themselves.

The political mobilization of the postwar era occurred under booming economic conditions, and the interests of those primary industries based upon the intensive use of natural resources were very much involved in the boom. The result was new intensity of demand from traditional claimants directed at both the Forest Service and BLM. Gregg details some of the major episodes of conflict that arose as these groups, whose legitimacy had long been established, were confronted with the arrival of relatively new members of the maturing interest group system, primarily the environmental lobbies. He discusses the "range wars" of the period that turned on the politics of grazing regulations, together with timber production and minerals development.

Gregg discusses the emergence of the "Sagebrush Rebellion" in the closing years of the Carter administration, a movement rooted in the reactions of the commodity interests to the inroads made by environmental groups, which for a time focused on the proposal that the lands over which controversy raged be returned by the federal government to the states as a matter of constitutional interpretation. Gregg points out that the "rebellion" was briefly a success in terms of media coverage and regional politics. It may have resulted in quickening the pace of commodity use in the short run, although no transfer of jurisdiction was ever seriously entertained.

In the longer perspective, however, things were very different, and reflected an important transformation of the underlying politics of public lands management. Since the vast majority of such lands are located in the West, the political leadership of that region has long been dominant in public lands issues. The salience of commodities production issues was assumed by many observers to be eternal. Gregg's account makes it clear, however, that the assumption was mistaken. The social and demographic changes of the postwar era have affected the West as much as they have affected life in other regions, with the result that production priorities have undergone erosion and lost their special status.

Gregg makes this case by considering the experience of the Reagan administration within this context. The administration, he notes, "candidly attempted to establish the primacy of commodity production in public lands policy." Despite the president's regional popularity and the strength of his administration's political resources, however, the attempts met routinely with failure. The reasons for failure, Gregg suggests, had to do with a misreading of the grassroots situation in the region. In keeping with the premises of its back-to-basics priorities for government and the economy, the administration attempted to generate a renaissance of the public lands politics of the old West, and spoke in the idiom of turn-of-the-century "production" ideas—about the need to build up civilization through increased mining, livestock production, timber harvesting, "taming the land and the rivers for settlement and for economic opportunity."

Gregg notes, however, that President Reagan's appointees were speaking not to the West of myth and symbol, but to the real West, increasingly an urban region in its patterns of population distribution, increasingly less dependent upon the primary industries of extraction and more dependent upon growth in the high technology and services sector. These changes in social demography were reflected in a new sense of wariness on the part of the region's governors, who, as Gregg points out, have played an important role in stabilizing, moderating, and diversifying the pace of regional development.

Gregg closes his essay with a brief consideration of a sample of current kinds of policy thinking that have developed in connection with discussions about the future of the multiple-use land systems. These range from various types of "privatization" ideas, premised on the convergence of private interests and public virtue, through recent "mainstream" thought on how principles of economic efficiency might more effectively be combined with new types of leasing arrangements under government supervision, to ideas that would attack the dilemmas of multiple-use politics head-on by reversing the historical priorities of production and substituting new-style environmental priorities in their place. Gregg cites in connection with the latter a 1985 speech given before the Sierra Club by former Arizona governor Bruce Babbitt that called for an overhaul of the basic statutes involved to accommodate "the new reality that the highest, best and most productive use of Western land will usually be for public purposes—watershed, wildlife, and recreation."

The reach of the federal government extends far beyond life on the public lands. As the modern state has come to be held politically responsible for prevailing economic conditions, governmental programs to shore up the evolving economic infrastructure (via transportation, public works, and stimulation of the housing industry, for example) have grown apace and generated many effects on private land use. In the mid-1970s the federal government's recently formed Council on Environmental Quality, modeled on the Council of Economic Advisers and intended to pull together the knowledge base for environmental affairs within government, estimated that twenty-five federal agencies were involved in these "growth-shaping" activities, their operations governed by more than seventy statutes.

Malcolm Baldwin, a long-time senior staff attorney at the Council on Environmental Quality and acting chairman of the Council during the transition between the Carter and Reagan administrations, focuses upon the federal government's changing role in the management of private rural lands in the postwar period. He contrasts the prewar heritage of programs and policy thought—marked primarily by the New Deal period's major innovations, the Tennessee Valley Authority's program for integrated regional development and the Soil Conservation Services's efforts to influence rural land use—with the changing agenda of the postwar era.

Baldwin groups the key events of the postwar era into three distinct periods: 1945 to 1961, when population and urban areas grew, economic development programs flourished, and government paid little attention to the conservation aspects of the boom; 1961 to 1975, when many new ideas for aesthetic, ecological, and land use programs were translated into the workings of the federal programs; and 1975 to the present, when energy and economic concerns again rose to the top of the agenda, and recession and political reaction set in.

From a constitutional and planning standpoint, the most ambitious policy initiative of the period was one which never came to fruition, although it was given extensive congressional and White House attention in the early 1970s. Baldwin examines the proposal of that time for a national land use policy act, which would have altered relationships between the federal government, the states, and private landowners in the interests of more balanced national development.

While the movement for a national land use act was politically unsuccessful (Baldwin notes that environmental groups were not enthusiastic supporters of it), a number of environmentally oriented extensions of the federal presence on privately owned lands were successful. The principal ones were the management of wetlands, coastal regions (which were put under unprecedented developmental pressure in the rising culture of mass leisure and abundance), and coal surface mining areas, mainly in Appalachia. Baldwin details the political, legal, and bureaucratic context within which these new areas of responsibility for the federal government were brought under an environmentally oriented system of management.

The nature and extent of government protection for wildlife, particularly at the federal level, is still another area that has changed dramatically since the mid-1960s. Thomas R. Dunlap in his essay considers both the intellectual and political history of the changes. Between 1966 and 1973 Congress established a program to identify, protect, and restore all species in danger of extinction. It asserted an unprecedented degree of federal control over wildlife policy (traditionally a state-level responsibility), expanded the definition of wildlife from certain species of mammals and birds to include almost all forms of life, provided protection for plants, and explicitly sacrificed a degree of economic development in some areas to preserve endangered species.

Dunlap argues that although comprehensive wildlife protection is ultimately attributable to America's transition from a rural and agricultural nation to an urban and industrial (even postindustrial) one, it is more immediately the product of a new attitude toward nature and wild animals formed by modern science, particularly ecology, together with the pressure of a well-educated, affluent social group greatly enlarged by the prosperity of the postwar era.

In tracing the evolution of the change, Dunlap surveys the wildlife policies of the prewar period, which laid the legal and administrative foundations for federal policy, and then considers the intellectual and social bases of the movement to save all species in danger of extinction. Dunlap discusses how these came to fruition in the passage of the Land and Water Conservation Act (1964), the Marine Mammal Protection Act (1973), and the Endangered Species Act (1973). He concludes with a short discussion of the institutionalization of the postwar impulse to save nature.

Still another area of change related to the politics of energy production, and one segment of those politics circled around the introduction of the postwar era's new nuclear technology. Long before the incidents at Three Mile Island and Chernobyl, opposition to nuclear energy production had a special place on the agenda of many environmental activists. In no other area of policy controversy were there more deep-seated suspicions of an unhealthy special relationship between government officials and industry executives, a distrust that played as important a role in retarding the diffusion of nuclear technology as did the technical problems of waste disposal, decommissioning facilities, insurance costs, the difficulties of emergency planning, and others with which all observers are now familiar. Michael L. Smith in his essay makes it clear why the early suspicions of environmentalists were justified, and shows how their persistence contributed to an important reworking of the relationship between government and industry.

Smith recalls those days in the midst of the postwar boom when the Atomic Energy Commission (AEC) was empowered to develop, regulate, and promote commercial nuclear power. Between 1954 and 1969, he says, the AEC conducted a successful promotional campaign while failing almost entirely to anticipate the regulatory crises and environmental problems of the emerging nuclear industry. Smith shows that among other things AEC officials pioneered in the abuse of "national security" arguments to keep information from the public, and suppressed troubling safety assessments generated by its own scientists (Friends of the Earth would later get access to internal AEC documents and expose the cover-up). The public relations strategies of the AEC and the nuclear power industry, and the promotional materials they generated (typified by the Disney production "Our Friend the Atom") provide insight into both the public and the official perceptions of the environmental crisis since 1969, Smith says.

Project Plowshare, an AEC effort to find nonmilitary uses for nuclear explosives (1956–1971), coincided roughly with the early promotional years. Smith uses it as a case study to show the relationship of regulatory standards to promotion in AEC policymaking.

The rise of environmental activism, of course, collided at many points with the developmental strategies of government and the nuclear energy industry. Growing pressure, applied through Congress and the courts, arrested the momentum of development, and the environmental impact statement requirements of the National Environmental Policy Act of 1970 helped to seal the fate of Project Plowshare's ideas for planetary engineering. By 1974 the regulatory practices of the AEC had been so discredited as to result in the demand for total overhaul. Once again there was a new "organic act," which reflected changes occurring in the political order.

The Energy Reorganization Act of 1974 split the AEC into two new agencies: a Nuclear Regulatory Commission to oversee licensing and regulation of nuclear power plants, and an Energy Research and Development Administration (ERDA) to cover the development and promotion of nuclear energy and the production of nuclear weapons.

Smith examines the environmentalists' challenge alongside the response of government and industry, and assesses the current state of federal nuclear power policy and of the nuclear power industry. He concludes that the most dangerous threat to an environmentally sound nuclear policy today may be the inability of any regulatory process to extract both safety and productivity from current nuclear technology.

In the closing essay in the volume, Christopher Schroeder considers the federal regulation of toxic substances, a set of problems that lies at the frontier of environmental politics and makes the new-style environmental regulation of the postwar era so much more conceptually complex and technically difficult than earlier types of public regulation. Prewar government experience with regulatory activity, typified by the operations of the Federal Reserve System or the Interstate Commerce Commission, for example, turned mainly on problems of economic stabilization and the conditions of competition within the economy. Such institutions drew upon a rich tradition of social thought about the proper roles of government, and a heritage of technical arguments about how such roles could be given practical effect. In the postwar era, on the other hand, government was confronted with problems of air and water pollution, the ecological and health effects of new materials circulating through the economy, and a host of other issues for which the inherited traditions of thought about the aims of public regulation were inadequate as guides, particularly at the level of practical implementation.

The regulation of toxic substances is a leading example of the difficulties of the new regulation and the continuing search for adequate grounds, both constitutional and scientific-technical, upon which to lay its foundations. The toxics problem is generated in large part by one of the most important sources of innovation in the postwar economy, the chemical industries and those dependent upon them. Schroeder cites an Environmental Protection Agency (EPA) estimate that over seventy thousand chemicals are now in production and use within the United States. Of these, EPA speculates 20 percent may be carcinogenic. He points out that while a few slender prewar precedents existed, the toxics problem is essentially a postwar phenomenon, resulting from the rise of this peculiarly modern industry. As the scope and scale of the problem itself have grown up in the period, so gradually has knowledge about it.

Schroeder discusses in this connection the discovery of the "low-level exposure dilemma": certain toxics can cause serious injury, even death, at extremely low levels of exposure; yet important areas of social and economic life can be disrupted by the attempts to eliminate toxics altogether.

The pattern of the federal government's response to the growth of knowledge in this area has been understandably haphazard. Schroeder points to over thirty federal statutes that regulate some aspect of the toxics problem, and seven different agencies that administer them, chief among them the EPA.

In Schroeder's analysis close study of the statutes reveals two main features of evolving toxics policy: the redefinition of "acceptable risk" downward from the level that an unregulated market economy, together with lawsuits between private parties, would produce, and the supplementation of market and private court actions with centralized regulation of risk on the part of new federal agencies.

Schroeder outlines the complex toxics regulatory structure and defines the toxics problem; he then discusses the history of federal responses to the problem as part of the environmental movement in the postwar era. He shows how the statutes regulating one or more aspects of the problem were written by various congressional committees responding to the demands of different issues and constituencies and reflecting varying philosophies and ambitions of key congressional leaders.

Now, Schroeder suggests, the controversies have moved inside the regulatory system: the role and independence of science advisory boards, the quantity and quality of specific scientific theory and evidence, the fixing of burdens of proof on participants, and the decisionmaking procedures to be followed by the agencies in publicly "making up their minds" are cited as some of the issues that constitute the new battlegrounds. He notes that throughout the deregulation tide of the 1980s, the machinery of toxics regulation was left largely untouched, and this despite the occasional feeling of environmentalists that their aims might be furthered more efficiently at state and local levels of jurisdiction.

Schroeder closes with the observation that centralized regulation is in a condition of near-paralysis, attempting to discharge the backlog of its current burdens. Given the history of development to date, he sees it as unlikely that the bottleneck will be broken unless Congress, the White House, and the interest groups can agree on reforms to simplify the regulatory process, an exercise that would require more mutual trust and confidence than is in evidence. In the interim, Schroeder points out, federal administrative proceedings will remain nearly the only forums for closely focused adversarial debate on the duties of the state in this evolving new area of the public interest. His essay, together with the others in the volume, suggests the ways in which the historically informed study of postwar environmental politics can serve as an index to the changing contours of American social thought in its attempts to keep abreast of the modern situation.

NOTE

1. Gianfranco Poggi, "The Modern State and the Idea of Progress," in *Progress and Its Discontents*, edited by Gabriel Almond, Marvin Chodorow, and Roy Pearce (Berkeley: University of California Press, 1982), 351.

1

THREE DECADES OF ENVIRONMENTAL POLITICS: THE HISTORICAL CONTEXT

Samuel P. Hays

The nation's political intelligentsia—those institutional leaders who think about, write about, and try to explain contemporary public life—have been more than puzzled about the rise of environmental affairs to a rank of national importance over the past thirty years. Although they have come to accept the reality of the political drive the rise entails, they have been persistently skeptical about the wisdom of its objectives. Rarely have they taken the role of leading the nation into new environmental frontiers. Their mixed emotions have led to a curious mixture of ideas about what this rise means. Many are tempted to accept the rather crude analyses of William Tucker and Ron Arnold that it is a plot of the "privileged classes" to turn the nation into untrustworthy paths and that the instruments of that plot are the environmental organizations whose leaders delight in manipulating an unwary public.[1] Those who cannot accept such polemical writings have taken more seriously a version of the same argument by Mary Douglas and Aaron Wildavsky, two scholars of high repute. They also argue that there is something here that is divorced from reality and must be explained away by the cultural peculiarities that make people subject to irrational fear and manipulation by others. "Primitive ways of thought," they have called it.[2]

It is a formidable task to challenge all this sophistry, which is so attractive to those who express skepticism about the wisdom and justification of environmental objectives. But that is precisely the purpose of this essay. From the vantage point of historical analysis, these efforts to explain away a social and political phenomenon that has been with us for three decades are the "ways of thought" that seem strange, remote from the real world, subject themselves to being explained through sociological and cultural analyses as "curious twists of mind." For environmental affairs are deeply rooted in the forces of history, in changes in society and demography, in the new world that emerged after World War II. They are but one aspect of massive social changes that include television and the computer, new realms of production beyond manufacturing, new attitudes about the relationships between men and women, new levels of income and education, and new intensities of information acquisition and exchange. Our task is not to be tempted constantly with the hope that it will all go away because it is an aberration of some sort, but to understand—seriously—what it is all about as a deeply rooted historical phenomenon. It is time we quit looking superficially at the environmental movement and instead looked at it straight. It is here and it is here to stay. What is it all about?

It may be of some help to provide an idea of the particular part of the beast I will be examining—the sources and the evidence. There is the vast product of media reporting that brings into focus the large issues of political controversy. But soon one finds that all that is superficial and is, in fact, a source of much of the confusion among our political intelligentsia. To go beyond the media requires an examination of the record of political give-and-take embedded in the documents created by administrative agencies, the courts, and the legislatures. Also a part of the recorded history are the newsletters dealing with air and water, land and the coastal zone, chemicals and occupational health, wetlands and wildlife, and even periscopic views into the inside of the Environmental Protection Agency (EPA).[3] There are also publications of the actors, the chemical, coal, and nuclear industries on the one hand, and the environmental organizations on the other: *Chemical Week*, *Coal Age*, *Nucleonics Week*, *Not Man Apart*, *National News Report* of the Sierra Club, *Exposure*, and *Nutrition Action*. And innumerable books have appeared in recent years.

In my review of this literature and in my personal inquiries, I have been especially interested in analysis at the deeper levels of social demography and values. I believe along with John Naisbett that one cannot really understand this country in the midst of vast social change if one's attention is limited to the media level of national politics.[4] I have been particularly interested in those studies, many commissioned by industry, that describe geographical variations in environmental attitudes.[5] I have also sought to construct regional patterns from legislative votes compiled by the League of Conservation Voters and a number of state environmental organizations. I have attempted to keep an eye on live environmental matters in more than a dozen states, from Maine and New Hampshire to Pennsylvania and Michigan, to Florida and Tennessee, and on to the Rocky Mountains, the Pacific Coast, and Alaska. I have consulted a wide range of materials, from state documents to newspapers to the publications of state and local environmental organizations, that provide clues as to the meaning of environmental values and actions.[6] I have sought out evidence concerning the environmental thrust from the bottom up as well as from the top down.

Let me also provide, at the outset, a warning: my analysis is in the style of a historian and not that of a policy analyst. My question is not one of which policy has been desirable or undesirable, has worked or not; I do not identify a problem and describe the resulting governmental action and outcomes of policy.[7] Rather, my concern is with historical evolution, with change and response to change. I am interested in problems, but at the level of analyzing how problems came to be defined in the particular way they were; I am interested in science and technology, but primarily in terms of why one path rather than another was taken; I am interested in management and its associated technical expertise, but primarily in terms of why managers and experts took up their own peculiar ways of thinking and acting. Above all, I am interested in the relationship between society and politics, between social change and political change. What are the roots of the environmental impulse, and how was it translated into public objectives? What interplay took place between these new public objectives and the older institutions with older values, which accepted the new objectives only with such difficulty? The drama unfolds the give-and-take in the political order.

FROM CONSERVATION TO ENVIRONMENT

The new interest in environmental objectives grew out of the vast social and economic changes that took place in the United States after World War II. Although some beginnings could be seen in earlier years, in the rising interest in outdoor recreation in the 1930s, for example, or in the few cases of concern for urban air and water pollution in the late nineteenth century, these trends are little more than precedents. It was the advanced consumer economy that came into existence following World War II that gave rise to a wide range of new public needs and wants. Incomes and standards of living rose; values changed amid rising levels of education; demands persisted that government supplement the private market to advance the new aims of the expanding and changing middle class. The roots of environmental interest lay in social demography, which was undergoing a fundamental historical transformation. The American people of the 1950s and 1960s differed markedly from those of a half-century before, and one among many of those differences lay in the new aspirations of personal, community, regional, and national life that were embodied in environmental affairs.[8]

It is customary for historians to link the environmental movement with the earlier conservation movement, yet they were quite dissimilar. While the conservation movement was concerned with efficiency in the use and development of material resources, the environmental movement was concerned with amenities and quality of life. The first was a part of the history of production; the second, of the history of consumption. The conservation movement arose not from public demand but from the strivings of professionals, scientists, and administrators to turn the wasteful use of resources into more efficient production. Efforts by these leaders to fashion a sustained political base in the form of organized activity failed miserably; for example, they continually rejected the policy initiatives of the most extensive conservation public of the time—the women's clubs.[9] The environmental movement, on the other hand, arose from the wider public to set new goals and demands about which the nation's administrative, technical, and professional leadership was usually skeptical. Environmental impulses, fueled from the public at large, constantly pressed the nation's leaders to go further than they felt desirable. Many of these reluctant participants in environmental affairs were, in fact, direct professional and institutional descendants of the early conservation leaders.

The conservation movement arose amid the concern for waste in the use of waters and forests.[10] Its classic elements were the use of engineering works to manage rivers and sustained-yield forest management to establish systems of continuous cropping. Water conservation originated in western irrigation, which reached a milestone with the Reclamation Act of 1902 and spread to the construction of reservoirs to enhance electric power production, navigation, and flood control, that panoply of multiple uses which reached its highest expression in the Tennessee Valley Authority.[11] The spiritual leader of forest conservation was Gifford Pinchot, who emphasized the primacy of wood production in forest management, expressed disdain for amenity values in woodlands, and spawned a host of public lands activities and professionalizing measures to establish himself firmly as the founding father of the entire forestry movement.[12] All these ideas moved forward rapidly during the administrations of Theodore Roosevelt. Conservation entered a later phase during the 1930s with the establishment of a soil conservation program that emphasized the efficient use and management of the nation's soil resources to undergird a sustainable agriculture.[13] This spirit was reflected even more in the many and varied resource development activities of the New Deal in public works projects and the Civilian Conservation Corps.[14] The New Deal fostered a massive drive for development in the nation's rivers and wildlands.[15]

The environmental movement turned the nation from this historical stage to a new direction that was not at all to the liking of conservation leaders. Environmentalists began to redefine "natural resources" according to not their physical, but their environmental, character. They emphasized not just the material resource itself but the human relationship to it. Rivers could be sources of water as a commodity to be used for agriculture, human consumption, electric power production, or navigation, but they could also be environments for human enjoyment. Forests could be sources of wood, but they could also be settings for home, work, and play that enhanced the quality of human life. The new environmental interest set off a widespread and fundamental debate over just what natural resources were in terms of their relationships with people. That meaning had been assumed in the earlier conservation movement; now it was subject to explicit interpretation and disagreement. While environmentalists affirmed the value of rivers as free-flowing waters and hence set off a series of controversies from Echo Park and Hells Canyon down to a host of smaller streams throughout the nation, those interested in more traditional conservation objectives affirmed the vital importance of rivers as sources of water as a developed commodity.[16] There was an equally vast gulf between those who affirmed the importance of wilderness and those who wanted to produce lumber from the nation's forests. The meaning and purpose of forests in human life was being redefined, and the older conservationists of the Pinchot vintage simply could not accept the change.[17] In the 1950s and the 1960s, the older conservationists and the newer environmentalists clashed repeatedly over the use of land and water resources.

Two conservation innovations of the 1930s played very different roles in this transition. The soil conservation movement, closely akin to the earlier water and forest conservation activities, was concerned with production, not environmental, resources. Hence, it was relatively easy in the 1950s for the Soil Conservation Service to take on the new task of land and water development by building upstream reservoirs and draining wetlands. Both of these tasks ran headlong into the newly developing environmental interests, which looked on both tributaries and wetlands as valuable natural environments.[18] The wildlife movement, in contrast, bridged the transition between the old and the new. Interest in sustainable game populations had brought the game management movement into the orbit of "natural resources" activities in the spirit of conservation; the task of sustained-yield game management did not differ markedly from that of sustained-yield forest management. To the wildlife community, however, water, land, and forests were important not in their own right but as habitat for animals. As the perception of wildlife as game broadened into the larger interest in appreciating wildlife for its own sake, that context was strengthened. Concern over habitat emphasized the environment surrounding animals, on which they depend for propagation and survival. Hence, while the water, forest, and soils components of the conservation movement clashed persistently with the emerging environmental interests, the wildlife component found common ground with it and served as a bridge between old attitudes and new ones.

The new interest in pollution also served to distinguish environment from conservation. In this case, the precedents one might be tempted to use to establish firm historical roots are tenuous. There was, of course, the successful public health drive to reduce infectious disease, with triumphs such as the purification of drinking water and the use of antibiotics. There were fitful efforts to deal with urban environmental problems such as air pollution and noise, sewage disposal and industrial wastewater,[19] but it was not until the 1960s that these efforts gained widespread support and led to equally widespread action. One need not go into detail to establish the point that all these efforts, including the later interest in toxic chemicals, were associated with postwar change and had no significant roots in the early twentieth century. It is a history of change, not continuity, that unfolds as we examine U.S. environmental issues in the twentieth century.

While the public interest in natural environments rested on aesthetic objectives, the concern for pollution had its roots in new attitudes toward the biological environment and human health. The emerging interest in biology and the environment was expressed in terms of "ecology," the functioning of biological systems and how pollution interfered with it. The developing water quality program acquired an ecological focus that appeared explicitly in the statement of purpose of the 1972 Clean Water Act: to "restore and maintain the . . . integrity of the nation's waters." This viewpoint came to be articulated in the early years particularly by the U.S. Fish and Wildlife Service.[20] Ideas about environmental threats to health, conversely, arose from new health aspirations that came after the control of infectious disease and were expressed in words such as "preventive medicine" and "wellness." Increasingly, control of air and water pollution came to be thought of as an aspect of advance in human health protection.

In both these expressions of concern about pollution, one could detect a sense of heightened goals and aspirations associated with a desire for a higher standard of living and "quality of life." While one could easily focus on the crisis aspect of pollution problems (and this usually was the tone of media coverage), on a more fundamental level the notion of pollution as a problem arose far more from new attitudes that valued both smoothly functioning ecosystems and higher levels of human health.

The roots of these new environmental interests lay in the changing demography and social values of the post–World War II years. Many studies describe those changes: of public attitudes and membership in environmental organizations;[21] of the values of outdoor recreationists individually and in groups;[22] of values sought by purchasers of woodlands, the "non-farm, non-industrial woodlot owners."[23] To these can be added more qualitative evidence in the form of expressions of environmental interest and values in environmental publications. These sources identify values associated with leading rather than lagging sectors of demographic change, those that emerge in greater scope and intensity over time, rather than diminish. They are associated with younger people and the more educated, and the changing values that they express. With age, environmental interest tends to decline, and with education it tends to rise. Environmental values do not seem to be distinctively associated with rising occupational or income levels and do not seem to increase markedly after the attainment of a college degree. They are associated with value changes at work primarily between elementary school and the fourth year of college.[24]

Environmental interest is a feature of the rise of the new mass-middle class, which has come to play an increasing role in American society and politics. From the production side, that new middle class is associated with the new information economy; it is also associated with the new consumer economy, a fact that defines its direction of values and interest more precisely. One can formulate a sequence of historical change in patterns of consumption. In the late nineteenth century, necessities dominated consumption; in the 1920s a new interest in conveniences emerged, with a major focus on consumer durables. After World War II, the growing emphasis on consumption lay in amenities, both as new items of consumption and as integral parts of both necessities and conveniences. Now consumers had considerable discretionary income, which could be spent in many ways to make life more enjoyable. Environmental consumption was an integral part of this new direction of the economy.

Some of these environmental goods and services could be purchased in the private market. One could buy property in areas of cleaner air and water and amid more natural surroundings, either on the urban fringe or in the wider countryside, and one could enhance the environmental quality of one's own home. Many environmental amenities, in contrast, involved collective goods, such as air, water, or public lands, in which the private market did not determine the allocation of use. These amenities hence became the subject of public action and of public controversy and debate. Although environmental amenities had long been sought out by the more affluent, in the twentieth century that desire became a mass movement. The very rich could still escape to distant islands and retreats,[25] but within the United States, the environmental movement emphasized environmental equity, a broader sharing of environmental amenities than the private market could provide.[26] Public land management stressed such equity, and the process by which nineteenth-century private estates were transferred to public ownership and management symbolized the thrust of action.[27] Public management of air and water involved similar objectives of broad public benefit. What at one time in history had been thought of as a luxury available only to a few—one had to be able to travel to the "north woods" to enjoy more salubrious air—came to be thought of as an integral aspect of the mass consumption economy.

Several twists in the social analysis of environmental values help to indicate more clearly their roots. One was the examination of population shifts undertaken by geographer Ronald Briggs for the decade of the 1970s.[28] Briggs went beyond the customary analysis of state population trends and examined county-level data. Population shifts, he argued, were not confined to flows from the North to the South and West, but took place also within each region. Analysis revealed a dominant pattern of movement from amenity-poor to amenity-rich areas. The presence or absence of natural amenities was the major factor in people's choices to leave one area and to go to another. The second analytical twist distinguishes regional variations in environmental values.[29] Change did not come uniformly to the entire nation, and a few analyses went beyond the nationwide samples to disaggregate by region. There were regions of great strength in environmental values: New England, Florida, the Upper Great Lakes, the Pacific Coast, and the Mountain West. There were also regions that lagged markedly, such as the mid-South and the Gulf states, and the midsection of the nation from the Dakotas down through Texas. The old factory belt from Pennsylvania through Illinois lay in between. These variations seemed to be consistent with variations in the spread of demographic change—and especially values associated with education—throughout the nation.

Some further insight comes from recent attempts to associate environmental values with psychological factors beyond the more traditional demography. The 1982 Continental Group study of environmental attitudes is especially useful in extending the analysis in this way.[30] That study ranged attitude expressions in a spectrum from resource development on one end to resource preservation on the other and sought to associate that range of opinion with other values. Two such relationships are of interest. In one, environmental values were associated with drives for higher levels of personal and occupational achievement, and developmental values with job security and more modest levels of aspiration. In the other, values were associated with the question of whether in the midst of difficult problems one turned to religion or to science. Those associated with developmental values tended to turn to religion; and those with environmental values, to science. These tendencies prompt one even more to associate evolving environmental values with the leading edge of change, rather than with older and more traditional ways of thought.

From the point of view of historical analysis, environmental politics displays a fairly typical case of change and response to change. At many times in history, major innovations occur that transform the economy, society, and politics, and these in turn create tensions between the new and the old, as the new challenges more accustomed patterns of thinking and acting. Such was the case in the transformation of an agricultural society in the nineteenth century to an industrial and manufacturing one. Such is also the case in the latter half of the twentieth century, as a manufacturing economy is transformed into an information and service one. A wide range of ways of thinking, of attitudes and values, of private and public institutions rooted in the manufacturing economy now are challenged by new modes of production and new modes of consumption. More strikingly, the role of consumption in the economy, subordinated in both economic thought and public policy in earlier eras, has come to assume a more formative influence in shaping personal choice and institutional patterns. From the broadest point of view, environmental controversy arises from the impact of new environmental consumer impulses on older institutions of material production rooted in earlier manufacturing and agricultural eras.

THE ENVIRONMENTAL IMPULSES

The environmental impulses that stemmed from these changes in values can be sorted out into different strands, each of which arose at a different point in time to take its place alongside those that had evolved earlier. The first to appear, dominating the years from 1958 to 1965, was the drive to manage resources as natural environments for human enjoyment. This drive arose out of the expansion of outdoor recreation, which had grown steadily between the two world wars and then expanded even more rapidly after 1945. First the National Park Service and then the U.S. Forest Service extended their outdoor recreation programs, and then the national Outdoor Recreation Resources Review Commission laid out a massive national plan to promote outdoor recreation. A host of proposals arose to identify areas of value as natural environments so as to manage them permanently in a natural state. The most widely debated feature of this thrust was wilderness, which led to the Wilderness Act of 1964; but the Wild and Scenic Rivers Act and the National Trails Act, both in 1968, reflected similar objectives.[31]

These more dramatic cases of interest in natural environments should not obscure its many and varied forms. In the urban environment, there were attempts to carve out beachheads in the midst of overwhelming pressures for development: the open space acquisition program of the Department of Housing and Urban Development, the interest in urban forestry and urban wildlife, the slow and persistent efforts to recapture the urban waterfront for low-density use and open space.[32] Far more was done in the countryside between the nation's cities and its wildlands, where the use of land, air, and water was in the balance, "up for grabs," so to speak, and where the contest between natural and developed environments was intensely keen. Here, a host of resources were identified as valuable: wetlands, natural rivers, barrier islands, pine barrens, tall-grass prairies, swamps, and those remnants of endangered flora and fauna called natural areas. All came to be the subject of identification, description, allocation, and management as natural environments.[33]

The drive also came to include the protection and management of high-quality air and water. Environmentalists obtained statutory authority that would protect air and water from degradation in programs centering on nondegradation and Prevention of Significant Deterioration (PSD).[34] The implementation of both went slowly. They were most effective in areas where the land was publicly owned and had already been classified as park or wilderness. Although the legal machinery led to little immediate action, a slowly rising trend in public interest in such matters could be followed, for example, through the growing prominence of visibility as a problem. Visibility standards were first applied in the Clean Air Act of 1977 to the national parks, but soon the standards came to be thought of as having wider relevance and were given special focus in the debate over acid deposition.[35]

Especially significant in extending an interest in natural environments was the growing appreciative interest in wildlife. What in earlier years had been thought of as "game" now came to be extended to include "nongame" animals under the more general rubric of wildlife. In 1973, the American Game Policy published in 1930 was revised under the new title of the North American Wildlife Policy.[36] The appreciative use of wildlife for observation came to exceed that of hunting; the majority of those who used even public game lands sought to shoot with a camera rather than a gun.[37] Out of all this interest came new nongame or appreciative wildlife programs, such as those for endangered species and ocean mammals, and new forms of financing, such as the state income-tax checkoffs. The U.S. Fish and Wildlife Service responded to these new interests with its programmatic environmental impact analysis, published in 1976, and its extensive studies of the attitudes of Americans toward wildlife, conducted by Stephen Kellert of the Yale Forestry School.[38] Appreciative wildlife attitudes undergirded almost every program to establish an aesthetic interest in natural environments.

All these actions represented not a throwback to some preindustrial longing, as detractors were inclined to argue, but a desire for an advanced standard of living that included a far larger share of natural environment amid developed environment than had been thought appropriate in years past. Such natural environments formerly had been looked upon as "useless," waiting only to be developed. Now they were thought of as "useful" for filling human wants and needs. They played a role as significant in the advanced consumer society as such material goods as hi-fi sets or indoor gardens. Architects and corporate business sought to express such values by incorporating nature into the design of hotels, office buildings, convention centers, and shopping malls in the built-up environment. This drive for natural environments was the most extensive and enduring feature of the organized environmental movement. For the largest environmental organizations, that interest was uppermost, and over the years its political base expanded from Washington, D.C., to almost every nook and cranny of the nation.[39] By 1984 the wilderness movement, which in the 1950s had been dominated by the Wilderness Society in Washington, had come to be organized in every western state, with local constituencies fashioning logrolling coalitions to put together state packages of proposals for wilderness designation, each unit of which was looked upon as someone's backyard. Reflecting this continuing public interest in natural environments, two organizations with such concerns, the Wilderness Society and the Nature Conservancy, each with very different political strategies, grew rapidly in the late 1970s and early 1980s.[40]

Between 1965 and 1972, air and water pollution came to exercise a formative influence in environmental affairs alongside the search for natural environments. The legislative landmarks were the Clean Air Acts of 1963, 1967, and 1970, the Clean Water Acts of 1965, 1970, and 1972 and the new pesticide law of 1972. All this activity had a twist, distinctive for the times, which emphasized ecological change and which later events tended to obscure. The concern for ecology or the "integrity of biological systems" seemed to structure ideas about pollution in this first stage of public policy toward it. The concern for ecology was closely related to the earlier concern for the aesthetic natural environment out of which it evolved. The new concern for pollution was not primarily focused on human health. Instead, it emphasized the role of pollution in the functioning of ecological systems, a degraded ecology as an undesirable human context, and a concern for protection of natural ecological processes. One heard of an overload in carrying capacity, the way in which animal populations outran food supplies, biological simplification under stress, disturbances in aquatic ecosystems under acidification, and reduced forest growth due to air pollution.[41]

The initial concern in this vein seemed to come from those who protested the destruction of wildlife habitat by development. During the 1960s there were many objections to the construction of dams and roads, the release of waste heat from power plants into rivers and lakes, and the destruction of wetland habitat by residential construction.[42] The U.S. Fish and Wildlife Service came to be the major source of objection to these threats and, within the inner realms of the federal government, a defender of ecological systems and values. It was that concern from which arose the initial concept of "adverse environmental impacts," and it was from the House Subcommittee on Fish and Wildlife that many new initiatives along this line came, including both the National Environmental Policy Act (NEPA) and the Council on Environmental Quality (CEQ).[43] The fact that the appropriation for staff activities of the CEQ rests with the Fish and Wildlife Subcommittee is no historical accident, and the initial context that shaped the NEPA, namely interagency review rather than public review, was influenced heavily by attempts by the Fish and Wildlife Service and the Water Pollution Control Administration (WPCA) to protect fish and wildlife habitat.

Several other features of these years reflect this ecological context. The new interest in water pollution emphasized streams and lakes as ecological systems; it was this new drive that led to the removal of the water pollution control program from the Public Health Service (PHS), first to an independent agency within the Department of Health, Education and Welfare (HEW) and then to the Department of the Interior. Fishing interests argued that the PHS cared little for water pollution beyond chlorination. The initial concern over atomic energy in the late 1960s emphasized thermal pollution rather than radioactive release or reactor accidents.[44] And the new interest in pesticides emphasized heavily, as did *Silent Spring*, their detrimental effects on wildlife.[45] By the 1970s much of this emphasis on pollution matters had changed to focus more exclusively on human health, but between 1965 and 1972, ecology provided a transitional context of the new interest in pollution.

Several observations mark the decline of the ecological focus. One was the shift in the interests of the Conservation Foundation, which in its origins and its role in environmental leadership in the 1960s thought and acted from an ecological point of view.[46] By the end of the 1970s, it was difficult to detect much of that past in the work of the foundation. Equally remarkable was the way in which the formulation of ideas about an extensive federal "ecological service" was soon forgotten. Amid debates about heightened federal environmental action in the late 1960s, the proposal for a major ecological research program to undergird environmental policy, coming in no small part from the Subcommittee on Fish and Wildlife, was cast aside in favor of impact analysis.[47] The idea survived with some success in the form of the Office of Biological Services in the Fish and Wildlife Service until the arrival of the Reagan administration, but not with the formative influence anticipated in the 1960s.[48] Amid the growing attention to human health in environmental matters in the 1970s, ecology took a back seat.

The ecological focus reappeared late in the 1970s from the entirely different quarter of the new interest in acid deposition. The main significance of that new debate lay in the renewal of interest in welfare effects of air pollution. In the 1960s, these forces played as important a role in air pollution matters as health concerns. The first version of the sulfur oxides criteria document issued in the spring of 1967 contained a major section on visibility as well as material on agricultural crops, reflecting the technical background of John Middleton, first director of the National Air Pollution Control Administration.[49] But implementation of the Clean Air Act veered strongly toward a dominant focus on health. In the early 1970s, the EPA lost the court case concerning secondary annual standards for sulfur dioxide, and it did not pursue the matter further. The issue of acid rain, however, along with increased scientific knowledge about transport, deposition, and effects, revived the concern with welfare impacts, and by extension, ecological disturbances. The major significance of the acid precipitation issue concerned the role of these secondary effects in public regulation. This perception of the matter tended to restore the broad context in which air pollution had been considered in the 1960s.[50]

New Environmental Impulses

During the 1970s, three new environmental impulses emerged to take their place alongside previous ones. One of these, emphasizing resource shortages and the "limits to growth," seemed to arouse little public interest. The rise in energy prices after 1973 led to public action to reduce the burden of increases in the cost of living. Environmentalists carried the concern further, discussing energy efficiency, least-cost supplies, and alternative sources.[51] But even this concern generated little analysis of the inflation that seemed to be a result of rising real energy costs, the decline in lower real-cost supplies, and the movement toward higher real-cost levels. Real-cost inflation was merely the way in which resource limits made themselves felt in the economy. By the beginning of the 1980s, the declining pressure on energy prices reduced even this limited public interest in the larger problem.

Interest in resource shortages and limits to growth was expressed not so much by the public as by technical and professional leaders. It was in these circles that studies were made, articles written, warnings of the future announced, and influence applied to public policy. These were debates among the nation's political intelligentsia without significant public action. In the late 1960s, there had been some interest in the "population problem,"[52] but even here it seemed that public interest was shaped far more by family size and the "limits to growth" within the family context than by any broader global concern.[53] The report of the Commission on Population and the Nation's Future was shelved as quickly by the public as by President Nixon, and despite continued action on the part of environmental leaders, the matter was not revived as a broad public issue outside the context of immigration.[54] The low public interest in such matters was reflected in the very limited public attention displayed for both the Global 2000 Report and the World Conservation Strategy.[55] Similar relative indifference was expressed concerning farmland conversion and soil erosion, which the American Land Forum sought to foster as issues in the early 1980s.[56]

Far more extensive in terms of popular involvement was a second new twist to environmental affairs in the 1970s, the practice and ideology of personal and community autonomy and decentralization. Limits-to-growth issues stirred public debate without stirring the public; the decentralist movement stirred public interest in the form of personal and community action without stirring much public debate. Evidence concerning wide public involvement in such matters was abundant. *Mother Earth News* and *Organic Gardening* were extremely popular magazines; ideas concerning the wisdom of "human scale" provided one of the few ventures into well-formulated ideology in the environmental movement.[57] There seemed to be a major demographic trend associated with self-help and self-reliance in which people in the city, the suburbs, and the countryside sought to pursue life with more independence.[58] Modern technology furthered the trend by making self-reliance less burdensome in time and human energy. The community focus on the enhancement of natural environment values and protection against threats from large-scale development and pollution helped to enhance this tendency. So did efforts to make personal choices about lifestyle that might protect one against chemical hazards.

There was the great popularity of new habits of eating, with an emphasis on organically grown foods. Natural food stores witnessed a major growth in the 1970s, and general food markets began to take on some of their products.[59] Many people sought to grow their own vegetables so that potential chemical threats could be lessened; by the 1980s it was estimated that more than 35 million American families had some sort of vegetable garden. The concern for toxic chemicals in the 1970s often took the form of efforts to protect supplies of drinking water by insulating sources from potential groundwater contamination, or by seeking to declare one's home and grounds off-limits to pesticide spray drift.[60] All these trends underlined both changes in personal lifestyles and a new interest in political autonomy and decentralization. One of its most striking expressions was the community energy movement and the promise of solar energy, which seemed to make possible a higher level of freedom from larger energy systems.[61]

Although both the limits-to-growth impulse and the personal-and-community-autonomy impulse, each in its own way, constituted new elements in environmental affairs in the 1970s, both were overshadowed in policy debate and action by new concerns for human health. These reflected a persistent transformation in human attitudes marked by the triumph over infectious disease and by the emergence of new health concerns, expressed by such words as "wellness" and embracing concepts such as physical fitness and optimum health. Most people were no longer worried about imminent death or uncontrolled infectious diseases. Major chemical threat episodes, such as the Donora smog, kepone in the James River, and PPBs in Michigan and Love Canal, all dramatized the concern.[62] Underlying it was a change in attitudes and values. Increasingly Americans expressed their interest in health in terms of a capacity to engage in daily affairs at an optimum level of physical and mental health. They began to change their personal habits of eating, smoking, drinking, and exercise. They came to look on chemical agents in the environment as having an adverse effect on their aspirations.

This new interest in human health seemed to be at variance with the dominant trends and capabilities of American science and medicine. Leaders in those fields tended to identify advances in health in terms of mortality. Periodic reports from the surgeon general marked progress with data about reduced death rates in various age categories.[63] Medical practitioners, moreover, tended to be concerned more with curing sickness than preventing it, and hence Americans experienced a limited ability by physicians to cope with knotty problems of optimum health.[64] While preventive medicine had made impressive accomplishments in vaccination against infectious diseases, it now seemed to be less interested in the limitations on optimum health that might come from environmental causes. Pollution issues seemed to be wrapped up in these contradictory tendencies—the high level of aspiration by the public for greater wellness on the one hand, and the relatively limited capacity of science and medicine to respond on the other. Hence, a major aspect of the public's concern for chemical pollution was either to take matters into their own hands and avoid contaminants by means of new personal lifestyles, or to demand public action to prevent exposure.[65] In such affairs the medical profession often followed rather than led public attitudes.

The range of health effects at issue seemed to expand steadily. Most attention was given to cancer, but soon a wider range emerged: genetic and reproductive disorders, fetal and infant malformations, neurological deficits and modified enzyme systems, lowered immunity and premature aging.[66] Interest in chemical pollution as a possible cause of such health effects seemed to widen. Attempts to prevent exposure to carcinogens led to many battles over specific chemicals and to unsuccessful efforts to establish generic cancer regulation, first by the EPA in its formulation of "cancer principles" in the mid-1970s, and then by the Occupational Safety and Health Administration (OSHA) in its abortive "generic cancer" policy a few years later.[67] Cancer took up most of the debate, but expanding scientific knowledge tended to expand the range of concerns that were being taken more seriously with each passing year. There were massive implications. If it could be demonstrated, for example, that blood-lead levels common to most Americans as "normal" reduced IQ scores by three points, the cost when widely shared by large numbers of Americans could be extensive—it was estimated that for each point of IQ deficit one's lifetime earnings were reduced by 1 percent.[68] Potentially more dramatic was the research linking toxic chemicals to reduced male and female fertility. In February 1984, *Chemical Week* reported soberly on the implications for the industry of the fact that three and a half million American families were not able to have children.[69]

Occupational health provided an especially sharp focus for these concerns.[70] In earlier years, occupational hazards had been thought of primarily in terms of physical injury, but after World War II, occupational illness received increasing attention. The use of antibiotics made other problems such as cancer and reproductive defects more visible. Workers were reluctant to take up such issues, often because it might well mean removal from the job and hence financial loss. Nevertheless, several labor unions pressed forward, lobbied for the Occupational Health and Safety Act, hired their own industrial hygiene professionals to tackle the issues, and pressed OSHA to move toward stricter regulation. The format of the resulting controversies was typical: the affected workers expanded the frontiers of scientific knowledge and sought to reduce exposures in the workplace; the industries demanded higher levels of proof, resisted emission controls, and opted for protective devices worn by workers. Change often came most readily as a result of liability suits. In the case of asbestos, for example, the courts, rather than the administrative agencies, pinpointed responsibility and assessed damages for injury. Such action often caused controls to be tightened voluntarily so that industries could avoid such costs in the future.

Here was the crux of the political interplay: the public was demanding that frontiers of knowledge and action with respect to health be expanded more rapidly than was possible, given the limited capabilities of science and medicine. Public controversy revealed how little was known about the presence of chemicals of potential harm, either in the environment or in humans.[71] The first comprehensive measurements of blood lead levels for the entire population were made only after 1975, and such monitoring was rare.[72] Occupational health experts continually stressed the degree to which the lack of exposure records as a part of medical case histories inhibited diagnosis. The problem was one of data. The collection of mortality data had long been standard operating procedure, and information was readily available about the "reportable" infectious diseases. Modern medicine now focused on a much different type of statistic; even cancer incidence, in contrast with mortality, proved difficult to determine. On a wide range of subjects, public demand for knowledge and action outran capability.

The historic change that all this implied for health science and medicine was the task of retooling from concern with acute effects of high-level exposures to the more subtle chronic effects of persistent low-level ones. Scientific method and analysis appropriate for the first were found to be far less able to identify the second. Thus, epidemiological studies that traditionally relied on the 5 percent confidence level used by biostatisticians might not be appropriate for smaller effects, because these were almost invariably, by their very size, beyond such limits. Such was the relationship found between lead and child behavior.[73] Environmental demands had contributed in no small part to the extension of chemical measurement capabilities from parts per million to parts per trillion in concentration. But the cost of such measurement made it all but impossible to chart the presence of chemicals, such as dioxin, in more than a few cases. A new test for blood lead developed in the late 1970s greatly lowered the cost of surveillance, but for most of the new health effects, diagnostic cost greatly limited the public demand for knowledge. The public desire to know was thwarted by the increasing "real cost" of knowing.

Underlying the new concern for chemical harm was the public perception of a chemical world out of control. In a series of episodes, from the early experience with atomic testing and pesticides to later cases of toxic chemicals in air and water and on land, the public gradually formulated fairly clear notions about the chemical universe: chemicals were persistent, not biodegradable, lasting in the environment for long periods of time; they were ubiquitous, transported through air and water to places far distant from their source; they "biomagnified" in the food chain so as to become more highly concentrated in the higher orders of mammals; and they were mysterious in that they could suddenly appear in ways not previously known or suspected. Chemicals dispersed into the environment affected biological life, and humans in particular, in such a way that their effects could be even partially controlled only with the greatest difficulty. The nuclear reactor episode at Three Mile Island conveyed the image of a technology that was out of control; the sudden realization that hazardous waste had been permitted to pervade the environment and now seemed impossible to contain reinforced that perception. How could an individual person bring under control a potential harm which the nation's prevailing institutions could not?

THE IMPACT ON PUBLIC AFFAIRS

Environmental impulses had varied effects on public affairs and shaped varied debates. One of the first of these involved the question of "how much," the level of public benefit that should be sought. What was the balance that should be struck between natural and developed environment? How much wilderness; how many miles of free-flowing rivers; how much in the way of wetlands, estuaries, barrier islands, and natural areas; how much open space and parkland within the metropolitan region? Each case involved conflict between those who wanted to develop land and water and those who wanted to manage them as a natural environment. No formula seemed to be available or offered as to "how much" in some ultimate sense; the issues were fought over case by case, as environmentalists pressed their claims as to the social desirability of more natural environments. These competing claims constituted some of the most celebrated cases in the politics of the environmental era.

National goals with respect to environmental pollution were far more focused. Early in the 1960s, the notion that social objectives should be thought of in terms of standards of environmental quality was agreed on. The debate over the precise level of ambient air or water or biological burden was another matter; the establishment and reestablishment of standards continued year after year as the contending forces debated the issue of "how much." Those responsible for emissions persistently sought to increase the allowable levels, and those who felt harmed equally sought to reduce them. These debates over standards took many forms. One was the threshold margin-of-safety formula, which in standard setting sought to add a cushion that took into account the plausible unknowns on top of known scientific evidence about harm. Another was cost-benefit analysis, which sought to establish a common denominator for the value of adverse effects, which, through control, could become a social benefit and hence could be compared with costs.[74] Still a third approach, which became more popular with industries responsible for pollution and with environmental managers toward the end of the 1970s, was risk analysis. Analysts were heavily influenced by the concern with cancer, which sought to establish some mathematical statement about risk of premature mortality. This strategy was limited to firmly known effects and did not incorporate the plausible unknowns; and it seemed to focus heavily on mortality rather than morbidity.[75]

Although the policymaking strategy of setting standards and implementing them seemed to give a degree of precise discipline to the issue of social goals, it did so only by excluding the more debatable frontiers of knowledge. Were secondary welfare standards to be given as much consideration as human health? Environmentalists sought to bring a wide range of effects into standard setting, but those responsible for emissions sought to narrow the range so as to limit their responsibility. Industry vigorously opposed the expansion of the visibility standard to integral vistas in the national parks, for example, and ridiculed the gradual extension of visibility as a policy objective throughout the East, all because it would justify tighter controls over sulfur dioxide emissions. While environmentalists sought to extend health protection from premature mortality to temporary and reversible morbidity, such as the Montana objective of preventing air pollution that would interfere with normal daily activities, industries sought to contain that drive.[76]

One is struck also by the limited degree to which developmental agencies in government incorporated environmental objectives into their daily way of thinking and acting. What little transformation they underwent was associated with the mitigation of environmental impact rather than the enhancement of environmental objectives. To conform to requirements of the law, the environmental impact statement (EIS) became widely used, although its role in the work of developmental agencies would seem to be more one of providing an early warning signal as to potential opposition to the agency than of enhancing environmental goals. An approach that focused more directly on objectives was the dual-stream planning process of the Principles and Standards of the Water Resources Council.[77] This process called for as much attention to planning for environmental objectives as for developmental ones. Similar attention was given to environmental goals through the various "futures" strategies used by several state governments.[78] Nevertheless, agencies such as the U.S. Army Corps of Engineers, the U.S. Forest Service, the Bureau of Land Management, and the U.S. Department of Agriculture continued to conceive of their missions as developmental, to look upon environmental objectives as requirements to which they would adjust minimally in the face of legal and political necessity. One can detect little significant change in the values they sought to advance, little leadership to innovate and enhance environmental objectives.[79]

A second effect of the environmental impulse on public policy was its influence in expanding frontiers in science and technology. Contrary to much of the political debate, environmentalists did not reject science and technology, but instead sought to extend them. Controversies over such issues involved disputes over the use of scarce resources. In matters scientific, environmentalists sought to allocate resources to extend the frontiers of scientific knowledge about the functioning of the biological world and human health and the environmental effects on them. In matters technological, they sought to promote innovation to reduce emissions at the source, rather than rely on either treatment or dispersion. In both cases, the political significance lay in the fact that environmentalists placed such great demands on existing scientific and technical resources and institutions that they could not respond effectively.

Environmentalists pressed for more monitoring and research on the one hand and the incorporation of new knowledge more quickly into the scientific consensus on the other. Environmental monitoring was very limited. Air quality monitoring was confined to the areas of high pollution or the isolated "hot spots," rather than extended into a comprehensive system; there was little information about changing levels of air quality in the cleaner areas, those that Pennsylvania, for example, called its "non-air basins." Hence there was the intense controversy over measurements, which would chart degradation under the PSD program. Water quality monitoring was a bit better because of the wide interest in a fuller range of quality measurements desired by fishing interests, but even here the extension of parameters outran resources.[80] The weakness of baseline ecological measurement was underlined by the acid precipitation controversy, which focused on the Hubbard Brook experiment station as the only case with twenty years of significant monitoring. As groundwater increased as an issue, the limited data only underlined the weakness of knowledge about biogeochemical cycles. Perhaps the most extensive data involved lead, largely because of the persistent effort by Clair Patterson of the California Institute of Technology to chart lead in a wide range of environmental media throughout the world.[81] And a major result of the acid precipitation issue was to enhance a wide range of environmental monitoring.[82]

Environmental demands also stimulated experimental research. Knowledge about the effects of lower levels of human exposure was pursued. The draft lead criteria document issued in 1983 relied heavily on new research reported since the earlier 1977 version to detail an increasing range of health effects at lower levels of exposure.[83] Research on the effects of acidification on lakes, building materials, visibility, forests, and agricultural crops came about largely because of the intensity with which environmentalists pursued the acid rain issue. There was an explosion of environmental research after the 1960s, with a large number of new scientific journals appearing to accommodate publication of results. Much of this activity took on a regional focus, as reflected in a new journal, *Northeastern Environmental Science*, which began publication in 1982.[84] But in some realms of research, environmental influences were limited. The regional experiment stations of the U.S. Forest Service undertook research on environmental uses of the forests primarily in response to recreational and landscape demands expressed by users and not from its own internal initiatives. The composition of its research advisory committees and the choices they made in research strategies reflected limited environmental interest.[85] The same was true of the new sea grant institutions, where research seemed to be shaped by developmental rather than environmental objectives.

The series of controversies that emerged after the mid-1960s over the assessment of scientific studies about environmental effects identifies more precisely the role of the environmental impulse in matters scientific. These controversies followed a general pattern. At one end were those who required high levels of direct proof of harm before they would believe that action was justified; at the other were those who sought to bring frontier knowledge about adverse effects into assessments as a ground for action more quickly. Many views ranged in between. As new scientific data appeared, no matter what the total fund of knowledge, there was continual disagreement between those who hesitated to draw conclusions because not enough was known and those who sought to act on the basis of newly discovered frontier knowledge.[86] There were disagreements over levels of proof. Environmentalists associated themselves with frontier scientists who sought the more rapid use of new knowledge because it would speed up action. Sources of pollution associated themselves with high-proof scientists so as to retard such action.[87] Although some in the scientific world were prone to describe these debates as being between "good science" and "bad science," such a scheme does not seem to be workable in identifying the nature of the controversy. Far more significant were the predispositions of scientists in the face of expanding knowledge: their institutional attachments; their commitments to particular methods such as epidemiology and experimental biochemistry; their psychological values, which predisposed them in personal choices to play it safe or to be innovative. In these matters, environmentalists played an important role in advancing the frontiers of knowledge, in giving support to scientists so engaged, and in bringing that knowledge more quickly into the workable consensus building that provided political support for public decisions.[88]

The environmental impulse also stimulated innovation in technology. In pollution matters, environmentalists stressed source reduction rather than treatment or dispersion. The weight of their influence on such matters at any one time seemed minimal, but over the years the direction was clear. Once dispersion in the form of either dilution (in water pollution) or the use of tall stacks (in air pollution) had come to be rejected, then treatment technology was stimulated. Once the cost of treatment came to be accepted, third-party innovators were prompted to become inventive with respect to alternative and lower-cost treatment technologies. Higher treatment costs promoted source reduction technologies. This gave rise in some quarters to the view that "pollution prevention pays," because it both increased conventional productivity and reduced pollution.[89] Joseph Ling of the 3M Corporation sought to popularize this point of view, but it was not widely accepted in the United States. In Europe, Ling's views were more popular.[90]

The debates had to do with the pace of technological innovation.[91] Environmentalists pressed for it to proceed more rapidly; for example, to coal washing, fluid bed combustion, and limestone injection in controlling sulfur dioxide. But existing technologies carried with them a built-in conservatism, a demand that they be continued through their economic life rather than be discarded because of technological obsolescence. Within each line of industry there was a range of technical modernization, from old plants to newer ones. In the normal course of events the old would gradually give way to the new. Environmentalists believed, however, that the pace was too slow; it should be accelerated. They sought to focus on the presence of obsolescence in the face of rapid technological change. Hence came the idea of technology standards in water pollution treatment and the view that all plants should conform to the "average of the best" technology in the industry. The dispute focused on which "average" and which "best." When industry sought to define it as the median among all plants, the EPA argued that the law justified the average of the "best ten percent." The courts agreed. All this identified the resistance to new technological innovation as a form of diseconomy arising from previous investment commitments.

Environmentalists sought to press on with a number of frontier technologies: photovoltaic solar cells, the electronically controlled internal combustion engine, biological farming, integrated pest management, decentralized source-stream recycling of materials, closed-loop water process recycling, least-cost energy efficiencies.[92] Many of these innovations assumed a point of view that stressed decentralization rather than support the historical preference for large-scale technologies. In that new emphasis on light rather than heavy technologies, environmental ideas about such matters seemed to be more than slightly in tune with the flexibility and versatility of the electronic age. The innovations were viewed with scorn by those associated with older and heavier technologies.

An understanding of the historical role that environmental action played requires a focus on the rapid changes in both science and technology in the latter half of the twentieth century throughout American society, and not on the less-than-adequate analytical context of pro- and antiscience or pro- and antitechnology. If one sorts out the old and the new in such matters, the environmental impulse was strongly associated with the frontiers of discovery and innovation.

THE ENVIRONMENTAL OPPOSITION

The environmental impulses and their impact on public life gave rise to formidable and persistent opposition. While those engaged in the daily round of environmental politics well recognized that resistance, the forces arrayed in opposition to the environmental movement have received little systematic treatment as a major element of environmental history. They deserve analysis as careful and complete as the environmental drive itself. The historical drama is one of thrust and counterthrust, of new values and old, of new impulses seeking to turn public policy toward new goals in the face of the reluctance of the old order to modify customary ways of thought and action. The prevailing pattern of politics in the years before World War II involved contests among diverse segments of development—business, labor, and agriculture—each seeking to gain a larger share of public benefits. Now those groups faced a significant challenge from the consumer side. The environmental drive was a forceful part of that challenge. It is no surprise that developmentalists fought back.[93]

Much of the strength of the opposition lay in its own reinvigoration in post-World War II society. Developmental institutions advanced in their role in tandem with environmental objectives. The drama of the political scene lay in the intensity of debate over rival claimants for public policy: those who wanted to expand material production rapidly and those who wanted to expand the environmental benefits of the advanced consumer society. A striking case of that drama lay in the Rocky Mountain West, where the old and the new competed vigorously for a claim to turf.[94] The traditional developmental institutions there, such as stockraising, lumbering, and hard rock mining, were now augmented by a range of new energy activities. But that region now expressed some of the strongest environmental objectives in the nation.[95] The new order made claims of its own for wilderness, wildlife, environmental forestry, outdoor recreation, cleaner water and air, and in-stream water flows; it contributed much to the defeat from within the region of both the Sagebrush Rebellion and the Reagan administration's asset management program. The old order fought back with its own claims, seeking to use the power of the federal government to restrain the environmental forces from within the West.

One can chart the environmental opposition through confrontations in state and federal legislatures, administrative agencies, and the courts. The fight was intense and persistent. Many have sought to argue that the passage of federal environmental legislation reflected fundamental agreement on public environmental values and objectives and that disagreement was secondary and concerned with differences over implementation of goals rather than of goals themselves. This view is more of an argumentative contrivance than a faithful recording of the facts. Many decisions by administrative agencies and courts expressed fundamental disagreements over objectives. Often it was the debate within the agencies and the courts that revealed most fully the intensity of the controversy. The opposition to environmental objectives in each arena of politics was more than relentless; it constituted a strategy of maximum feasible resistance and minimum feasible retreat.

So many cases of opposition are on record that it would be foolhardy to do more than categorize some generic types. There was, for example, the attempt to shape the course of scientific assessment. This appeared with the first criteria document, the draft sulfur oxides report in the spring of 1967, which led to a storm of protest from the coal and utility industries over the scientific conclusions. The intensity of this opposition gave rise to concern by some state public health authorities; if the National Air Pollution Control Administration (NAPCA) buckled under to demands that the conclusions be altered, no state could withstand the ensuing pressure.[96] The answer came with an instruction from Congress through the work of Senator Jennings Randolph that the criteria document be revised under the guidance of an advisory committee that included representatives from the affected industries.[97] From that time forward, no scientific assessment, whether a criteria document or otherwise, remained independent of the real world of very high stakes involved in scientific assessment.[98] The vast importance of all this to the environmental opposition was reflected in the Reagan administration's strategy of replacing one group of scientific advisers with another more receptive to the views of the opposition. The resulting furor revealed environmental politics at its most basic level. That politics could be observed in the care, or lack thereof, with which EPA administrators chose members of their science advisory boards.[99]

A major strategy of containment by the environmental opposition was to restrict both in numbers and in geographical territory the identification of natural resources that were valuable environmental assets subject to potential management for environmental objectives. Mere listing of specific areas of high environmental value, such as wilderness, wetlands, barrier islands, estuaries, parks, and wildlife species, became crucial, because such listing could well be the first step toward action. Hence, classification of wilderness under the three wilderness inventory programs, or state inventories of "areas of critical environmental concern" in the coastal zone, became targets of contention. By the same token, pollution sources sought to restrict the identification of sources of harm, such as toxic waste dumps or chemical carcinogens, in order to minimize the range of environmental controls to which they might be subject. As was the case with standards, identification of environmental assets or environmental problems involved fundamental debates over goals. They were not simply issues of implementation of agreed-on policies; on the contrary, they constituted a continual set of cases, one by one displaying a repeated drama: different groups in American society differing over the wisdom of pursuing specific environmental goals.

It is important to emphasize the high degrees of success enjoyed by the environmental opposition. The main theme of recent environmental history has emphasized the environmental triumphs. A careful assessment calls for a mixed review. There were significant, though less publicized, failures, which in the customary legislative record are forgotten. Significant attempts to add to the range of publicly owned and managed natural environment areas were turned back: a program of national estuarine areas as extensive as the lakeshore and seashore program was transformed into a small set of research areas; [100] the drive to protect barrier islands in a similar manner, proposed by Congressman Philip Burton, was restricted to control of federal activities on such islands;[101] the marine sanctuary program advanced at a snail's pace; purchase of inholdings within the national parks moved equally slowly; the drive for urban national parks met little success beyond the initial choices in the early 1970s.

Equally noticeable was the modest, even slow, pace of pollution control activity. The Clean Air Act had anticipated a program extending far beyond the original six criteria pollutants, and NAPCA announced in 1968 that several dozen criteria documents were in the offing.[102] But lead alone has been added to the list, and this only by a series of citizen actions and court orders.[103] A research program in behavioral toxicology, which in the late 1960s seemed more than promising, could have greatly expanded air pollution problems to include temporary disability that led to lost time at work and in personal affairs.[104] This program did not go beyond the beginning stage. An initial effort in 1965 by Senator Abraham Ribicoff to control pesticide discharges into streams by direct inspection and supervision of manufacturing processes was turned back without difficulty, and the pesticide control program throughout its history has had to struggle for effectiveness. The control of toxic chemicals has been marked more by lack of action than by vigor; few chemicals have been tested for low-level chronic effects of low-level exposures; the promise of strategies to move beyond the acute effects of acute exposures has not yet been realized.

Two subtle but vastly important realms of success for the environmental opposition have been little noticed. One was the degree to which it was able to shape the terms of environmental debate. There was hardly a realm of public thought in which those who feared and struggled against environmental action did not take the initiative to dominate the definition of environmental issues. In so doing they described environmental affairs in terms of what they were not rather than what they were: antitechnology, bad science, single-issue politics, adversarial strategy, the environment versus the economy, no-risk philosophy, hostility to cost-benefit and cost-effective analyses, housewives' data, pollutant of the week, elitism, and populism. These words and phrases and the ideas they implied often structured the way in which the public media and the professional media defined the discussion. Most of the intellectual efforts of environmentalists, therefore, were channeled into acts of self-defense defined by their opponents rather than into positive initiatives they themselves had shaped. This gave the opposition considerable leverage with the nation's political intelligentsia.

The opposition's second success was a marked shift in the drift of scientific opinion and assessment toward an acceptance of the demand for higher levels of proof of harm. In the 1960s, scientific experts were more likely to talk approvingly about the need to act in the absence of full knowledge, to work in terms of "reasonable anticipation of harm," to bring frontier knowledge more quickly into the realm of public policy. Much of the early pollution control program in air and water was worked out under the assumption of a forceful role for this kind of scientific inference.[105] But over the years, pollution sources demanded higher levels of proof and singled out frontier scientists for special—and often massive—attack. Scientists have seen the professional careers of a number of their colleagues severely damaged, including John Gofman, Ernest Sternglass, Samuel Epstein, James Allen, Thomas Mancuso, Beverly Paigen, Dante Piciano, and Melvin Reuber. Although they have admired those few who have been able to remain strong professionally in the face of these attacks, such as Irving Selikoff, Clair Patterson, Edward Radford, Karl Morgan, and Herbert Needleman, most scientists have not been willing to take such professional risks.[106] In the face of such criticism, often sharpened by attempts from the environmental opposition to define all such issues in terms of "good science versus bad science," the self-images of scientists have worked a powerful influence on those less self-confident. A special focus of this growing influence of the demand for higher levels of proof of harm was the weakened role of the "margin of safety," that nebulous area of plausible inference that is the heart of every decision about standards.[107] While the "margin of safety" still remained by 1984, over the years it had become severely bruised and battered.

The response of traditional production sectors to the environmental impulse was not uniform. Agriculture, for example, had played a major role in the earlier conservation movement; when the issue now could be defined either as soil erosion or as the loss of productive cropland, some of the older players reached a common ground with environmentalists. Nevertheless, a host of issues about farming methods, from pesticides to commercial fertilizers, from the destruction of fence rows to the use of no-till agriculture, [108] from non-point water pollution to field burning, as well as land use planning and the use of the countryside for recreation—all served to divide farmers from environmentalists. Within the states, rural legislators provided the strongest environmental opposition, objecting both to carving out natural environment lands and programs within the countryside and to the imposition on rural areas of pollution controls really meant, they argued, for the cities.[109] Only when it came to siting large-scale industrial and waste facilities in rural areas, or the massive impacts of raw materials extraction, did farmers and environmentalists reach common ground. The various "resource councils" of the Dakotas, Montana, and Wyoming reflected that kind of cooperation.[110]

Cooperation between environmental groups and organized labor was more frequent. In this case, one must distinguish between the construction unions and the industrial unions.[111] The former were often at odds with environmentalists over siting and federal funding for large-scale projects. With industrial unions it was different; they had joined with environmentalists in the early 1950s in opposing the construction of the Echo Park dam in western Colorado. The United Steelworkers of America held the first nationwide citizen conference on air pollution in Washington, D.C., in 1969.[112] As the 1970s wore on, environmental ties with the Industrial Union Department of the AFL-CIO grew on such joint interests as occupational health, community air pollution, toxic chemicals, and the "right to know." There were major controversies between environmentalists and the steel workers over such issues as the bottle bill, and though coal miners joined the antinuclear drive, they still sided with the coal companies on sulfur dioxide control and (especially) acid rain. The relationships were mixed, but in general they provided more opportunities for cooperation than for conflict.

With industry, in contrast, such opportunities for cooperation were limited; industry provided most of the leadership and the resources for the environmental opposition. There was nothing mysterious about this; the two groups had mutually exclusive interests. Lands managed as wilderness were not available for mining and lumbering. Waste treatment added new production costs. Chemical companies wanted to increase the use of pesticides, and environmentalists wanted to reduce them. These normal conflicts between those who produce and those who consume, those who adversely affect others and those adversely affected by them, ran through much of the economy. Many such conflicts could be resolved by adjusting prices in the private market, but others inevitably led to public action. This was especially the case where the resource itself was widely shared or publicly owned, such as the public lands, water in streams or lakes, or air. Perhaps the most curious aspect of these relationships between industry and environmentalists was the failure to find common ground on the basis of a shared interest in technical process innovations which, as Joseph Ling pointed out, would kill several birds with one stone. While efforts to establish such a middle ground took place in Britain, they did not occur in the United States.[113]

All the actors in the environmental opposition came together on one objective: to slow down the pace of environmental advance. Attempts to enhance natural environment management should be restricted; there was already too much wilderness, there were too many parks, there were too many protected wetlands; those that existed were too restricted as to use. "Multiple use" became the battle cry of those who sought to enhance development on environmental lands.[114] The drive for pollution control had gone too far; more time should be spent on accepting the existing levels of pollution—they were not harmful at all—and on protecting human and biological life from the impact rather than on preventing emissions in the first place.[115] Light technologies should not be allowed to impede the growth of more important heavy ones. Notions about limits to physical resources were simply either a result of misinformation or a result of preoccupation with the future.[116] There was a distrust of the expanded influence of the public on environmental decision making, and especially on scientific and technical questions about which only the better-informed were capable of making sound judgments. As the environmental opposition grew in numbers and political strength throughout the 1970s and into the 1980s, it attacked on a wide front. Its influence rose in the latter years of the Carter administration, and with the Reagan presidency it succeeded beyond its wildest hopes.[117] The alacrity with which it took advantage of the new opening backfired, but even with the adjustments that the administration made in response to political protests from environmentalists, the opposition still scored high. Despite its growth in numbers and resources, by 1984 the environmental movement could maintain little more than a holding action against its opposition.[118]

THE POLITICS OF ENVIRONMENTAL MANAGEMENT

As environmental politics evolved, its context shifted from broader public debate to management. Increasingly one spoke of air quality management, water quality management, forest management, range management, the Bureau of Land Management, Coastal Zone Management, risk management, river management, and wilderness management. Hardly an environmental problem could be dealt with outside the terminology and conceptual focus of management, and, in turn, management played a powerful role in shaping the world of environmental choice. The influence of management grew because of its power and its authority to coordinate discordant elements in the "system" on its own terms, and even more because it constituted the persistent institution of government, with ongoing day-to-day capabilities for communication and action. Institutional power was the stuff of political power; it arose from a continuous presence requiring that others reckon with it day in and day out; it set the bounds of choice if not the actual agenda. While the larger ideological debates in environmental affairs came and went, management shaped the world of day-to-day political affairs.[119]

Frequently it is argued that administrators simply implement policies made elsewhere in the legislature. Hence, administrative choices are secondary and derivative. This distinction between lawmaking and law-implementing institutions makes sense in terms of the formal structure of government, but when one explores political controversy through the various governmental functions, the argument breaks down. It is in management that the fundamental choices are made. The legislature and the court establish some of the outer limits of possibility, but the choices that make a difference to those concerned with the outcome are made by the managers. As soon as a law is passed by Congress, the parties rush to the agency to influence the way it decides to carry out the law. That outcome can range all the way from closely following the legislative objectives to virtually nullifying them. Administrative choice is more final; hence, it is more critical. Management and administration, and the politics of management and administration, are at the heart of the American political system.

When environmental political choice shifted from the wider public arena to the realm of administration and management, it was transformed into a vast array of technical issues: allowable cut and non-declining even flow; visual corridors and residual pricing; animal unit months; discount rates; mixing zones and lethal concentration 50s; acentric chromosome fragments and erythrocyte protoporphyrin; case control epidemiology and levels of significance; benign and malignant tumors; minimum low flows; integral vistas; oxygen depletion; buffering capacity and aluminum mobilization; Class I, II, and III PSD areas and testing protocols. Such issues did not reduce the intensity of debate; they only specified it and structured it more precisely. Far more important, debate was shifted from options that both the public and the media understood to seemingly esoteric issues that neither sought to comprehend. Hence, decisions became more obscured from the public view, more private and less public. Management and administration shaped an arena of politics in which the range of actors diminished in number, and only the initiated could follow and participate. That changed neither the significance of the debate nor its stakes.

It is within this realm of technical politics, therefore, that one can observe and follow most precisely the course of give-and-take in environmental affairs. Movement forward or backward with respect to environmental objectives, success or defeat for one side or the other, or just stalemate, can be identified in this arena with great precision. Here the main competing pressures and options were worked out, such as the clash between high proof of harm and frontier science about adverse environmental effects, or the balance between developed and natural environments, or the degree to which government sought to influence the direction of technical innovation. Whether the arena of choice was extended to include the legislature or the courts on the one hand, or the wider public on the other, depended on managerial action. The extent to which management sought to be innovative or conservative with respect to the cutting edge of questions about values, science, and technology was the main point of the drama. The major structural achievement of the environmental era was the erection of this managerial and administrative apparatus of political action. The major historical problem is to observe the degree to which it succeeded in disciplining contending environmental and developmental forces in the wider society so as to facilitate action.

To develop an ability to engage in environmental politics in the arena of managerial choice was the major achievement of citizen environmental organizations. Although their success in mobilizing members and the public, and their accomplishments in legislative action and litigation, were more noticeable, the test of what they could accomplish lay in their ability to work within the managerial and administrative orbit. Long-standing tendencies in administrative politics emphasized the face-to-face relationships between the regulators and the regulated as each sought to modify choices amid the anticipatory power and influence of the other. Those relationships had hinged on technical matters of administrative arrangements, on science and technology. Because they shared a common perspective about efficiency in matters of production, they found it relatively easy to come to terms. The injection of sustained institutional influence from the consumer side, however, into this traditional pattern upset the balance and brought to the relationships between regulators and regulated an intrusion that led to far greater uncertainty. Environmentalists sought to shape the direction of values, science, and technology; their demands and their strategies were less than welcome.

Thus there was a constant debate over the "opening up" of administrative decisions so that the new directions could be given more weight. In this effort, environmentalists benefited from the judicial commitment to supervising the Administrative Procedures Act of 1946 within the doctrine of "fairness." Freedom of information strategies had come from sources far beyond the environmental impulse,[120] but the court's supervision of the National Environmental Policy Act was shaped by much the same concerns.[121] While the act had been conceived in its earliest days as concerned primarily with interagency review, it was the court that modified it into a procedure for public review.[122] Similarly, the court required full participation in, and "on the record" bases for, administrative choices.[123] It balked only at the thought that the Executive Office of the President should be bound by the same open procedures.[124] Nevertheless, environmentalists were faced with constant efforts by both the regulated and the administrators to limit openness, to manage it for their own ends, and to restrict the new influences.[125]

Three focal points of administrative environmental politics were science, economic analysis, and planning. In all three, environmental organizations sought to increase their abilities to meet their opponents effectively. This was difficult, for the technical resources essential to each were readily available to both administrators and developmentalists in the business community and relatively unavailable to environmentalists. In the early years of the environmental era, citizen organizations often relied on volunteer contributions from technical and professional experts, such as analysis of the costs and benefits of the Cross-Florida Barge Canal, the environmental effects of pesticides, the effects of clearcutting, or the ecological role of wetlands. Such volunteer contributions continued, but over the years, environmental organizations began to develop their own in-house skills. Much of this came with environmental litigation at the Environmental Defense Fund (EDF) or the Natural Resources Defense Council (NRDC).[126] The Wilderness Society developed its own capacity for economic analysis.[127] The National Clean Air Coalition (NCAC) was able to finance outside technical studies. Environmental citizen involvement in forest planning was assisted greatly by the technical expertise of Randy O'Toole and the Cascade Holistic Economic Consultants of Eugene, Oregon.[128]

The most extensive capabilities in such matters came with the review and analysis of existing literature so that the latest in scientific and technical knowledge could be brought to bear on nuclear energy, forest management, environmental and health effects of pollution, atmospheric transmission, and wetlands protection. Environmentalists applied the investigative skills they had learned in higher education to the environmental scene and contributed their own bit to the opening up of scientific and technical choice. The expansion of knowledge often went far beyond the capability of individuals or institutions to absorb and apply it. Often those who made assessments about such matters did so on the basis of selective analyses made by others rather than through a review of original literature.[129] On more than one occasion, environmentalists unclogged avenues of communication and transfer that were either sluggish or choked. The NRDC, for example, brought two frontier researchers, Sergio Piomelli and Herbert Needleman, into the scientific assessment proceedings on the first lead criteria document, and thereby worked their new findings more rapidly into decision making on lead.[130] But this was only an especially dramatic case. Among citizen environmentalists generally, there occurred an extensive search for the latest scientific data so that it could be used to bolster their case.[131] In so doing they considerably increased the level and speed at which information was transferred.

Administrative environmental politics involved a still broader realm beyond the immediate interface of management, the regulated, and environmentalists—the relationships between administrators and the general public. Over the years, these two sectors became estranged. On the one hand, there was persistent public pessimism about the degree of commitment of management to environmental progress; on the other, there was persistent management distrust of the ability of the public to comprehend wisely and act reasonably on environmental issues. To the public, environmental administrators moved slowly, temporized continually, followed the wishes of the regulated too frequently, and were generally characterized by lethargy and inaction. To administrators, the public became too emotional and excited about environmental concerns and was incapable of participating effectively in decisions about technical matters. Much of the blame, they argued, fell on the news media, which simply fanned the flames of public fear.[132] This mutual distrust reached new heights amid the toxic chemical issues of the late 1970s and the early 1980s.[133]

Much of the distrust arose from the persistent pressure on administrators by the environmental public. Resource agencies retained their older primary developmental missions, which were sustained not only by administrative loyalties but also by personal and professional commitments to commodity management; they found themselves constrained by these loyalties from playing a lead role in advancing values expressed by the environmental public.[134] By the 1970s, agencies managing land and water resources seemed to be trapped in persistent tension with public environmental demands. There was similar mutual distrust with respect to health and ecological objectives. As the public demanded that health frontiers be advanced to include new problems of genetic defects, reproductive effects, immune capabilities, and physiological and neurological disorders, as well as cancer, managerial agencies felt pressed beyond their capabilities and defended more limited action. Frequently they were tempted to affirm with confidence that a supposed threat presented little harm when the public was convinced that the unknowns and even the plausible unknowns could not justify such statements.[135] The public remained suspicious of talk and action that tended to diminish the importance or wisdom of higher health and environmental aspiration.

By the beginning of the 1980s, attitude studies had begun to reveal this vast difference in attitude between managerial leaders and the environmental public. Research was designed to compare public attitudes about environmental affairs with the attitudes of public leaders. One study conducted by the American Forest Institute found that the public was far more inclined to believe that trees in the national forests should be preserved rather than cut, in contrast to the views of "thought leaders" in Washington. The survey firm urged its clients to direct their public relations campaigns to those leaders, because they, so the firm argued, perceived the issue "more rationally and with greater expertise."[136] The EPA began to define its relationship to the public as one of public relations, rather than substance. The public was emotional, tended to accept "bad science," and was subject to erratic media influence. The agency's main task was to restore public confidence by persuading the public to change its views rather than by greater managerial effectiveness in response to public definition of problems. The EPA strategy did not satisfy the public. The image evolved and persisted: a public that sought more environmental progress and an agency that held back.

Environmental management came to exercise an influence that was restraining, more than formative, in the advance of environmental objectives. Those objectives, rooted in changing public values, had a role in public affairs largely because the public mobilized to influence the apparatus of government. On occasion, leading sector initiatives came from management, but these were few. The prevailing pattern was one of public thrust and managerial reaction, of public impulses pushing managers further and faster than they thought wise. Hence, agencies acted and reacted somewhat reluctantly, defensively, seeking to parry and hold off rather than to serve as vigorous agents of environmental change. The Forest Service, for example, did not seek out ways to establish a lead role in innovation in wilderness expansion; the National Park Service did not follow up on early efforts to reduce the pressures on parks of growing numbers of visitors; the EPA did not extend pollution control readily into new realms of toxic air contaminants and acid deposition. The incorporation of environmental impulses into management constituted a system of decision making and control that disciplined and retarded as much as it implemented, that held back public demands as much as it carried them out.

For environmentalists, management provided both luring opportunities and significant risks. If one established a system of environmental planning that relied heavily on managerial strategy, it could be used for either environmental or developmental purposes. Federal air quality standards could be used as a weapon either to prevent further action or to force advances.[137] Forest planning could be used to increase allowable cuts and endanger environmental forest values as well as to enhance them.[138] Coastal zone management could enhance identification of areas of critical environmental concern or focus on commodity development under the rubric of multiple use.[139] The requirement that criteria documents be updated regularly could support either action or inaction, depending upon whether the agency felt sufficiently strong politically in its relationship with the regulated industry.[140] Despite the fact that management could be used against environmental objectives, there seemed to be little way out; hence, environmentalists felt just as constrained as their opponents to accept the administrative machinery and to seek to influence it for their own ends.[141] The vast importance of administrative choice in the national drama of environmental affairs was therefore solidified.

During the later years of the 1970s, several environmental institutions sought to serve as a middle ground in the debate between environmentalists and developmentalists. These organizations did not directly represent citizens but were rooted in professional and technical institutions somewhat separate from the process of mobilizing and expressing public values. Foremost among them were the Conservation Foundation and the Environmental Law Institute. They devoted considerable time and effort to negotiating amid the conflicting sectors of administrative decision making. The Conservation Foundation stressed what came to be called environmental mediation, the intervention of third parties into disputes to seek out a middle ground of compromise.[142] The Environmental Law Institute veered more toward a context of legal action, in which all parties to disputes would be dealt with evenhandedly.[143] In these strategies, both organizations deliberately sought not to identify themselves with the cutting edge of environmental impulses, but to occupy a middle ground and hence to urge that environmental objectives be restrained sufficiently to meet similar initiatives for restraint on the developmental side. Their actions took them out of the mainstream of environmental action and identified them as brokers among competing interests. In their emphasis on independence and professionalism, they sought—not always successfully—to rise above the fray. They found common ground with still another "middle ground" force, the technical experts with a self-image of independence rather than advocacy. By 1980, this group of experts had formed the National Association of Environmental Professionals.[144]

The middle ground that these groups sought to establish was distinctive. Rather than expressing agreement with frontier actions to advance natural values, the frontiers of science about health and environmental effects, or advances in process technology, they tended to convey skepticism about pushing too fast and too far. They sought to work within the managerial context in which considerable skepticism was expressed about too-rapid environmental progress. They tended to share the views of environmental managers that such progress was either limited in potential, too costly, or unnecessary. The middle ground that these efforts sought to capture drew them away from citizen environmental action and into the orbit of managerial thinking. Hence, their strategies were often discounted by those seeking to advance environmental progress in leading-edge fashion as too closely associated with efforts to constrain environmental impulses.[145]

THE RESPONSE OF GOVERNING INSTITUTIONS

The environmental drive inevitably became intertwined with two sets of governing institutions. One can be described by the term "separation of powers" and the other by the term "federalism." We should extend each of these terms beyond its usual meaning. The varied "powers" that were separated involved not just the legislative, administrative, and judicial branches, but also a differentiation within the administrative branch between the agencies and the Executive Office of the President. In effect, there were four branches of the federal government. "Federalism" involved not just relationships between state and nation but the entire hierarchy of local, state, and national institutions. Here was a varied lot of independent and autonomous institutions of public decision making that provided diverse opportunities for political expression.

Those seeking to influence government find one institution more responsive to their demands and another less so; hence, they pick and choose where to call for action, and patterns relating particular impulses and particular institutions evolve. At the federal level, environmentalists found Congress and the courts more willing to reflect their views, and the agencies and the Executive Office of the President more reluctant. The initial step in environmental action was to work through Congress to establish policies that administrative agencies by themselves resisted. The drive for wilderness set this pattern, and it was repeated on many an occasion. Congress provided the arena in which public demands were expressed. Here environmentalists found leverage as legislators responded to their constituents, and a host of environmental laws resulted.

The courts were equally responsive, not in terms of offering policy choices but in establishing the legitimacy of the new public interest in environmental affairs. Courts approved of standing to defend an environmental interest as being of equal importance as the defense of person and property. They accepted environmental demands as a legitimate part of administrative choice, and in that spirit, readily took up the environmental impact statement. The courts emphasized procedure rather than substance. Their role was to decide whether environmental impulses were legitimate demands on government, not that they should win out in competition with other demands. In much the same manner, courts accepted environmental objectives as a legitimate use of the state's police powers, though a balance had to be struck with other objectives.

The administrative agencies and the Executive Office of the President sought to exercise restraint on environmental demands. The flow of ideas and actions that came to public life through these branches of government tended to arise not from broadly based public values but from the more limited concerns of technical, professional, and administrative cultures. That leadership remained skeptical about the wisdom of the environmental impulse. When environmentalists found the agencies to be open procedurally to them, their opposition began to view the Executive Office of the President as a more useful source of environmental restraint. The opposition obtained significant leverage at that level during the last two years of the Carter administration, and far more with the advent of the presidency of Ronald Reagan. The Reagan strategy was to bring the agencies under firmer control; and those who advocated a more powerful executive came to their views, in part, in opposition to the influence of environmental values in public affairs. No matter that to the "middle ground" environmental professionals the Reaganites went much too far and invited backlash; such discomfort did not entice them to embrace the innovative role of either Congress or the courts, but only to reform the executive strategy of restraint.[146]

Federalism provided opportunities either to enhance or to restrain authority at one level when action at another was not favorable. Many issues involved a choice to use either state or national action in cases that concerned the preemption of state powers by federal authority. There were also options as to state or local action, described by the word "override." Many state statutes provided that local governments could not establish legal conditions more stringent than those of the state.[147] The constitutional setting in this case was somewhat different. Whereas the division between state and federal authority required a sorting out of constitutional powers, the division between state and local authority did not. Local governments were created by state governments, and the two did not hold ultimate authority concurrently. For all practical purposes, however, this contest for authority was as intense as that between state and national levels. Many land use questions were framed by the question of whether the state should "recapture" powers it had given to municipalities through zoning legislation.

Environmental action often consisted of a defense of one's home and community against some developmental intrusion. Hence, that action often appealed to the right of localities to make decisions about their environmental quality. Perhaps it was a matter of siting an unwanted industrial plant or a waste disposal facility. A variety of strategies, from zoning to local referenda, were devised. To circumvent such local opposition, developers in turn sought an "override" from state government in which state agencies could make decisions that localities would be required to accept. Amid local protest, however, state agencies were reluctant to act. As the environmental era proceeded, an accumulation of such cases led to hesitation by state governments to force undesired siting into communities against their wishes. Disposal of hazardous and radioactive wastes sharply fixed the issue. Even though most legislatures had provided that in hazardous waste siting decisions, communities could advise but not veto, state agencies still hesitated to force decisions against overwhelming opposition.

In response to such state lethargy, developers sought out the power of the federal government to preempt the power of the states. Environmental impulses arising from the grass roots had gone too far and corrupted state policy. Only strong federal preemption would work. Such authority had arisen in certain cases, such as the federal noise control program in which railroads and airlines had sought federal regulations to prohibit communities from establishing noise control levels that they felt were too strict.[148] There were many other cases: local and state pesticide regulations stricter than federal ones;[149] action by communities to restrict the transport of radioactive waste through their jurisdictions;[150] actions to prevent states and communities from passing "right-to-know" laws stricter than federal ones;[151] federal action to prevent states from restricting siting of energy facilities on the coastal zone.[152] This repeated drive by industry to use federal power to override that of the states and localities was greatly aided by the new Reagan administration.

The public lands presented a different case. Federal lands inevitably created a contest over the use of federal power. Over the years, conservationists and environmentalists sought to enhance federal authority as friendly to their public land interest. Commodity users, on the other hand, had always relied on state authority, on occasion had fostered strategies to transfer federal lands to the states, and had upheld land management in the western states as an appropriate model. Such political positions underwent some modification in the environmental era. Throughout the West, environmental impulses steadily grew to exercise influence in public land management at the local and state levels. Support for environmental public land uses increased among state recreational and fish and wildlife agencies and in executive departments of state governments. There were indigenous demands for wilderness, against clearcutting, for visual corridors, for protection of streams against sedimentation, and for restrictions in the use of herbicides and pesticides.[153] As these demands began to restrain wood production in first one and then another national forest, sentiment began to emerge among those on both the private and public sides of wood production objectives for an upward flow of authority within the U.S. Forest Service. This appeared in the demand during the Reagan administration for marked increases in the allowable cut through Washington-based decisions. This, in turn, generated a vigorous opposition within the West and fostered a state environmental stance against national authority.[154]

There were many variations on these themes. Environmentalists usually sought federal action to further natural-environment objectives for managing air, water, and land, but at the same time they developed extensive pragmatic strategies when they found that federal planning could be used against them and that local and state action could be used in their behalf. Developmentalists continued to rely on local and state action when it could be associated with issues of jobs and economic development that had local support. That strategy, however, now had to compete in many places with local interest in environmental quality, and this fact tended to push developmentalists toward higher levels of authority. Hence, amid the complexities of federalism, one could discern a tendency for the consumer-oriented stance of environmental expression to emphasize local choice and the producer-oriented stance of large-scale production to emphasize national choice.

These alternatives in the use of governmental institutions were closely associated with the politics of technical information. The context of environmental decision making was heavily laden with the detail of science and technology, of future prediction and past record, of economic analysis, of finely tuned administrative mechanisms. Political give-and-take involved the ability to command, communicate, and apply detail to decisions. Hence, the ability to carry out effective information strategy played a critical role in political success or failure. Much of this depended on the resources available to environmentalists or developmentalists to command technical detail through their own experts. Much of it also depended on the ability to command government resources, either to turn technical knowledge to one's advantage or to prevent critical information from becoming available to one's opponents. It was their capability in these matters that often gave the environmental opposition a considerable edge. That opposition often sought government as an ally in restricting the flow of information to the environmental public. Hence, the politics of information became closely intertwined with the varied roles of governmental institutions.

Several cases illustrate the problem. In the 1960s, the U.S. Public Health Service (PHS) sought to obtain an inventory of industrial waste water as complete as the inventory already available for municipal sewage. During various river basin water quality conferences, industries had declined to provide this information. State public health authorities defended their secrecy, since what information they had obtained had been secured on the condition of confidentiality.[155] The PHS drew up a survey form and sought to submit it to industrial sources to acquire the information. The survey had to be cleared by the Bureau of the Budget (BOB) on the grounds that it involved public expenditures. But the BOB committee that had to approve the survey included several industry representatives, so it refused to approve the action. It took a congressional subcommittee report to bring the issue to a head, and by the time of the debate over revision of water quality legislation in 1972, a mechanism for requiring such information had been worked out.[156] In this case federal action was used to open up information that had been closed off at the state level.

There was also the case of the Lead Liaison Committee established by the Public Health Service in the late 1950s to provide continuing discussion with industry over the health effects of lead. The mechanism served as an opportunity for scientific and technical experts from both industry and government to share their views and work out conclusions about such matters among themselves. The meetings were not open to the public. By the early 1970s, however, in a new era of openness in government, the EPA decided that the committee's proceedings would be made public. In return the lead industry withdrew from the committee, explaining in later years that the publicity reduced the committee's effectiveness.[157] The regulation of lead involved intensely debated scientific and technical issues. The shift from private to public debate over these issues constituted a major change in the relationship between government, on the one hand, and environmentalists and industry, on the other. From that point on the inability to control the flow and assessment of information was the lead industry's most crucial weakness.[158]

The courts constituted an equally significant arena in the politics of information. The judiciary valued a "full record," both for its own decisions and in supervising decision making by federal agencies. The judicial record could be enhanced by the process of discovery, in which one litigant could obtain evidence from the files of its opponent, or by the court proceeding itself, in which information from documents and witnesses could be placed on the record. Many a proceeding hinged on the issue of whether the parties wanted to risk creating a record that in future actions as well as the one in question might be detrimental to their interests. Discovery greatly fostered the claims of injury by persons exposed to asbestos, for example, for it brought to light that industry had known about carcinogenic effects much earlier than had been thought before.[159] To forestall such a record of environmental harm was a prime consideration in environmental litigation, and led to many an out-of-court settlement in which the defendant refused to accept liability and prevented "record building" by agreeing to make payments to those who claimed injury.[160]

The courts insisted on openness in administrative proceedings as well. Decisions had to be made on the record that would be open to all interested parties, including legislators, other administrators, and the public, and no decision could be made on the basis of information obtained after the record had been formally closed. What of the tendency, however, for administrative decisions to be made not in the agency but in the Executive Office of the President, and especially the Office of Management and Budget? It was at this point that the court hesitated. Did the president have the authority to make decisions privately or not? The shift invited the use of the Executive Office of the President as a context of influence by the environmental opposition that did not have to be publicly exposed.[161] When the Reagan administration sought to reduce openness at the agency level through private meetings between the regulators and the regulated, the strategy backfired, and the most manifest procedural norms the courts had evolved were restored.[162] But this only enhanced the role of the Executive Office of the President as the agency of unpublicized political choice.

Within the realm of federalism, the most striking fact was the enormous variation in technical and information capabilities between the various levels of government. The air pollution program in the 1960s focused squarely on the problem. Given the superior resources that enabled industry to bring experts to bear on decision making, local governments seeking increased control of air pollution had little clout. They tried to enhance the technical role of the Public Health Service to increase their leverage. Congress provided funds for community air quality programs to build up local and state management capabilities, and fostered the criteria document assessment strategy to provide authoritative statements about environmental effects that localities and states could use. Environmentalists used the first draft criteria document on sulfur oxides in a number of local and state proceedings in the late 1960s; this action led to stricter standards than industry wanted. Industry, in turn, looked on the use of federal expertise as an "interference" in local affairs. Nevertheless, the use of nationally mobilized scientific and technical opinion remained highly influential in state and local action. Only a few states, such as California, developed a scientific and technical capability that sustained even stronger pollution control strategies than those maintained by the federal government. Most states simply followed the national lead.

Environmental implementation was a different matter. In this area the technical ability of industry often far outweighed that of state and local agencies. In issues of air quality emissions and water quality discharges, both involving the impacts of individual sources on ambient quality, modeling became the basis of regulatory choice. The industry would make its case with supporting technical detail and then the regulatory agency would not have sufficient time or resources to counter it.[163] Often such resources hinged on federal grants to the states, and when such funds declined, so did the states' ability to compete on a technical basis. During the 1960s and 1970s, federal funds considerably strengthened the capabilities of state governments, but the Reagan administration's strategy of transferring costs to the states set in motion an opposite tendency that weakened the role of state government. At times, states sought to shift the cost of environmental management to the regulated industries, but this strategy had little success. In a host of managerial actions ranging from monitoring to modeling, technical limitations of state and local environmental authorities reduced their effectiveness in the face of the far superior resources of the regulated industries.

One of the more significant features of the politics of information concerned the issue of the right of the public or public authorities to information about the practices of those responsible for harm. Public action depended on information, but private sources argued that the information demanded was confidential, proprietary, a trade secret. Such a political strategy stymied many an environmental action. Claims ranged from the identity of chemicals subject to toxic substances regulation,[164] to packaging (subject to scrutiny to determine whether resulting waste could be reduced), to the potential toxic properties of strip mine overburden. The stakes were focused especially by information about exposures and health effects in the workplace. The Occupational Safety and Health Administration promulgated regulations requiring that health and exposure records of workers be available for epidemiological study, but industry countered that such information was privileged and defended the "right" of workers to confidentiality. Then there was the similar protest by industry against the spate of right-to-know proposals concerning workers and communities exposed to chemicals.[165] Every case involved control of information that might be used either to protect industry against legal action or to aid those affected who sought protection. Workers initially looked for help at the federal level when they tried to exercise their "right to know," but industry blocked their efforts. When protective action shifted to states and localities, with some success, industry sought a federal override from a sympathetic Reagan administration. Although this case has had more than its share of political drama, it should not obscure the fact that a host of choices about disclosure of information became closely intertwined with choices about the level of government at which information control strategies were worked out.

VALUES AND LIMITS

Throughout these twists and turns of environmental politics one can observe two broad forces at work. One is the new set of values that emerged in the years after World War II, deeply rooted in changing demography, standards of health and living, and enhanced levels of human aspiration. Environmental politics has involved the working out of these historic changes in what people sought to think, be, and do. The other force has involved the private and public apparatus that constitutes the organizational society, the way in which some people fashion managerial institutions to shape the social and political order according to their views of desired arrangements. From the public came demands for environmental improvement and progress. From the managerial world of science, technology, economic analysis, and planning came the message that that aspiration was on the point of outrunning resources. It was ironic that in creating such doubts about environmental demands, the nation's institutional leaders were giving stark testimony to the "limits of growth" that many of them often denied. The world, they argued, simply could not be made as healthful or clean or safe or filled with natural beauty as the environmental public seemed to want. One would have to settle for less.

It was a curious twist that environmentalists were the conveyors of optimism about the possibilities of human achievement, while the administrative and technical leadership were consistently the bearers of bad news. In the media the roles were reversed: environmentalists warned of impending catastrophe, while the technical leadership exuded optimism. Such language was used even by the parties to the debate. To remain, however, on this level of media understanding seems not to fit the wide range of evidence about what people did and the values implicit in their actions. The driving public force behind environmental affairs was rooted in hope and confidence about possibilities for a better life; there was a constant search for the very latest in science and technology to harness to such aspiration. The opposition, which in the grand debates over "limits" radiated confidence about the future, was constantly warning environmentalists that their demands were extravagant and beyond the capability of existing resources.[166]

Such discrepancies between what people do and how they explain what they do are not unique; they are the stock-in-trade of sensitive historical analysis concerned with self-images as well as actions. So we need not pause to unravel inconsistencies inherent in daily human life. What is particularly interesting is the way in which the new mass-middle class sought to shape a newer world revolving around its values and conceptions about the good life, and in so doing gave expression to precisely the rising standard of living that the managerial and technological leaders professed to extol. Like Karl Marx's proletariat, the successful middle class turned out to be different from what its creators sought, and began to demand that the production system generate environmental goods and services for themselves and their children. One could go even a step further and point out that it was Adam Smith who said that the main purpose of production was consumption, and that when consumer wants were no longer filled by the existing mode of production, it was time for a change. Such was the challenge of the environmental impulse to America's prevailing managerial institutions.[167]

NOTES

1. William Tucker, *Progress and Privilege* (New York: Doubleday, 1982); Ron Arnold, *At the Eye of the Storm* (Chicago: Regnery Gateway, 1982).
2. Mary Douglas and Aaron Wildavsky, *Risk and Culture* (Berkeley and Los Angeles: University of California Press, 1982).
3. Newsletters I have found particularly useful are *Environment Reporter; Chemical*

Regulation Reporter; National Wetlands Newsletter; Land Letter; Coastal Zone Management; Occupational Health and Safety Letter; Environmental Health Letter; Inside EPA; Land Use Planning Report; Public Land News; Weekly Bulletin of the Environmental Study Institute.

4. John Naisbitt, *Megatrends* (New York: Warner Books, 1982).
5. Examples of such studies are Stephen R. Kellert, *American Attitudes, Knowledge and Behaviors Toward Wildlife and Natural Habitats* (Washington, D.C.: U.S. Fish and Wildlife Service, 1978–1980); Opinion Research Corporation, *The Public's Participation in Outdoor Activities and Attitudes Toward National Wilderness Areas* (Princeton, N.J., 1977); The Continental Group, *Toward Responsible Growth: Economic and Environmental Concern in the Balance* (Stamford, Conn., 1982).
6. Some state sources are the *Maine Times;* the *New Hampshire Times,* the *Deseret News* (Salt Lake City); the *Missoulian* (Missoula, Montana); *New York Environmental News*; the *North Woods Call* (Michigan); *Maine Environment; ENFO* (Florida); the *Plains Truth* (Northern Plains Resource Council); *Newsletter*, Tennessee Citizens for Wilderness Planning; *Our Wetlands* (Wisconsin); *Crossroads Monitor* (Wyoming); *Newsletter*, Idaho Environmental Council.
7. A useful, and contrasting, example of a policy approach is Norman J. Vig and Michael E. Kraft, *Environmental Policy in the 1980s: Reagan's New Agenda* (Washington, D.C.: CQ Press, 1984).
8. While the materials of popular debate often cast the environmental movement as negative, with a major focus on opposition to modern values, science, and technology, an approach followed by Tucker, Arnold, and Douglas and Wildavsky, this paper takes a quite different tack to identify the movement as an outgrowth of positive aspiration associated with an advanced industrial society, including advanced applications of science and technology. Douglas and Wildavsky briefly consider, but reject, this approach. See their *Risk and Culture*, 1–15. Their treatment seems to be strikingly devoid of empirical observation about the values of the advanced industrial society and environmental behavior. The values described here were characteristic of advanced industrial societies throughout the world. See Ronald Inglehart, *The Silent Revolution: Changing Values and Political Styles Among Western Publics* (Princeton, N.J.: Princeton University Press, 1977).
9. Carolyn Merchant, "Women of the Progressive Conservation Movement, 1900–1916," *Environmental Review*, 8, no. 1 (Spring 1984): 57–85.
10. For a more extended treatment, see Samuel P. Hays, *Conservation and the Gospel of Efficiency* (Cambridge, Mass.: Harvard University Press, 1958).
11. The classic work on the Tennessee Valley Authority is David Lilienthal, *TVA, Democracy on the March* (New York: Harper, 1944).
12. For Pinchot's own view, see Gifford Pinchot, *Breaking New Ground* (New York: Americana Library, 1947; Seattle: University of Washington Press, 1983).
13. D. Harper Simms, *The Soil Conservation Service* (New York: Praeger, 1970).
14. John A. Salmond, *The Civilian Conservation Corps, 1933–1942* (Durham, N.C.: Duke University Press, 1967).
15. It might well be argued that many aspects of the environmental movement, especially with respect to land and water resources, were a reaction against the extensive developmental projects of the New Deal. The wilderness movement, for example, was fueled by the rapid advance of roads constructed by the Civilian Conservation Corps into wilderness candidate areas.
16. For environmental river values, see *American Rivers*, quarterly publication of the American Rivers Conservation Council (Washington, D.C., 1973–); for river development views, see *Waterways Journal*, published in St. Louis, Mo., which covers events from the viewpoint of the eastern inland and coastal navigation industry.
17. Publications of the Wilderness Society, *The Living Wilderness*, and the Sierra Club, *Sierra Bulletin* (later *Sierra*), are useful in charting the evolution of environmental wildlands values.

18. The issue came to a dramatic head with channelization. See House Committee on Government Operations, Conservation and Natural Resources Subcommittee, *Stream Channelization* (4 parts), 92d Cong., 1st sess. (Washington, D.C., 1971). Wetlands issues can be followed in the *National Wetlands Newsletter* published by the Environmental Law Institute, Washington, D.C.; a useful, up-to-date report is U.S. Congress, Office of Technology Assessment, *Wetlands: Their Use and Regulation* (Washington, D.C., 1984).
19. The historical context of such issues can be followed in Martin V. Melosi, *Pollution and Reform in American Cities, 1870–1930*, (Austin, Tex.: University of Texas Press, 1980).
20. The Fish and Wildlife Service has received little focused attention from environmentalists or from academic writers. Hence, its role in the 1960s has been greatly underestimated. Some pieces of the puzzle are dealt with in Joseph V. Siry, *Marshes of the Ocean Shore: Development of an Ecological Ethic* (College Station, Tex.: Texas A&M University Press, 1984).
21. See, for example, Robert C. Mitchell, "The Public Speaks Again: A New Environmental Survey," *Resources* (Resources for the Future), no. 60 (Sept.-Nov. 1978); Council on Environmental Quality et al., *Public Opinion on Environmental Issues* (Washington, D.C., 1980); for later data see Robert C. Mitchell, "Public Opinion and Environmental Politics in the 1970s and 1980s," in Vig and Kraft, *Environmental Policy in the 1980s*. For a review of studies of environmental values, see Kent D. VanLiere and Riley E. Dunlap, "The Social Bases of Environmental Concern: A Review of Hypotheses, Explanations and Empirical Evidence," *Public Opinion Quarterly* 44, no. 2 (Summer 1980): 181–97.
22. For a survey of wilderness users, see John C. Hendee, George H. Stankey, and Robert C. Lucas, "Wilderness Use and Users: Trends and Projections," in their *Wilderness Management*, U.S. Department of Agriculture, Forest Service Miscellaneous Publication No. 1365 (Washington, D.C., 1978). See also George H. Stankey, "Myths in Wilderness Decision Making," *Journal of Soil and Water Conservation*, Sept.-Oct. 1971.
23. See, for example, Lawrence S. Hamilton and Terry Rader, "Suburban Forest Owners' Goals and Attitudes Toward Forest Practices," *Northern Logger and Timber Processor*, July 1974, 18–19. This is typical of a vast amount of literature.
24. "In general, evidence . . . provides very weak support for the assertion that social class is positively associated with environmental concern. What support there is rests primarily on the moderately strong relationship between environmental concern and education. The evidence for occupational prestige provides very weak support at best, while the overall evidence for income is highly ambiguous." See VanLiere and Dunlap, "The Social Bases of Environmental Concern," 190.
25. See, for example, the magazine *Islands* (1981–), Santa Barbara, Calif.
26. It is relevant to note the very rapid growth of visits to the national forests (from 7,132,000 to 14,332,000) and to the national parks (from 3,248,000 to 15,531,000) during the depression years from 1929 to 1939 to identify the important role of mass, rather than elite, consumption, or to contrast the enormous difficulties of the mass recreation movement in Europe in securing outdoor recreation opportunities amid the traditional exclusionary policies of private estate owners, in contrast with the openness of the American wildlands. For the data, see U.S. Dept. of Commerce, Bureau of the Census, *Historical Statistics of the United States* (Washington, D.C., 1974), Series H808: R104.
27. Significant public land purchases throughout the East, such as those in the Adirondacks, came through acquisition of former private estates or hunting grounds.
28. For a brief summary of Briggs's data, see "Amenity-Rich, Amenity-Poor," *Demographics* 3, no. 7 (July-Aug. 1981): 9.
29. For regional variations in attitude studies, see items in n. 5.
30. The Continental Group, *Toward Responsible Growth*, 57, 77.

31. For a general review of wilderness politics see Craig W. Allin, *The Politics of Wilderness Preservation* (Westport, Conn.: Greenwood Press, 1982).
32. The classic work on urban land use is William K. Reilly, ed., *The Use of Land* (New York: Crowell, 1973). For urban forestry, see Silas Little, ed., *Urban Foresters Notebook* (Northeast Forest Experiment Station, 1978), Forest Service General Technical Report NE–49. For a recent review of waterfront activities, see Patrick Barry, "The Last Urban Frontier," *Environmental Action* 15, no. 9 (May 1984): 14–17; see also Heritage Conservation and Recreation Service, U.S. Department of the Interior, *Urban Waterfront Revitalization; the Role of Recreation and Heritage* (Washington, D.C., 1979). For urban wildlife, see *Urban Wildlife News*, published for several years after 1977 by the Urban Wildlife Research Center, Ellicott City, Md.
33. See, for example, *Natural Areas Journal* (1981–), published by the Natural Areas Association. A wide range of these resources are described in various articles in *Nature Conservancy News*, published by the Nature Conservancy.
34. For PSD, see Thomas M. Disselhorst, "Sierra Club v. Ruckelshaus: 'On a Clear Day. . . . ' ", *Ecology Law Quarterly* 4, no. 3 (1975): 739–80.
35. Rudolph Husar of Washington University, St. Louis, refocused attention sharply on visibility by associating reduced visibility with sulfate particles and by using airport visibility data reaching back to the 1930s to identify trends and patterns in regional variations. The issue was closely associated with that of acid precipitation. Husar's then-recent work was included in the National Research Council report *Sulfur Oxides*, published by the National Academy of Sciences in 1978; see pp. 29–37.
36. Wildlife Management Institute, "The North American Wildlife Policy, 1973," which includes a copy of the "American Game Policy, 1930" (Washington, D.C., n.d.).
37. In 1975, for example, 21 million Americans participated in hunting and 50 million in wildlife observation.
38. U.S. Fish and Wildlife Service, *Operation of the National Wildlife Refuge System, Final Environmental Impact Statement* (Washington, D.C., 1976); Stephen R. Kellert, *American Attitudes, Knowledge and Behaviors.*
39. The National Wildlife Federation, Sierra Club, Wilderness Society, Friends of the Earth, National Parks and Conservation Association, and Defenders of Wildlife all had firm roots in natural environment issues. By 1984 there were state wilderness organizations throughout the West, and the Nature Conservancy was organized in thirty-five states. By that time wilderness politics had come to focus on separate state bills rather than national bills, reflecting the position of strength within each state from which wilderness advocates were then negotiating.
40. The Wilderness Society increased in membership from 50,000 to 100,000 between 1978 and 1983, the largest rate of growth in its history; the Nature Conservancy had 36,000 members in 1976 and 193,000 in 1983. See the relevant annual reports.
41. For such concepts see, for example, two textbooks, Kenneth E. F. Watt, *Understanding the Environment* (Boston: Allyn and Bacon, 1982); and Penelope ReVelle and Charles ReVelle, *The Environment: Issues and Choices for Society* (Boston: Willard Grant Press, 1981). They indicate the way in which new ecological ideas had come into general thinking.
42. For relevant events, see National Wildlife Federation, *Conservation Report*, 1966, May 27 (geothermal steam leases); July 28 (pollution of estuaries); Sept. 30 (thermal pollution of waters); and Oct. 7 (highway construction).
43. The first serious proposal in the legislative history of the Environmental Impact Statement (EIS) came from Rep. John Dingell of the Committee on Merchant Marine and Fisheries on March 23, 1967. The bill reflected the committee's concern for the impact of development on fish and wildlife. This and similar

measures frequently took the form of amendments to the Fish and Wildlife Coordination Act.

44. Senate Public Works Committee, *Thermal Pollution*, 90th Cong., 2d sess., 1968, 1969. Hearings on the extent to which environmental factors are considered in selecting power plant sites, with particular emphasis on ecological effects of discharge of waste heat into rivers, lakes, estuaries, and coastal waters.
45. See, for example, Robert L. Rudd, *Pesticides and the Living Landscape* (Madison, Wis.: University of Wisconsin Press, 1964); Thomas R. Dunlap, *DDT: Scientists, Citizens and Public Policy* (Princeton, N.J.: Princeton University Press, 1981).
46. For changes in the approach of the Conservation Foundation, see its annual reports, which chart a transition from its earlier origins in natural history to its later concerns for land use, environmental mediation, and economic limitations to environmental objectives, among others.
47. A number of bills proposed in 1969 anticipated a comprehensive program for "research on natural systems"; see H.R. 952 proposed by Rep. Charles Bennett of Florida; H.R. 7923 by Rep. James Howard of New Jersey; H.R. 12,900 by Rep. John Saylor of Pennsylvania; and S. 1075 by Senators Henry Jackson and Ted Stevens. These bills spoke of research on "ecological systems, natural resources, and environmental quality."
48. U.S. Department of the Interior, Fish and Wildlife Service, *Biological Services Program* (Fiscal Year 1975–).
49. U.S. Department of Health, Education and Welfare, Public Health Service, *Air Quality Criteria for Sulfur Oxides* (Washington, D.C., 1967), Public Health Service Publication No. 1619.
50. For the broad range of relevant effects see EPA, Office of Research and Development, *The Acidic Deposition Phenomenon and Its Effects: Critical Assessment Review Papers, Public Review Draft* (Washington, D.C., May 1983). For a collection of papers concerning ecological effects, see Frank M. D'Itri, *Acid Precipitation: Effects on Ecological Systems* (Ann Arbor, Mich.: Ann Arbor Science, 1982). An excellent and brief review of effects from a European perspective is Environmental Resources Limited, *Acid Rain: A Review of the Phenomenon in the EEC and Europe* (London, 1983).
51. One of the first major environmental statements on energy appeared in Gerald O. Barney, ed., *The Unfinished Agenda* (New York: Crowell, 1977), 50–68. An even earlier, though brief, statement is The Georgia Conservancy, *The Wolfcreek Statement: Toward a Sustainable Energy Society* (Atlanta, Ga., 1976).
52. One of the two groups to emerge from the organization shaping Earth Day in 1970 was Zero Population Growth, which thereafter worked exclusively on population problems. See its publication, *National Reporter* (1969–).
53. See Samuel P. Hays, "The Limits-to-Growth Issue: An Historical Perspective," in *Growth in America*, ed. Chester L. Cooper (Westport, Conn.: Greenwood Press, 1976).
54. The Commission on Population Growth and the American Future, *Population and the American Future* (Washington, D.C.: U.S. Government Printing Office, 1972).
55. Council on Environmental Quality and U.S. Department of State, *The Global 2000 Report to the President*, 3 vols. (Gland, Switzerland: International Union for Conservation of Nature and Natural Resources, 1980); World Wildlife Fund, *World Conservation Strategy* (n.p., 1980). The Year 2000 Committee, established to implement the World Conservation Strategy in the United States, was confined to prominent institutional leaders. The Global Tomorrow Coalition was a coalition of citizen environmental groups; however, its work was not widely publicized, known about, or reflected in grassroots citizen environmental activity. For the work of the coalition, see its publication, *Interaction* (Washington, D.C., 1981–).
56. For the American Land Forum, see its publication, *American Land Forum Magazine* (1980–); also *American Farm Land*, newsletter of the American Farmland Trust

(1981–). See also W. Wendell Fletcher and Charles E. Little, *The American Cropland Crisis* (Bethesda, Md.: American Land Forum, 1982).

57. See, for example, Kirkpatrick Sale, *Human Scale* (New York: Coward, McCann, Geoghegan, 1982).
58. Subscribers to *Mother Earth News*, a publication that reflected these views, were about evenly divided among the central city, the suburbs, and the countryside. For self-help ideas, see John Lobell, *The Little Green Book: A Guide to Self-Reliant Living in the 80s* (Boulder, Colo.: Shambhala Publications, 1981).
59. For information about natural food stores, see *Whole Foods*, "the natural foods business journal" (Irvine, Calif., 1977–). This publisher also compiles an annual *Source Directory*. For a convenient collection of articles from its pages, see *Whole Food Natural Foods Guide* (Berkeley, Calif.: And/Or Press, 1979); see especially "First Annual Report on the Industry" in the *Guide*, 268–74, which concluded that in 1978 there were 6,400 natural food stores in the nation, with sales of $1,152,000,000.
60. The issue of protection against spray drift arose in many forest areas. See, for example, the Oregon case in Carol Van Strum, *A Bitter Fog: Herbicides and Human Rights* (San Francisco: Sierra Club Books, 1983).
61. Alan Okagaki, Albert J. Fritsch, and C. J. Swet, *Solar Energy: One Way to Citizen Control* (Washington, D.C.: Center for Science in the Public Interest, 1976).
62. Relevant sources include Michael Brown, *Laying Waste* (New York: Washington Square Press, 1979); Edwin Chen, *PBB: An American Tragedy* (Englewood Cliffs, N.J.: Prentice-Hall, 1979); Joyce Egginton, *The Poisoning of Michigan* (New York: Norton, 1980); Adeline Gordon Levine, *Love Canal: Science, Politics and People* (Lexington, Mass.: Lexington Books, 1982); Samuel S. Epstein, Lester O. Brown, and Carl Pope, *Hazardous Waste in America* (San Francisco: Sierra Club Books, 1982).
63. See charts in U.S. Department of Health, Education and Welfare, *Healthy People: The Surgeon General's Report on Health Promotion and Disease Prevention, 1979* (Washington, D.C., 1979), 22–23, 34–35, 44–45, 54–55, 72–73, in which improvements in health for five age groups are charted in terms of reduction in death. The report reflects many concerns for "better health" in each group, which imply notions beyond reduction in deaths, but has difficulty in translating these into clearly defined goals beyond reducing "premature death."
64. This rising public interest is reflected in the growing popularity of *Prevention* magazine. Its circulation rose from 50,000 in 1950 to 270,000 in 1960 to 2,434,017 in 1981, and now stands at over 4,000,000.
65. A persistent, but somewhat subordinated, theme in such matters was the relationship between health and natural environments. Horticultural therapy began to be used in hospitals. A study at Paoli Memorial Hospital near Philadelphia concluded that patients recovered faster if they were in rooms with views of trees and grass than if they saw brick walls. *Pittsburgh Press*, May 22, 1984. Connections were drawn between clearer skies and psychological mood; see Michael R. Cunningham, "Weather, Mood and Helping Behavior: Quasi Experiments with the Sunshine Samaritan," *Journal of Personality and Social Psychology*, 37, no. 11 (1979): 1947–56.
66. For a nontechnical account of infant malformations, see Christopher Norwood, *At Highest Risk: Environmental Hazards to Young and Unborn Children* (New York: McGraw-Hill, 1980). See also Bernard Rimland and Gerald E. Larson, "The Manpower Quality Decline: An Ecological Perspective," *Armed Forces and Society*, Autumn 1981, 21–78.
67. For EPA's "cancer principles," see Nathan J. Karch, "Explicit Criteria and Principles for Identifying Carcinogens: A Focus of the Controversy at the Environmental Protection Agency," *Analytical Studies for the U.S. Environmental Protection Agency*; vol. IIa, *Decision Making in the Environmental Protection Agency: Case Studies* (Washington, D.C.: National Research Council, Committee on Environmental

Decision Making, 1977). OSHA's attempt can be followed through the pages of *Chemical Regulation Reporter* and *Occupational Health and Safety Reporter*. See also "Industry Raps OSHA's Proposed Cancer Policy," *Chemical and Engineering News*, July 3, 1978, 14–15; "OSHA's War on Cancer: What it Means to Plastics," *Plastics World*, June-Sept. 1978.

68. See Hugh M. Pitcher, "Comments on Issues Raised in the Analysis of the Neuropsychological Effects of Low Level Lead Exposure" (Washington, D.C.: Office of Policy Analysis, U.S. Environmental Protection Agency, 1984). "In reviewing the Draft Lead Criteria Document we realized that, if substantiated and found to be causal, the cognitive effects of low level lead exposure would generate large social costs. As an indication of the size of this cost, several studies indicate that an IQ difference of one point is associated with a 1% change in lifetime earnings (corrected for education and other socioeconomic characteristics)."
69. "Probing Chemical Causes of Infertility," *Chemical Week*, Feb. 15, 1984, 26, 29.
70. For occupational health issues, see Nicholas A. Ashford, *Crisis in the Workplace: Occupational Disease and Injury* (Cambridge, Mass.: MIT Press, 1976). Events can be followed in *Occupational Health and Safety Letter* (Washington, D.C., 1970–) and *Occupational Health and Safety Reporter* (Washington, D.C., 1970–). See also Daniel M. Berman, *Death on the Job: Occupational Health and Safety Struggles in the United States* (New York: Monthly Review Press, 1979).
71. "Of tens of thousands of commercially important chemicals only a few have been subjected to extensive toxicity testing and most have scarcely been tested at all." So concluded *Toxicity Testing: Strategies to Determine Needs and Priorities* (Washington, D.C.: National Academy of Sciences, 1984).
72. J. L. Annest, J. L. Pirkle, D. Makuc, J. W. Neese, D. D. Bayse, M. G. Kovar, "Chronological Trend in Blood Lead Levels between 1976 and 1980," *New England Journal of Medicine*, 308 (1983): 1373–77.
73. The best statements on this problem have been made by British child psychologist Michael Rutter with respect to the neurological effects of lead on children. See his analysis, "Low Level Lead Exposure: Sources, Effects and Implications," in *Lead versus Health: Sources and Effects of Low Level Lead Exposure*, ed. Michael Rutter and Robin Russell Jones (London: Wiley-Interscience, 1983), 333–70. The British debate on this issue was closely followed in the United States, and British lead researchers were brought into the review of the lead criteria document by the EPA Science Advisory Board in 1984. See also Michael Rutter, "The Relationship Between Science and Policy Making: The Case of Lead," *Clean Air* 31, no. 1 (1983): 17–32.
74. A good account of the politics of cost-benefit analysis is Mark Green and Norman Waitzman, *Business War on the Law* (Washington, D.C.: Corporate Accountability Research Group, 1981).
75. For a statement favorable to risk analysis, see the Business Roundtable Air Quality Project, "National Ambient Air Quality Standards," (Cambridge, Mass., 1980).
76. When it revised its air quality regulations in 1980, Montana included the provision that the program's objective was to prevent air pollution that "interfered with normal daily activities," and specifically identified lowered lung function in schoolchildren as an "adverse effect" that should be prevented. See clippings on this issue in the *Missoulian* (Missoula, Mont.) in author file; *Down to Earth* (Helena, Mont.: Montana Environmental Information Center, Mar.-Apr., May-June, and Sept.Oct. 1980). For a business-sponsored view, see Benjamin G. Ferris, Jr., and Frank E. Speizer, "Criteria for Establishing Standards for Air Pollutants," *The Business Roundtable Air Quality Project* (Boston, 1980): "We define an adverse effect as medically significant physiologic or pathologic changes generally evidenced by permanent damage or incapacitating illness to the individual."
77. For the issues in the principles and standards problem, see U.S. Water Resources Council, *Summary Analysis of Public Response to the Proposed Principles and Standards*

for Planning Water and Related Land Resources and Draft Environmental Statement (Washington, D.C., 1972).

78. See, for example, Commission on Maine's Future, Final Report, Dec. 1, 1977.
79. Two reports that are more sanguine concerning value change in the Corps of Engineers are Daniel A. Mazmanian and Jeanne Nienaber, *Can Organizations Change? Environmental Protection, Citizen Participation, and the Corps of Engineers* (Washington, D.C.: Brookings Institution, 1979); Martin Reuss, *Shaping Environmental Awareness: The United States Army Corps of Engineers Environmental Advisory Board, 1970–1980* (Washington, D.C.: Historical Division, Office of Administrative Services, Office of the Chief of Engineers, 1983).
80. For a brief statement of the problem, see Gene E. Likens, "A Priority for Ecological Research," *Bulletin of the Ecological Society of America* 64, no. 4 (Dec. 1983): 234–43; see also James T. Callahan, "Long-Term Ecological Research," *BioScience* 34, no. 6 (June 1984): 363–67.
81. Clair C. Patterson, "Natural Levels of Lead in Humans" (Chapel Hill: Institute for Environmental Studies, University of North Carolina, 1982), Carolina Environmental Essay Series 3. This is a summary of data, much of which was developed by Patterson and his associates.
82. For a brief look at the range of research, including monitoring, stimulated in the private sector, see John D. Kinsman, Joe Wisniewski, and Jimmie Nelson, "Acid Deposition Research in the Private Sector," *Journal of the Air Pollution Control Association* 31, no. 2 (Feb. 1984): 119–23.
83. See U.S. Environmental Protection Agency, *Review Draft Air Quality Criteria for Lead*, vols. 3 and 4 (Research Triangle Park, N.C., 1983). See also Environmental Protection Agency, Office of Policy Analysis, *Costs and Benefits of Reducing Lead in Gasoline, Draft Final Report* (Washington, D.C.: Office of Planning Analysis, Office of Policy, Planning, and Evaluation, EPA, 1984), chaps. 5–6.
84. See *Northeastern Environmental Science* (Troy, N.Y.: Northeastern Science Foundation, 1982–).
85. See Joseph E. DeSteiguer, "Public Participation in Forestry Research Planning" (Ph.D. thesis, Texas A&M University, 1979), which analyzes the public participation process carried out by the U.S. Forest Service in its review of its research program; the author charts significant distinctions among the choices as to desired research expressed by each of five groups. My own review of the relevant documents substantiates DeSteiguer's description and identifies the limited role of research based on enhancing environmental objectives in contrast to that enhancing commodity objectives.
86. The controversy over lead is a useful case for analysis. Covering some three decades of dispute, the debate continued in much the same manner despite a considerable increase in scientific data and a lowering of the consensus view on threshold levels from 80 μg/dl to 30 μg/dl of blood. At each stage of this threshold reduction, the lead industry scientists argued that the accepted level provided ample protection, and public health scientists argued that frontier knowledge indicated the desirability of allowing for effects at lower levels. In air pollution matters generally, sources of pollution argued that more research was needed both in the 1960s when the laws were first being developed and later in the 1980s when the focus was acid deposition.
87. Two major centers of "high-proof" analysis and political action, both heavily supported by industrial sources of pollution, were the Council on Agricultural Science and Technology and the American Council on Science and Health. Their activities can be followed in their publications, *News from CAST* (Ames, Iowa) and *ASCH News and Views* (New York).
88. A useful review of these controversies is contained in Earon S. Davis and Valerie Wilk, *Toxic Chemicals: The Interface Between Law and Science*, published by the Farmworker Justice Fund, Inc., (n.p., 1982). A useful exchange that sharpens some of the issue is "Examining the Role of Science in the Regulatory Process: A

Roundtable Discussion about Science at EPA," *Environment* 25, no. 5 (June 1983): 6–14, 33–41. Frontier activities with respect to environmental and occupational health can be followed in an organization that brought such scientists together, the Society for Occupational and Environmental Health. See its publication, *Letter* (Washington, D.C., 1975–).

89. Michael G. Royston, *Pollution Prevention Pays* (Oxford: Pergamon Press, 1979); Donald Huisingh and Vicki Bailey, eds., *Making Pollution Pay: Ecology With Economy as Policy* (New York: Pergamon Press, 1982); Monica E. Campbell and William M. Glenn, *Profit from Pollution Prevention: A Guide to Industrial Waste Reduction and Recycling* (Ontario: Pollution Probe Foundation, 1982).
90. Joseph T. Ling, "Industry's Environmental Challenge: Prevention" (Paper delivered at the International Conference on the Environment, Stockholm, Sweden, April 12–31, 1982). See also "Low- or Non-Pollution Technology Through Pollution Prevention," prepared by 3M Company for the United Nations Environment Programme, Office of Industry and the Environment (n.p., 1982).
91. This issue was referred to as "technology forcing." See John E. Bonine, "The Evolution of 'Technology-Forcing' in the Clean Air Act," *Environment Reporter*, Monograph no. 21, July 25, 1975.
92. For integrated pest management, see the newsletter, *IPM Practitioner* (Berkeley, Calif., 1978–); for biological farming, see the newsletter *Alternative Agriculture News* (Beltsville, Md., 1983–); for recycling, see *BioCycle* (formerly *Compost Science/Land Utilization*) (Emmaus, Pa., 1963–); for solar energy, see *Solar Age* (Harrisville, N.H., 1975–). No general center in the United States focuses directly on innovative environmental technologies with an emphasis on process change, comparable to Environmental Data Services which does so in England. See its publication, *ENDS Report* (London, 1978–).
93. For a general source of publications by the environmental opposition, see National Council for Environmental Balance, Inc., Louisville, Ky.
94. *High Country News* is a good source for these developments. See, for example, Ed Marston, "Fighting for a Land Base," *High Country News* (Paonia, Colo.), Jan. 23, 1984, 15.
95. See the Continental Group, *Toward Responsible Growth* (n. 5), 97. The study contrasted attitudes on a four-point scale from resource utilization to resource preservation. It identified the Rocky Mountain West as the strongest region to support its end-point category, "resource preservation." The regional scores on this point were Rocky Mountain, 35 percent; New England, 32 percent; Pacific, 28 percent; Middle Atlantic, 28 percent; North Central, 26 percent; Middle West, 20 percent; West South Central, 25 percent; South Atlantic, 18 percent; and Middle South, 17 percent.
96. Dr. William J. Peeples, Commissioner, Department of Health, State of Maryland, to Senator Edmund Muskie, Nov. 15, 1968, in Senate Committee on Public Works, Subcommittee on Air and Water Pollution, *Air Pollution: Hearings on Air Quality Criteria*, 90th Cong., 2d sess., July 1968.
97. For a brief summary of the hearings and the controversy, see *Conservation Foundation Letter*, July 29, 1967. The resulting National Air Quality Criteria Advisory Committee of fifteen members included representatives from the paper, coal, petroleum, and automobile industries; see *Environmental Science and Technology* 2, no. 6 (June 1968): 400.
98. These controversies can be followed most readily in the pages of *Environment Reporter*, especially with respect to lead, sulfur dioxide, particulates, and nitrogen oxide.
99. As the 1970s advanced, appointments to and activities of the EPA Science Advisory Board, which can be followed in *Environmental Reporter*, seemed to follow a deliberate policy to make scientific decisions more open and reflect a wider range of opinion. Under the Reagan administration, that policy was reversed. See, for example, the formaldehyde case in House Subcommittee on Investiga-

tions and Oversight of the Committee on Science and Technology, *Formaldehyde: Review of Scientific Basis of EPA's Carcinogenic Risk Assessment*, 97th Cong., 2d sess., 1983. A more open policy was established with the appointment of William Ruckelshaus as EPA administrator.

100. The estuarine proposals are included in hearings of the 1960s: House Subcommittee on Fisheries and Wildlife Conservation, *Estuarine and Wetlands Legislation*, 89th Cong., 2d sess., 1966; and *Estuarine Areas*, 90th Cong., 1st sess., 1967.
101. For the larger proposal, see House Subcommittee on National Parks and Insular Affairs, *To Establish a Barrier Islands Protection System: Hearings*, 96th Cong., 2d sess., 1980.
102. See statement by John Middleton, administrator, the National Air Pollution Control Administration, in Senate Committee on Public Works, *Air Pollution: Hearings on Air Quality Criteria*, 90th Cong., 2d sess., 1968.
103. For brief accounts of this issue, see two items, Stanton Coerr, "EPA's Air Standard for Lead," and David Schoenbrod, "Why Regulation of Lead Has Failed," in *Low Level Lead Exposure: The Clinical Implications of Current Research*, ed. Herbert L. Needleman (New York: Raven Press, 1980), 253–57, 258–66. See also Gregory S. Wetstone and Jan Goldman, "Chronology of Events Surrounding the *Ethyl* Decision," (Draft, Washington, D.C.: Environmental Law Institute, 1981).
104. "Behavioral Toxicology Looks at Air Pollutants," *Environmental Science and Technology*, 2, no. 10 (Oct. 1968): 731–33, which describes the work of the NAPCA behavioral toxicology unit in Cincinnati under the direction of Dr. Charles Xintaras.
105. These views of the 1960s were reflected in Senate Subcommittee on Air and Water Pollution, *Air Quality Criteria*, 90th Cong., 2d sess., 1968.
106. Few of these cases have received systematic treatment. The attempt by the lead industry to undermine the credibility of Dr. Herbert Needleman, a prominent frontier researcher on the neurological effects of low-level lead exposures on children, is well known. At times their attacks appear on the record. See description of Dr. Needleman as representing "what is at best a minority view that adverse health effects occur at blood-lead levels below 40 μg/dl," in House Subcommittee on Health and the Environment, *Oversight—Clean Air Act Amendments of 1977*, 96th Cong., 1st sess., 1980 (testimony of Jerome Cole of the International Lead Zinc Research Organization), 344. An indirectly related analysis of the roles of dissenting scientists that stresses more the controversial nature of their public role is Rae Goodell, *The Visible Scientists* (Boston: Little, Brown, 1975). For some aspects of the James Allen case, see William J. Broad, "Court Upholds Privacy of Unpublished Data," *Science* 216 (April 2, 1982): 34, 36.
107. See statement concerning the role of the margin of safety in the recent proposed revisions to the particulate standard to control fine particulates in *Environment Reporter*, Mar. 23, 1984, 2121.
108. See, for example, for no-till agriculture, Maureen K. Hinkle, "Problems With Conservation Tillage," *Journal of Soil and Water Conservation* 38, no. 3 (May-June 1983): 201–6.
109. A summary of voting tabulations for environmental issues in the Michigan House of Representatives for several years in the 1970s is as follows:

	Percent Environmental Vote		
Percent Urban District	1973–74	1975–76	1977–78
75–100	57.3 (69)*	62.6 (69)	62.8 (69)
50–75	47.7 (12)	48.7 (12)	41.3 (12)
25–50	33.6 (15)	38.4 (15)	35.1 (15)
1–25	30.7 (10)	38.1 (10)	21.3 (10)
0	19.3 (4)	27.0 (4)	18.3 (4)

*Number of legislative districts in each population group.

110. See, for example, the activities of the Northern Plains Resource Council, which can be followed in its publication *The Plains Truth.* Cooperation on these issues can be followed in regional publications such as *High Country News* as well as *Not Man Apart,* a monthly publication of Friends of the Earth.
111. For examples of labor-environmental disagreement as well as cooperation, see Richard Kazis and Richard L. Grossman, *Fear at Work: Job Blackmail, Labor and the Environment* (New York: Pilgrim Press, 1982). See also activities of the OSHA/Environmental Network (author file).
112. United Steelworkers of America, *Poison in our Air: Air Pollution Conference* (Washington, D.C., 1969).
113. See *ENDS Report,* (n. 92).
114. The Oregon Coastal Zone Management Association was formed by development-oriented local governments in western Oregon to counter environmental coastal programs. Its general ideological focus was to stress a policy of "multiple use" on the coast. See its publication *The Oregon Coast,* which was issued first in 1978 and in 1981 was incorporated as a regular feature in a new journal, *Oregon Coast,* which attempted to link aesthetic coastal values with expanded tourist economy activity.
115. The iron and steel industry argued that the particulate standards could be doubled with no adverse health effects. Its attorneys, through the American Iron and Steel Institute, commissioned a study by British scientist W. W. Holland to support its case. See W. W. Holland et al., "Health Effects of Particulate Pollution: Reappraising the Evidence," *American Journal of Epidemiology* 110, no. 5 (Nov. 1979), with subsequent reply by Carl M. Shy, "Epidemiologic Evidence and the United States Air Quality Standards," *ibid.* 110, no. 6 (Dec. 1979): 661–71.
116. Julian Simon, *The Ultimate Resource* (Princeton, N.J.: Princeton University Press, 1981).
117. These activities can be followed in industry publications such as *Chemical Week* and, in a more neutral source, *Environment Reporter,* during the Reagan years. See also an environmental critique, Friends of the Earth et al., *Ronald Reagan and the American Environment* (Andover, Mass.: Brick House Publishing Company, 1982).
118. As of this writing it appears that passage of a large number of state wilderness bills might be the most successful positive action by environmentalists during the Reagan administration.
119. "The bureaucracy and its organized clienteles are surely the most durable components of the policy process." See John Edward Chubb, "Interest Groups and the Bureaucracy: The Politics of Energy" (Ph.D. thesis, University of Minnesota, 1979), 21.
120. Andrew C. Gordon and John P. Heinz, eds., *Public Access to Information* (New Brunswick, N.J.: Transaction Books, 1973), esp. 184–222.
121. For NEPA and the courts, see Lettie M. Wenner, *The Environmental Decade in Court* (Bloomington, Ind.: Indiana University Press, 1982); Frederick R. Anderson and Robert H. Daniels, *NEPA in the Courts; a Legal Analysis of the National Environmental Policy Act* (Washington, D.C.: Resources for the Future and Johns Hopkins University Press, 1973).
122. Initially the environmental impact reports were to "accompany the proposal through existing agency review processes." Tim Atkeson, legal officer of the Council on Environmental Quality, said, "As we read the law and the legislative history, the public's involvement comes by disclosure of the thing (report) at the end of the process. The public gets a retrospective look, and their impact comes largely as some comment about the same decision in the future." See *Water Resources Newsletter* 5, no. 6 (Dec. 1970): 4. Court decisions modified this approach considerably.
123. A celebrated "on the record" case concerned regulations proposed by the Office of Surface Mining under the 1977 Surface Mining Act. The President's Council of Economic Advisers (CEA), via its Council on Wage and Price Stability (CWPS),

objected to the economic impact of the new regulations. See the CWPS "Report of the Regulatory Analysis Review Group Submitted by the Council on Wage and Price Stability, Nov. 27, 1978, Concerning Proposed Surface Coal Mining and Reclamation" (author file). The Department of the Interior objected to the use of "post record closure" information in the decision making on the regulation which CEA "interference" involved. Its action forced that post-record data into the record to be made available for comment by others, but did not change policy beyond this case. See "Compilation of Conversations and Correspondence on the Permanent Regulatory Program Implementing Section 501(b) of the Surface Mining Control and Reclamation Act of 1977" (author file).

124. For a discussion of this problem with respect to lead, see Gregory S. Wetstone, ed., "Meeting Record from Resolution of Scientific Issues and the Judicial Process: *Ethyl Corporation v. EPA*" (meeting held Oct. 21, 1977, under the auspices of the Environmental Law Institute, Washington, D.C., 1981), discussion on 84ff., and especially between Judges Skelly Wright and Harold Leventhal of the U.S. Court of Appeals, District of Columbia.
125. One recent case indicated that, while the National Oceanic and Atmospheric Administration sought comments on its state program evaluations, it did not feel obligated to respond to them; it rejected a petition by Friends of the Earth to change its policy. See *Environment Reporter*, Dec. 30, 1983, 1497–98. In another case, the Natural Resources Defense Council argued that EPA used negotiation with industry to test chemicals rather than work out agreed-on test rules in order to avoid public participation. See Senate Committee on Environment and Public Works, Subcommittee on Toxic Substances and Environmental Oversight, *Toxic Substances Control Act Oversight: Hearings*, July 29, 30, Aug. 1, 1983 (Washington, D.C., 1984).
126. The works of these two organizations can be followed in their publications, *EDF Letter* and *NRDC Newsletter*. For a specific example of their analytical capabilities, see David G. Hawkins, "A Review of Air Pollution Control Actions under the Reagan Administration as of July 1982," (Washington, D.C.: NRDC, July 1982).
127. See publications of its Economic Policy Department, such as Peter M. Emerson and Gloria E. Helfand, *Wilderness and Timber Production in the National Forests of California*, (Washington, D.C., 1983); and Gloria E. Helfand, *Timber Economics and Other Resource Values: The Bighorn-Weitas Roadless Area, Idaho* (Washington, D.C., 1983).
128. See its publication, *Forest Planning* (Eugene, Oreg., 1980–).
129. Members of the EPA Science Advisory Board, in reviewing the draft criteria document for lead, April 26–27, 1984, for example, had available for their evaluation the summaries produced by EPA but not the original literature on which the review was based. The give-and-take of the meeting indicated that only a few members of the board had first-hand knowledge of the research, and these were relied on heavily by other members for that information. (Interview with Dr. Herbert Needleman, who made a presentation at the meeting but was not a member of the board, May 5, 1984.) A similar role was played, for the same reasons, by Gordon J. Stopps of the Haskell Laboratory of Dupont in the proceedings of the first National Academy of Sciences study on lead; committee members deferred to him because of his knowledge of airborne lead and because they were far more familiar with ingested lead. See Philip Boffey, *The Brain Bank of America* (New York: McGraw-Hill, 1973).
130. NRDC, in cooperation with the National Air Conservation Commission of the American Lung Association, persuaded Needleman and Piomelli to publish a summary of frontier findings, "The Effects of Low-Level Lead Exposure," which was incorporated into proceedings on the criteria document. The issue can be followed in *Environment Reporter*, 1977–78; see especially July 7, 1978, 427–28. Needleman later testified at the House Subcommittee on Health and the Environ-

ment, *Oversight—Clean Air Act Amendments of 1977*, 96th Cong., 1st sess., 1980, 372–86, concerning the proceedings.

131. The Nuclear Information and Research Service, Washington, D.C., provided offprints of scientific and technical articles; see its publication, *Groundswell* (1979–). The Center for Science in the Public Interest communicated to the readers of *Nutrition Action* (Washington, D.C., 1976–) information concerning food hazards gathered by a staff of professional scientists.
132. For an EPA report highly critical of the role of the media in toxic chemical cases, with case studies on kepone, lead (Dallas, Tex.), Love Canal, and dioxin, see *Inside EPA, Weekly Report*, Jan. 13, 1984, 12–14.
133. See, for example the publications of the Citizen's Clearinghouse for Hazardous Wastes, Inc., Arlington, Va., critical of the EPA for its criticism of citizen action with respect to hazardous wastes.
134. A case study of this problem is Ben W. Twight, *Organizational Values and Political Power: The Forest Service Versus the Olympic National Park* (University Park, Pa.: Pennsylvania State University Press, 1983).
135. See the interplay of scientific debate in the Love Canal case as described in Adeline Gordon Levine, *Love Canal: Science, Politics, and People* (Lexington, Mass.: Lexington Books, 1982).
136. American Forest Institute, *Research Recap*, Nov. 10, 1977, 3–4.
137. The *Conservation Foundation Letter*, Nov. 1969, reported that citizen successes in setting stricter standards had been so striking that "there are indications that some industries are wondering if they might not fare better under federal standards rather than state standards." For industry reaction to state standards and report on testimony at Senate hearings by Fred E. Tucker, manager of pollution control services for the National Steel Corporation, favoring national air quality standards, see *Environmental Science and Technology*, May 1970, 4–5. Tucker was critical of "the people who appear to be playing a numbers game with air quality standards by setting lower and lower allowable pollutant levels in state standards."
138. The issue can be followed in *Forest Planning* (n. 128).
139. See *Conference Proceedings, Coastal Zone Management, Today and Tomorrow; the Necessity for Multiple Use, Economic Considerations of Coastal Zone Management*, sponsored by the Oregon Coastal Zone Management Association, 5 vols. (Newport, Oreg., 1980).
140. A case in point was the EPA action to advance the date for removal of lead in gasoline in 1984. Although revision of the lead criteria document reflected advances in scientific data since the earlier version in 1976, and although that new knowledge was reflected in the cost-benefit analysis produced by the EPA Policy Analysis Division, which took up action on the lead-in-gasoline issue, it declined to use the new health effects data as a basis for its decision on the grounds that the data were too controversial. One can rightly interpret this as a decision in which, in spite of the weight of scientific opinion about the matter, EPA did not feel that it could withstand the political opposition of the lead industry.
141. This mixed potential was implicit in the strong environmental support for planning in both the Forest Management Act of 1976 and the Federal Land Planning and Management Act of 1976.
142. For environmental mediation, see *Resolve*, which was published beginning in 1978 by the Center for Environmental Conflict Resolution in Palo Alto, California, and in early 1982 moved to the Conservation Foundation.
143. The approach of the Environmental Law Institute (ELI) is best reflected in its journal, *Environmental Forum*, which began publication in May 1982; it seemed to be governed by a policy of balancing opinion between the various parties in controversy rather than advancing a "leading-edge" environmental position. This could be understood as an attempt by ELI to service its legal professional clientele

no matter what side of legal controversy it was on, and to foster an opinion forum of the same political stance.

144. The political position of the National Association of Environmental Professionals can be followed in its journal, *The Environmental Professional* (Elmsford, N.Y.), which began publication in 1979.
145. Citizen environmental organizations considered that both the Conservation Foundation and the Environmental Law Institute were useful within their own limited "middle ground" spheres, but not reliable in advancing environmental objectives in leading-edge fashion. Thus, when the foundation worked with industry in the Clean Sites program in 1984, a joint venture to take private action to clean up hazardous waste sites, most environmental organizations felt that the venture would be used to undermine legislative strategy intended to make the hazardous waste program more effective.
146. This stance is reflected in Vig and Kraft, *Environmental Policy in the 1980s.*
147. In Minnesota, the air pollution statute provides that "no local government unit shall set standards of air quality which are more stringent than those set by the pollution control agency," *Nucleonics Week*, Feb. 26, 1970. In Illinois, the mayor of Catlin, near Danville, objecting to a proposed six-thousand-acre strip mine by Amax Coal Company on prime agricultural land in his township, complained, "the state pre-empted our rights to regulate coal mining locally," reported by Harold Henderson, "Caving in on Coal," *Illinois Ties*, Sept. 5–11, 1980, 81. In 1980, Connecticut established a Hazardous Waste Facility Siting Board with power to override local zoning laws, *Environmental Science and Technology*, Aug. 1980, 894. In 1980, the Wisconsin Department of Natural Resources revised its wetlands regulation, NR 115, with legislative approval, which prohibits counties from forming regulations more strict than those of the state, *Our Wetlands*, published by the Wisconsin Wetlands Association, Aug.-Sept. 1980.
148. Noise control issues were defined in 1981 by congressional staff: "Industry generally supports federal noise regulations as preferable to myriad local rules which would differ from place to place. But local governments . . . want the authority to establish rules that are stricter than the federal regulations. Currently the noise act pre-empts stricter state and local noise regulations." See Environmental Study Conference, U.S. Congress, *Weekly Bulletin*, Feb. 23, 1981, C7.
149. Industry sought a preemption provision in the 1972 extension of the federal pesticide law, but failed. See *Farm Chemicals*, Oct. 1970, 70; Nov. 1971, 12; Dec. 1971, 12. For environmental opposition to preemption, see Senate Subcommittee on Environment, Hearings on the 1972 Pesticide Act, 130–31 (testimony of Cynthia Wilson).
150. Radioactive waste transport was one of a wide range of nuclear issues involving preemption of state authority; see *Groundswell* 4, no. 5 (Sept.-Oct. 1981): 1–3, 5–6, for a listing of state laws subject to potential preemption.
151. Right-to-know controversy can be followed in the publication *Exposure*, issued by the Environmental Action Foundation, Washington, D.C., and in *Chemical Week* (New York), as well as *Chemical Regulation Reporter* (Washington, D.C.).
152. For the main case, from California, see *Coastal Zone Management*, Jan. 2, 1980, 1; subsequent events can be found in later issues, Feb. 25, 1981, 1; May 6, 1981, 1; July 15, 1981, 1; Aug. 5, 1981, 1; Aug. 12, 1981, 2; Sept. 16, 1981, 1–2; Sept. 23, 1981, 4–5; Oct. 7, 1981, 1–2.
153. Evidence from western sources throughout the 1970s and thereafter reflects a wide range of such activity. The best single source is *High Country News*, but it is confined to the Rocky Mountain states. State sources include *Alert*, publication of the Washington Environmental Council; the *Conservator*, the Colorado Open Space Council; *Earthwatch*, Montana Environmental Council; and *Crossroads Monitor*, Wyoming Outdoor Council.
154. A survey of western adverse reaction to the drive to increase allowable cuts during the Reagan administration is in an eight-part series of articles, "Taming the

Forests," by R. H. Ring, in the *Arizona Star*, Feb. 5 to Feb. 12, 1984. Especially useful is the last installment, which summarizes responses on the matter from state park and recreation, and fish and game agencies in the West.

155. See, for example, U.S. Department of Health, Education and Welfare, *Proceedings, Conference in the Matter of Pollution of the Interstate Waters of the Mahoning River and its Tributaries*, vol. 1, Feb. 16–17, 1965 (Washington, D.C.: the department, 1965), 195–213. In this proceeding the State of Ohio maintained that its pollution discharge information could not be divulged without express permission of the discharger.
156. House Committee on Government Operations, Natural Resources and Power Subcommittee, *House Report 1579*, 90th Cong., 2d sess., 1968. This report was given renewed publicity, with extended analysis, by Senator Lee Metcalf in his investigations of advisory committees in 1970.
157. Gregory S. Wetstone and Jan Goldman, "Chronology of Events Surrounding the *Ethyl* Decision" (Washington, D.C.: Environmental Law Institute, 1981), especially 2–3, 13.
158. Frustration of the lead industry with the course of events in the first lead criteria document is reflected in testimony of Jerome F. Cole, Director of Environmental Health, Lead Industries Association, in Subcommittee on Health and the Environment, *Oversight—Clean Air Act Amendments of 1977, Hearings*, 1980, 341–72. In the revision of the lead criteria document in 1983–84, considerable evidence was presented concerning health effects of blood lead levels below 30 μg/dl; this was taken seriously by EPA staff and the EPA Clean Air Science Advisory Committee despite the continued position of the Lead Industries Association (LIA) that there was no evidence of adverse effect below 40 μg/dl. See summary of the health effects evidence in EPA, Office of Policy Analysis, *Costs and Benefits of Reducing Lead in Gasoline* (Washington, D.C., 1984, Draft Final Report), chaps. 5 and 6. The LIA still disagreed strongly with the conclusion by the Center for Disease Control (CDC), the American Academy of Pediatrics, and EPA that the "threshold point" was 30 μg/dl; hence, when the CDC took action in 1983 to issue an advisory to public health professionals in the nation that the accepted threshold should be lowered, the LIA took up legal action to thwart it.
159. See newspaper clippings following the case in *The New York Times* and *The Washington Post* (author file).
160. The recent Agent Orange case was only one of the more dramatic instances. See the Tyler, Texas, asbestos suit, which was settled for $20 million "without trial and verdict." *The New York Times*, Dec. 20, 1977.
161. See Council of Economic Advisors, "Compilation of Conversations and Correspondence on the Permanent Regulatory Program Implementing Section 501(b) of the Surface Mining Control and Reclamation Act of 1977" (author file), which indicated considerable reliance by the CEA on the regulated industry for the formulation of its views.
162. The regulation of formaldehyde was the most widely debated case; see n. 99. It was not, however, an isolated case; for others, see Robert Nelson, *A World of Preference: Business Access to Reagan's Regulators*, Democracy Project Reports—No. 5 (n.p., Oct. 1983).
163. A useful case of the politics of modeling with respect to air pollution is Phyllis Austin, "Keeping the Air Clean: In a New Period of Regulatory Accommodation the Emphasis is on Granting Licenses Quickly," *Maine Times*, Nov. 11, 1983, 2–5. Another case involved the impact of electric power generation on the Hudson River striped bass population; see L. W. Barnthouse et al., "Population Biology in the Courtroom: The Hudson River Controversy," *BioScience* 34, no. 1 (Jan. 1984): 14–19.
164. A host of issues involved confidentiality. They can be observed, for example, in the first round of premanufacture notices for chemical registration under the 1976 Toxic Substances Control Act and the development of regulations pertaining

to Section 5 of that act. See *Chemical Regulation Reporter*, Apr. 27, 1979, 82; May 11, 1979, 147–49; May 18, 1979, 218–19.

165. An excellent source through which to follow this issue is *Chemical Week*. For its response to the Philadelphia ordinance, see Feb. 4, 1981, 50; see also "Philadelphia Sets Toxic Chemical Rules," *Chemical and Engineering News*, Feb. 2, 1981, 4–5. For California action see *Chemical Week*, Sept. 28, 1983, 29, 32.
166. See, for example, the speech by EPA administrator William Ruckelshaus before the Detroit Economic Club in April 1984, cited in *Coastal Zone Management* 15, no. 21 (May 24, 1984): 5.
167. "Consumption is the sole end and purpose of all production; and the interest of the producer ought to be attended to, only so far as it may be necessary for promoting that of the consumer. The maxim is so perfectly self-evident, that it would be absurd to attempt to prove it." See Adam Smith, *An Inquiry into the Nature and Causes of the Wealth of Nations* (New York: Modern Library Edition, 1937), 625.

2

FROM CONSERVATION TO ENVIRONMENTAL MOVEMENT: THE DEVELOPMENT OF THE MODERN ENVIRONMENTAL LOBBIES

Robert Cameron Mitchell

For almost two decades, the environmental movement has been one of the major actors in national environmental politics. It encompasses a wide range of interests from land use planning to the regulation of toxic wastes, and an equally wide range of organizations at all levels from national to regional, state, and local. The movement maintains a strong sense of identity based on a concern for the integrity of natural systems and a belief in the desirability of public policies that promote environmental quality. This chapter presents an account of the development of the twelve national groups whose activities include significant lobbying or litigating components. These organizations, which include the National Audubon Society and the Sierra Club, have several million dues-paying adherents; command multimillion-dollar budgets; employ corps of full-time lobbyists, lawyers, and scientists; and enjoy widespread public support. Several have local chapters numbering in the hundreds across the country or work with independent local environmental organizations on issues of common concern.[1] These groups have sought to influence public policy by publicizing environmental issues, lobbying Congress, monitoring administrative agencies, and, where necessary, litigating. Endorsing candidates for public office, once practiced by only a handful of groups, has recently become more widespread.

In this chapter I address two questions. First, how have these groups mobilized the resources needed to maintain the organizational strength on which they base their successful advocacy? This is a question basic to all social movements.[2] Second, what factors enabled them to play a significant and enduring role in environmental politics? The first part of the chapter is an account of the development of environmental groups over the past thirty years. In the second part, I discuss the groups' success in mobilizing public support for their efforts to influence environmental policy. The evolution of their tactics is discussed in the third part, with an emphasis on how the movement overcame the key institutional obstacles to their advocacy.

THE RISE OF THE ENVIRONMENTAL MOVEMENT

Three decades ago, it would have been difficult to predict the environmental movement's emergence as an effective lobby. In 1954, Grant McConnell described the current state of the conservation movement, its predecessor, as "small, divided, and frequently uncertain."[3] According to McConnell, conservationists lacked an ideology that commanded widespread public support. The movement's "wise use" wing was no longer animated by the unifying Progressive vision of resources for the people and had come to represent particular interests, "some as selfish as the special interests which were so denounced in the Progressive era."[4] The preservationist wing, in contrast, elevated certain values such as wilderness preservation above material values, but did so in a way that excluded most people.

Even as McConnell wrote those words, however, changes were occurring that would facilitate the transformation of the conservation movement into the modern environmental movement. Samuel Hays has described in some detail in chapter 1 the broad social changes and the emerging environmental impulse that enlarged the potential constituency for the environmental groups. Particularly important was the shift from the complacency of the 1950s to the rising social consciousness of the 1960s, with its willingness to acknowledge, and to challenge, the darker side of the American dream. The challenges facing conservationists also changed. Huge postwar development projects that menaced the remaining natural areas of outstanding quality and high symbolic value, such as Dinosaur National Monument and the Grand Canyon, were proposed. There was also increasing evidence that the side effects of modern technology posed a new set of environmental problems with serious implications for both nature and human health.

These changes together created the conditions that facilitated the emergence of the national environmental movement. Of the twelve organizations considered here, the seven groups founded before 1960 constitute the conservation movement's important organizational contribution to the environmental movement. During the early years of the environmental era the seven were joined by five new groups. Although all the groups now share a strong common identity and frequently cooperate in coalitions or lawsuits, their unity is tempered by a diversity of heritage, organizational structure, issue agenda, constituency, and tactics. The groups also compete with each other for the staples of their existence—publicity and funding. Overall, however, the movement's diversity is one of its strengths.

The three oldest groups are products of the Progressive Era. The first to be founded was the Sierra Club (1892), followed by the National Audubon Society (1905) and the National Parks and Conservation Association (1919). Today both the Sierra Club and the National Audubon Society have hundreds of local chapters, and both rank among the top three environmental groups in size and influence. Between the two world wars three more national organizations were added to the conservation movement. Two of them were founded by sportsmen—the Izaak Walton League (1922) and the National Wildlife Federation (NWF) (1936). Although both are built on local groups, the Izaak Walton League's chapters are under the direct control of the national group, whereas the NWF was founded as a federation of state wildlife federations whose autonomous member groups are principally sportsmen's clubs.

The third of the interwar conservationist organizations was founded in 1935 by a small group of men, primarily easterners. Led by Robert Marshall, a Forest Service civil servant and a man of independent wealth who personally supported the organization during its early years, the Wilderness Society attempted to fill what it regarded as a gap in the goals of the existing groups, the preservation of the remaining wildlands outside the national parks system. Defenders of Wildlife (1959) came into being during the waning years of the conservation era. Although it was initially more interested in individual animals than in wildlife species—the mistreatment of wild animals in roadside zoos was one of its founder's primary concerns—Defenders' interests have broadened to include wildlife habitat and endangered species, thus placing it squarely in the environmental rather than the animal protection camp.

Until the 1960s, the predominant concerns of both wings of the conservation movement involved land and wildlife. A common feature of these "first generation" environmental issues is that they involve threats to particular sites or species. Gifford Pinchot and his allies worked tirelessly to ensure the "wise use" of the country's rangeland, forests, and wildlife through careful government planning. Preservationists, such as John Muir of the Sierra Club and Robert Marshall of the Wilderness Society, battled to preserve wild and scenic natural areas from commercial exploitation,[5] whereas those associated with the Audubon Society focused on endangered birds. The Izaak Walton League added water pollution to the conservationist agenda, because its fishermen founders were concerned about the rapid loss of sportsfish-bearing streams and lakes to development and pollution.

An awareness of a new generation of environmental problems first began to develop in the 1950s. Air pollution, for example, was becoming more and more bothersome in California despite the strict controls imposed on the factories and refineries. Naturalists began to notice that peculiar things were happening to wildlife, especially birds, that seemed to be traceable to modern pesticides. It was becoming clear that modern technology posed a serious threat to the environment. Stewart Udall, a sensitive conservationist and President Kennedy's secretary of the interior, called this "the quiet crisis of conservation" in a book he published in 1963.[6] A key event that clarified the nature of the problem and helped to transform the quiet crisis into the not-so-quiet ecological crisis was the appearance in 1962 of *Silent Spring*, Rachel Carson's best-selling work that documented the threat posed by DDT to wildlife and to humans.[7] Carson did more than call attention to the second-generation problems; she also popularized a powerful intellectual framework that helped to explain the occurrence of these problems: the ecological perspective.

Carson's success as the environmentalist Paul Revere in awakening people to the potential threats posed by pesticides was a significant achievement, because these second-generation problems are inherently more difficult to communicate to the public than the threats posed to wildlands and "critters" by development projects. Table 2.1 compares the characteristics of the first- and second-generation issues. Traditional conservation issues typically involve a specific place or species that faces a clear threat whose cause, such as a dam or the draining of marshland, is unambiguous. In the case of the second-

Table 2.1
CHARACTERISTICS OF FIRST- AND SECOND-GENERATION ENVIRONMENTAL ISSUES

	First-Generation	*Second-Generation*
Locus	Typically specific locus in the natural environment such as a scenic river valley, endangered species, or forest area	Often not tied to a specific place or species
Consequences	Relatively specific, immediate, and apparent. If harvested, a forest loses its mature growth; if submerged in a reservoir, a valley's natural state is irretrievably and unmistakably lost.	Relatively far-reaching, delayed, and subtle. Increase in CO_2 in upper atmosphere may cause gradual increase in levels in the seas over a long period.
	No impact on human health.	Potentially significant impact on human health; for example, air pollution affects humans as well as forests; exposure to toxic chemicals can cause cancer in humans.
Causes	Direct and unambiguous, such as sheepherders overgrazing pastures and sport hunters slaughtering wildlife species to extinction.	Indirect and difficult to prove. Often requires imaginative scientific detective work to determine the causative agent, and even then the evidence is often open to debate owing to uncertainties. For example, the causes of acid rain have been the topic of debate for several years.
Cost of Solution	Relatively small. Some were even solved by government management, which paid for itself out of sales (timber from national forests) and fees (for wildlife hunting).	Can be very substantial, such as the cost to consumers of lead-free gasoline and the various automated devices required to meet air pollution emission requirements. Costs of toxic waste cleanup can be astronomical.
Opposing Interests	Those threatened by conservationist victories were either local economic interests, such as loggers, hunters, or fishermen, whose wasteful behavior was becoming obsolete with the advent of growing resource scarcity, or relatively small industrial corporations that lack the capacity to avoid wasteful competition.	The amelioration of these problems can threaten entire industries, such as commercial agriculture, or major patterns of social life, such as Americans' reliance upon the automobile for personal mobility.

generation issues, the locus is not necessarily site- or species-specific; the consequences are often delayed or subtle; and the cause is typically difficult to prove with certainty. These characteristics leave room for argument, and the industries that stand to lose if the environmentalist position prevails are both powerful and strongly motivated to fight.

Carson's perspective prefigured the politically reformist and culturally radical approach that environmentalists would take in the years to come on issues as diverse as acid rain and trihalomethanes in drinking water. First, she endeavored to make the problem of pesticide use as meaningful as possible to the ordinary person; hence her use of the silent spring metaphor and her emphasis on DDT's potential effects on human health, despite the fact that the evidence for these effects was rather scanty. Second, she cast her argument in scientific terms, developing the technical case for her assertion that DDT was behind the natural disasters she chronicled in considerable detail. Third, she took great pains to propose an alternative to chemical pesticides in the form of biological methods of insect control and went to some length to show that this approach was consistent with modern agriculture. Thus, while her critique challenged the hegemony of agribusiness on pesticide policy, it by no means implied revolutionary changes in American society. Culturally, however, her work was more radical. She called into question the prevailing assumptions about the relationship between humans and nature by pitting a control-of-nature perspective against an ecological approach based on the "complex, precise, and highly integrated system of relationships between living things which cannot be safely ignored."[8] The need to adopt the latter perspective was urgent, according to Carson, because of the potentially catastrophic long-term effects of continued DDT use. In this she prefigured the survival theme that was to be so prominent in Earth Day discourse.

By the end of the 1960s, the conservation groups had adopted both of the defining characteristics of the environmental movement: the ecological perspective and an agenda that included the second-generation issues. National Audubon, for example, began to campaign for goals not exclusively related to bird protection.[9] The National Parks Association, for its part, changed the name of its magazine from *National Parks Magazine* to *National Parks and Conservation Magazine: The Environmental Journal* to reflect "the enlargement of our knowledge and functions, and our basic activities."[10]

No group epitomizes the evolution from conservation to environmental group better than the venerable Sierra Club, which, by 1970, was treated by the press as the quintessential expression of activist environmentalism. The club was founded in 1892 by John Muir and a small group of nature-oriented people, many of them mountaineers from the San Francisco area, who wanted to preserve from development Yosemite Valley and the Sierra Nevada, where they hiked and climbed. For more than half a century, the organization was small and largely Californian. It was also a club, and every new member had to be sponsored by two members and approved by the membership committee before admission. Its militantly preservationist members pursued several campaigns to preserve wild areas in the Sierra Nevada, the most famous of which was their losing battle to keep the beautiful Hetch Hetchy Valley from being flooded to serve as a reservoir for San Francisco.

The Sierra Club entered a period of change in the 1950s, as its leaders began to address the new threats to America's natural heritage. It expanded concerns beyond its California base by establishing an Atlantic chapter in 1951,[11] and joining with the Wilderness Society in the campaign against construction of the proposed Echo Park Dam, which threatened the Dinosaur National Monument in Utah and Colorado. Then, in 1952, the club took the momentous step of hiring David Brower as its first executive director, after he convinced the board members that he could raise money for the club and "sell conservation."[12] Brower, the club's first full-time professional-level employee, aggressively pursued an activist program in defense of wild areas, leading the club in national campaigns such as the one to save the Grand Canyon from being flooded by proposed downstream dams. These successful battles, which were among the major conservation issues of the 1950s and early 1960s, united most conservation groups behind preservationist goals and vastly increased public awareness of public land issues.[13] They also established the club's reputation as a hard-hitting adversary. In the late 1960s, an influential congressman who opposed their stand on the Grand Canyon dams grumbled, "No group in this country has had more power in the last eight years."[14]

In 1966, the club moved beyond its historic concern with the preservation of natural areas, when its board of directors voted to add population and "urban amenities" to its action agenda.[15] Within a year or two, the second-generation environmental issues were squarely on its agenda, the ecological perspective was integral to its world view, and the club regarded itself as part of the new environmental movement. In an Earth Day editorial in the club's magazine, in which he emphatically rejected the idea that the Sierra Club represented the "old conservation," Edgar Wayburn, its volunteer vice president, proudly declared, "our priority projects range from expansion of the National Park System to the survival issues of population, pollution and pesticides."[16] The set of environmental issues was further expandable. Within a few years, the Sierra Club and its sister organizations found themselves grappling with the issues raised by the Arab oil embargo and the expansion of nuclear power. A decade later, the club added nuclear weapons and the possibility of a "nuclear winter" to its official list of concerns.

The Sierra Club and the other older groups did not in any way abandon their commitment to the first-generation issues when they embraced environmentalism. The ecological perspective provided an intellectual framework for linking the first- and second-generation issues, which enabled the older groups to work on both kinds of issues and to cooperate with the newer groups, which tended to concentrate only on the second-generation issues. In the editorial cited earlier, Wayburn further argued that the "so-called old issues" are not narrow, but involve the "full spectrum of environmental values." As an example he cited the National Timber Supply Act, which the club successfully opposed.[17] In addition to eliminating many potential wilderness areas, the stepped-up logging and increased industrial processes resulting from the Timber Supply Act would have caused more outpouring of carbon dioxide into the atmosphere, as well as more smog and more water pollution. The conservationists' national campaigns on the first-generation issues in the 1960s demonstrated to the environmentalists of the 1970s that hard-hitting tactics in a national campaign could attract a good deal of media attention, win them a following, and gain results on Capitol Hill. The Sierra Club's loss of its tax-exempt status in 1966 for its role in the Grand Canyon fight also taught the potential costs of political action.

When did the conservation movement become the environmental movement? A convenient, though necessarily arbitrary demarcation for the transition from one to the other is 1967, the founding year of the Environmental Defense Fund (EDF), the first of the new breed of national environmental groups. The creation of these new groups was greatly assisted by the availability of start-up money from foundations and individual patrons.[18]

The origins of the Environmental Defense Fund lie in the battle to ban DDT. Despite Rachel Carson's efforts, progress toward banning DDT was not made until the matter was brought to court. Victor Yannecone, an attorney who lived on Long Island, brought suit in

1966 against the Suffolk County Mosquito Control Commission's local use of DDT for mosquito control.[19] Although he received volunteer scientific assistance from several concerned biologists affiliated with local educational and research institutions when he began his crusade, Yannecone had no relationship with any organization. A couple of years later, Yannecone and his colleagues obtained some financial support from the Audubon Society's Rachel Carson Fund.[20]

In a series of lawsuits on Long Island and in Michigan, Yannecone persuaded state courts to hear the case. These hearings provided him and his colleagues with public forums where the scientific case against DDT could be aired in depth through an adversarial proceeding. In 1967, Yannecone and the scientists formalized their association by founding the Environmental Defense Fund, the first environmental law group. For all his brilliance at litigation, Yannecone was unpredictable and difficult to work with, which led to his departure from EDF a year or two later. The scientists then approached the Ford Foundation for funding to hire a legal staff so they could pursue litigation on DDT and similar hazards.

The Ford Foundation, by this time, had become more openly aggressive in directly funding social reform projects and was particularly interested in the possibility of funding environmental law groups. Thanks to the DDT litigation and other legal developments, the courts were a promising arena for environmental advocacy, but this course was still new and potentially controversial. The Ford staff did not leap at the chance to fund EDF, because they were uncertain about whether the nascent organization was viable in the long run. No doubt its connection with the unpredictable Yannecone gave them pause as well. As it pondered these matters, two other groups of people independently approached Ford for funding as environmental law firms. One group included several prominent New York attorneys and a younger lawyer named John Adams; the other was composed of seven well-recommended Yale law school students who were eager to dedicate their careers to the environmental cause. Ford officials informally brokered a merger between the two—as the responsible foundation official described the situation, "one group had prestige and no staff and the other had staff and no prestige"—and the National Resources Defense Council (NRDC) was the result.[21] An initial grant of $100,000 in 1970 was followed by a series of further grants that totaled $2.6 million by 1977. The success of the NRDC grant emboldened the Ford officials to make a similar grant to EDF in 1971,[22] which enabled them to recruit more staff and conduct a membership drive. The Ford Foundation's role in institutionalizing environmental law advocacy was aptly summarized by this official: "We did not create it, but we have had a lot to do with its size, quality, and rapid development."[23]

EDF has tended to specialize in issues dealing with toxic chemicals, wetlands, water quality, and power generation, whereas NRDC's specialties include air pollution, nuclear power, and solid waste. In comparison with EDF, which at its founding was a membership group, dominated by scientists, which engaged in litigation, NRDC was originally an environmental law firm run by lawyers.[24] However, within a few years the two groups became much more alike: NRDC developed a membership and hired some scientists, while EDF's lawyers gained a greater role in the organization. Both then expanded their advocacy to include administrative law and lobbying. By the end of the 1970s, after Ford gradually phased out its support, the groups developed sufficient alternative sources of funding to be financially independent.

Friends of the Earth (FOE), another of the new groups, is an offshoot of the Sierra Club. Founded by David Brower, shortly after his ouster as the club's executive director for being too freewheeling an activist and acting without consulting the club's elected leadership (his actions cost the club its nonprofit status), FOE espouses a radical environmentalism deeply suspicious of growth and technological change because of their consequences for already overburdened natural systems. It has lobbied aggressively on a broad range of issues from wilderness to recombinant DNA. Amory Lovins undertook his influential espousal of the "soft path" approach to energy production under FOE's auspices.[25] FOE's periodic financial problems ultimately led to Brower's ouster toward the end of the 1980s and to a prospective merger between FOE and the next group to be considered.

Another schism, this time from the schismatic, resulted in a fourth new group, the Environmental Policy Institute (EPI). EPI was founded in 1972 by a group of FOE's Washington lobbyists who tired of what they believed to be the inefficiencies of the FOE leadership and organization. These dissidents acquired some wealthy patrons whose largesse permitted them to eschew the distractions of having to serve a membership or publish a magazine and to concentrate on lobbying. EPI pursues a deliberately narrow agenda of land, wild and scenic rivers, and nuclear energy issues with dedication and considerable political savvy. Its leaders were the principal environmental force in Washington pressing for passage of the Surface Mining Control and Reclamation Act of 1975. In the absence of a membership of its own, EPI's modus operandi is to work with coalitions of grassroots groups, which enables EPI lobbyists to speak for a significant constituency on any given issue.

The last of the new environmental groups is Environmental Action, the only national organization that is a direct product of Earth Day in 1970.[26] It is notable that neither the older nor the newer environmental groups played major roles in organizing Earth Day, although they welcomed the event, gave it their active support, and benefited greatly from it. The idea originated with Senator Gaylord Nelson, one of several outspoken environmental advocates in Congress, who originally conceived the event as an environmental teach-in along the lines of the successful antiwar teach-ins of the day. Nelson recruited Denis Hayes, a recent Stanford graduate and campus antiwar activist, to head the effort and helped him raise money from labor (the United Auto Workers in particular), foundations, and a few individuals. Once Hayes began his labors in late fall 1969, the campus response was so fervent that older sponsors wisely decided to step back and let the students run their show. Hayes ultimately hired thirty-five paid full-time organizers, in addition to enlisting the volunteer services of hundreds of students across the country.[27]

Following the success of Earth Day, which enlisted widespread mainstream grassroots support and received a remarkable amount of attention from the news media, Hayes and some of the student organizers used their Earth Day mailing list to found an organization to continue their crusade. In keeping with the ideals of the 1960s student movement, decision-making authority was vested in the staff collective, each of whom received the same (low) salary regardless of position. This organization, Environmental Action (EA) pursues a strategy of publicizing issues through its magazine and press campaigns and also by lobbying. Unlike most of the other environmental groups, both old and new, EA directly focuses its fire on big business (the "Filthy Five") and on certain members of Congress regarded as being unduly attentive to business views on environmental issues (the "Dirty Dozen"). This mixture of anticorporatism, media campaigns, low-budget operation, and legislative activity bears a strong resemblance to Ralph Nader's approach to consumer issues, although EA's nonhierarchical structure is the antithesis of the Nader organizational structure.

As the nation entered the environmental decade of the 1970s, the twelve groups described earlier dominated the movement's Washington presence.[28] Their lobbyists share the mistrust for both business and government common to other contemporary public interest groups.[29] They tend to believe that, without their efforts, environmental policy would be controlled by unrepresentative elites who act to further their own special interests to the detriment of the public's interest in environmental quality.[30] They also share a worst-case view of environmental problems and environmental regulation, assuming that the problems are very likely to be even more serious than the available evidence shows and that environmental laws will be ineffective if administrative discretion is granted to the relevant administrative agencies. This, and the need to maintain room to bargain, motivates them to seek the most stringently protective laws possible and to favor laws and regulations that minimize the exercise of administrative discretion.

Although they frequently band together to coordinate and lobby on issues of common concern, the national environmental groups are diverse in their ideological outlook.[31] FOE and Environmental Action may be said to constitute the radical wing of the mainstream environmental movement. Unlike Greenpeace, until recently a nonlobbying group founded in 1971 to oppose nuclear weapons testing by nonviolent "direct action" tactics, FOE's and EA's Washington lobbyists are committed to democratic change through legislation and litigation. Their radicalism lies in the depth of their commitment to fundamental environmental reform born of a sense of crisis, their sensitivity to the equity implications of environmental issues, and the degree of their uneasiness about legislative compromise and cooperation with business and industry.[32] The National Wildlife Federation is widely regarded as the most moderate of the groups because of its less confrontational stance on some issues. The other groups stand squarely in the American reform tradition. Contrary to the predictions of some that the movement would shift toward "increasingly coordinated strategies for class politics and the reorganization of production,"[33] these groups remain firmly placed not too far to the left of the political center.[34] They have neither embraced the emphasis on restructuring American society along direct participation lines advocated in the late 1970s by antinuclear protesters belonging to Clamshell Alliance and its clones,[35] nor the program for radical decentralization and community control advocated by some appropriate technology theorists.[36] In this they reflect the views of their members, the vast bulk of whom define themselves as "moderate liberals," rather than "strong liberals" or "radicals."[37]

MOBILIZING CITIZEN SUPPORT

A crucial source of the environmental groups' political strength is the popularity of their cause. This is demonstrated in poll findings showing that strong majorities of the public support environmental protection, even when economic trade-offs are involved, that most Americans are sympathetic to the environmental movement, and that millions identify themselves as environmentalists.[38] Particularly striking is the fact that this support has remained steady since Earth Day, notwithstanding the energy crisis, the escalating cost of environmental protection, and the economy's ups and downs. It reflects what appears to be a fundamental shift in public awareness since the 1960s during which an understanding of important aspects of the ecological paradigm has spread from environmental activists to the public at large.[39] Several times over the past fifteen years, new polls confirming the issue's continued popularity have defused attacks on the movement by opponents who claimed it was no longer representative of a broad public interest.[40]

The popularity of a cause is not a sufficient condition for political success, especially when those opposed to reforms are strongly motivated and have deep pockets. Those in favor of the reforms need to be organized so that their resources, either time or money or both, can be mobilized to work for the groups' goals. Before the rise of the environmental movement, the conservation groups had achieved some success in organizing local activists into Sierra Club, National Audubon, or Izaak Walton chapters. Many of these activists were strongly motivated by their recreational involvement with the natural areas they worked to protect and deeply committed to their groups. The rise of the second-generation issues, which are national in scope, and the increasingly important role played by Congress in environmental policy made the recruitment of a mass membership imperative.[41] Such members might be less personally involved in the groups, but they could provide money to support Washington lobbyists, write letters, and add legitimacy to the groups' claim to speak for an otherwise underrepresented interest.[42]

In seeking to mobilize a constituency, environmentalists faced a problem common to all advocacy groups whose goals involve nonexcludable amenities, such as "good government," lower taxes, or cleaner air, which benefit large numbers of people. How could they reach such a diffuse constituency with their membership appeals? How could the groups encourage contributions from people who will still enjoy the fruits of any environmentalist victories whether they contribute or not, since such benefits as improved water quality or reduced SO_2 in the air are nonexcludable?[43] Without a low-cost method of bringing their appeal to the attention of potential members and providing them with a convenient way to make a contribution, the groups could not have mobilized a mass membership. Direct mail made this possible.[44]

Direct mail is so ubiquitous today that we have to remind ourselves that it was not widely used to recruit new members for public interest groups until the 1960s and that many direct mail techniques were not developed until the 1970s.[45] The importance of this technology is that it provides virtually the only efficient way for people to learn about the groups' goals and make their contributions.[46] The major focus of the appeal letters is the need for action to combat a particular threat to the environment. The group describes its activities to relieve the threat and asks for the prospective member's monetary contribution. In addition to purposive incentives, most groups also offer a material incentive, such as a subscription to their magazine. By making the contribution, via the convenient stamped, self-addressed envelope enclosed in the "package," the member engages in what might be called "delegated activism," in effect hiring surrogate lawyers who litigate and surrogate lobbyists who lobby on his or her behalf. Numerous small contributions can add up to impressive amounts of money, even after the costs of recruiting and serving the members are taken into account.[47]

The ability of groups to recruit mass memberships by direct mail rests on a number of factors. One is the credibility of the group making the appeal, and here the environmental groups have fared well, thanks to their early successes, sympathetic media coverage, and the willingness of prominent people to lend their names to their direct mail appeals. Another is the appeal of the group's grievances. Social movement theorists of the resource mobilization persuasion tend to deemphasize this factor as a reason why some groups are more successful than others in mobilizing recruitment.[48] As far as mobilization by direct mail is concerned, no matter how professionally grievances are merchandised, not every set of grievances is capable of mobilizing a viable constituency. Despite being founded by well-known figures (including Margaret Mead) who were able to obtain sympathetic press coverage and raise sufficient start-up money to mount a professional direct mail campaign, New Directions—a group founded with some fanfare in 1976 to lobby on international issues of energy, environment, and peace—was so unsuccessful in acquiring members that it was forced to close its doors a few years later. At that time, at least, international issues were not salient enough to the "liberal giver" constituency to support a lobby devoted to their solution.

Environmental issues, by contrast, are sufficiently appealing to provide significant amounts of financial support for not one but eleven national groups.[49] There are several reasons for this. First, the first-generation issues have a powerful appeal to the sympathetic public and to activists who identify with scenic wildlands and endangered species and are motivated by the possibility of their irreversible loss. The vivid and picturesque nature of these issues makes them easy to sell through the groups' direct mail and publications programs. Second, the potential threat to health and life posed by many of the pollution issues allows the groups to motivate contributors by portraying the potential losses if these conditions are allowed to persist. The fact that the threat is not confined to a particular group but potentially affects everyone gives these issues a broad appeal. Third, the multiplicity of environmental issues allows the groups to appeal to numerous segments of the population. People concerned about toxic wastes might not necessarily be motivated by wildlife appeals, and vice versa. Fourth, the continued emergence of new environmental issues such as toxic waste dumps keeps the environmental cause in the news, particularly when dramatic events occur, such as the contamination of Love Canal and the accident at Three Mile Island.

The groups' success in mobilizing public support is shown in their membership growth between 1960 and 1983. The groups count as members everyone who has contributed a minimum (usually tax deductible) membership fee that in 1984 ranged from $12 for the National Wildlife Federation to $30 for Friends of the Earth.[50] The first year for which we have reasonably accurate membership data for all the groups is 1960. At this time, just before the publication of *Silent Spring*, the seven national conservation groups were small, tended to be regional rather than national in scope, and counted a total of only 124,000 members. The Sierra Club recruited its members primarily from California, whereas most of the National Audubon Society's members were easterners. The largest group by far at that time was the Izaak Walton League, whose membership consisted primarily of conservation-minded fishermen. The National Wildlife Federation had not yet begun to recruit members directly, and Defenders was just getting formed. Over the next twenty-three years, as shown in Table 2.2, the number of groups almost doubled, the number of people who contributed to all the groups, both old and new, expanded rapidly, and the balance between sportsmen and nonsportsmen shifted decisively in favor of the latter.

The movement's growth occurred in three spurts, each attributable to a period of heightened public concern that environmental ameni-

Table 2.2
MEMBERSHIP IN TWELVE NATIONAL ENVIRONMENTAL GROUPS
SELECTED YEARS, 1960–1983

	Year Founded	*Membership (in thousands)*			
		1960	*1969*	*1972*	*1983*
Progressive Era					
Sierra Club	1892	15	83	136	346
National Audubon Society	1905	32	120	232	498
National Parks & Conservation Association (NPCA)	1919	15	43	50	38
Between the Wars					
Izaak Walton League (IWL)	1922	51	52	56	47
The Wilderness Society (TWS)	1935	(10)	(44)	(51)	100
National Wildlife Federation (NWF)*	1936	—	465	525	758
Post-World War II					
Defenders of Wildlife	1959	—	(12)	(15)	63
Environmental Era					
Environmental Defense Fund (EDF)	1967	—	—	30	50
Friends of the Earth (FOE)	1969	—	—	8	29
National Resources Defense Council (NRDC)	1970	—	—	6	45
Environmental Action (EA)	1970	—	—	8	20
Environmental Policy Institute (EPI)	1972	Not a membership group			
TOTAL		124	819	1,127	1,994

Note: Numbers in parentheses are estimates made by the group in lieu of actual membership data.

*NWF had not yet created its Associate Member Category in 1960. (Its older affiliate member category is not comparable with the other groups' membership figures because NWF's affiliate members do not contribute directly to the national organization.)

ties were threatened unless the groups' goals were realized.[51] Significantly, the first occurred in the years before Earth Day, demonstrating that that event was the culmination of a process rather than its cause. During this period, 1960 to 1969, the former conservation groups' total membership increased almost sevenfold, from 124,000 to 819,000, as they became more aggressive and outspoken about threats to scenic national treasures and wildlife. The Sierra Club, the Wilderness Society, and the National Audubon Society each quadrupled its membership. They became genuinely national groups by recruiting large numbers of members who lived outside their traditional geographic bases. The Sierra Club also abolished its archaic membership requirements in favor of an open membership. Finally, a giant newcomer, the National Wildlife Federation, emerged on the scene. In 1962, the federation inaugurated an extremely successful direct mail campaign to recruit associate members. For a low membership fee it offered these members an attractive publication and the opportunity to support wildlife protection. The influx of new associate members—465,000 strong by 1969—quickly made the federation the largest environmental group of all.[52] Sportspeople were now a minority component of the environmental constituency; both Audubon and the Sierra Club dwarfed the no-growth Izaak Walton League, and the formerly tiny Wilderness Society almost matched it in size.

The next membership surge was due to the publicity surrounding Earth Day, when environmental issues were firmly placed on the national agenda. In just three years, 1969 to 1972, 300,000 new members were added to the environmental ranks, a 38 percent overall increase. The older groups were in the best position to reap the Earth Day harvest, because they had the best name recognition, commanded the financial resources and expertise to conduct large direct mail campaigns, and could make appeals based on both first- and second-generation issues. As a result, older groups accounted for five out of every six new environmental organization members after Earth Day.

The post-Earth Day period was a time of testing for the movement. A number of observers believed environmentalism would soon fade, as the public realized how much a thorough environmental cleanup would cost. The groups, in contrast, assumed their rapid growth rate would continue and budgeted accordingly. Neither view was correct. Although membership growth slowed almost to a standstill after 1972, causing budget overruns and staff cutbacks in several groups, the number of members did not decline overall, confounding those who labeled the movement a fad. Any lingering doubts about the continuing viability of the environmental cause were erased when

Reagan administration attacks on environmentalist gains stimulated a great new influx of members.[53] Although both the first- and second-generation issues were affected by the Reagan administration's policies, the groups most closely identified with the first-generation issues reaped the greatest harvest, thanks to the motivational appeal of those issues and the high visibility of James Watt's alleged misstewardship of the nation's resources as President Reagan's first secretary of the interior. The Wilderness Society grew by a phenomenal 144 percent between 1980 and 1983, the Sierra Club by 90 percent, and Defenders of Wildlife and Friends of the Earth by about 40 percent. Subsequent growth has been strong, so that by 1989 these groups counted a total membership of more than 2,500,000.[54]

Those reached by environmentalist direct mail are well able to afford the relatively nominal cost of membership, and many give additional donations. With the notable exception of the National Wildlife Federation's associate members, many of whom are blue-collar high school graduates, environmental contributors are disproportionately well educated and economically well off. In 1978, for example, over half the members of Environmental Action, the Environmental Defense Fund, the Sierra Club, and the Wilderness Society had completed one or more years of graduate school, a level of educational attainment achieved by only 7 percent of the general population. The differences in income, though less, were still quite striking. More than half of the general public had annual household incomes below $14,000 in 1978, but only 22 percent of these groups' members were at that level.[55]

Although the size and growth of the environmental groups' membership helps ensure their viability as a force in environmental politics, the "upscale" character of their membership—in part due to the fact that direct mail reaches certain segments of the population better than others—makes the movement vulnerable to the charge that it is elitist and therefore not genuinely representative of a broad public interest. In the late 1970s and early 1980s, several critics accused the groups of opposing classical liberal values (for example, economic growth, individualism)[56] and claimed they represented the views of a "new class," whose interests were at odds with working-class people and the poor.[57] Others raised questions about the degree to which the public interest groups advocate policies that give them increased access to the state and whether those policies benefit anyone other than a few professional "public citizens."[58] A somewhat different criticism was made by James Watt, who charged that the environmental group leaders who criticized him were out of touch with their members, who, he claimed, supported his policies.[59]

These attacks struck at the heart of the groups' legitimacy as advocates of the public's interest in governmental reform and, if sustained, could have seriously weakened their political clout. Criticism was defused, however, by the findings of national polls showing that even during the depressed economic conditions of the early 1980s, majorities supported the proenvironmentalist position on numerous questions, including those that posed sharp economic-environmental protection tradeoffs.[60] Particularly damaging to the argument that environmentalists are elitists was the finding that public support for strong environmental protection was remarkably consistent across income, educational, and age categories.[61] As to the legitimacy of the movement leadership's claim to speak for the members, a National Wildlife Federation membership poll showed extremely strong support for the federation's position on a number of policy issues on which it disagreed with Watt.[62]

THE DEVELOPMENT OF MODERN ENVIRONMENTAL ADVOCACY

I have described the institutionalization of the environmental movement, its issue orientation, and the expansion of its active constituency. The movement's role in environmental politics has turned on its ability to translate these advantages into effective tactics. Three basic tactical options were available to the environmental groups: education, direct action, and policy reform. By the late 1960s, an exclusively educational approach, as practiced by National Audubon from the 1930s through the 1950s, was ruled out as insufficiently aggressive. Direct action, conversely, was regarded as too aggressive, although Environmental Action was sympathetic with this approach in the early 1970s.[63]

The national environmental groups adopted the third option and attempted to influence policy through traditional channels of political power. Their efforts are concentrated in Washington, for reasons stated by Brock Evans, formerly the Sierra Club's chief lobbyist:

> We must be there, because it is Congress that ultimately decides which areas shall be logged and which shall remain wild; it is the EPA that promulgates the vital air and water pollution regulations; and it is the President himself and his aides who, by a phone call, can often determine the fate of a bill in Congress or a policy in the bureaucracy.[64]

Depending on the particular legislative status of an issue and the resources the groups are willing to commit to it, contemporary environmental advocacy involves the use of one or more of the following tactics to further the cause: informational campaigns, congressional and White House lobbying, grassroots letter-writing campaigns, participation in administrative agency proceedings, and litigation. Organizations frequently practice the politics of ad hoc coalitions, where several groups (including nonenvironmental lobbies whenever possible) unite in a formal or informal coalition to work together on a given issue.[65] Recently, many of the groups entered electoral politics by establishing political action committees to fund favored candidates, some of whom also received campaign aid from local environmentalist workers.[66] With the exception of litigation and possibly candidate support, none of these tactics is new, having been employed by the conservation movement at one time or another. What is new is the scale of the environmental movement's advocacy, its sophistication, and its continuity in the sense that the groups now have the capacity to follow key issues from legislative enactment to the typically protracted, yet crucial, implementation stage, where an agency develops and enforces regulations designed to make the law work.

This strategy is compatible with the groups' mass-membership form of mobilization in which the members delegate authority to the group and provide it with the money necessary to lobby and litigate actively. Three developments—one legal, one legislative-administrative, and one organizational—have made this form of advocacy possible. These are as follows: changes in the rules of standing, which gave the groups access to the courts; changes in the tax laws and regulations, which made it possible for the groups to engage in noneducational advocacy without fear of losing their nonprofit tax status; and the professionalization of the groups' advocacy, which permitted a much greater degree of continuity and staff expertise on issues than the groups had ever had before.

After its inception as an environmentalist tactic in the mid-1960s, litigation quickly became one of the most important weapons in the environmentalist arsenal. The courts offered the Environmental Defense Fund a unique opportunity to debate the merits of DDT and enabled the fund to gain access to administrative decision making on the issue. These kinds of opportunities expanded greatly during the environmental era. The spate of environmental laws passed by Congress during the 1970s shifted responsibility for pollution control from the states to the federal government, thus enabling the groups' overworked lawyers to gain their objectives by filing a single suit in the federal courts instead of fighting legal battles on a state-by-state basis. The National Environmental Policy Act (NEPA), which was passed in 1969 and signed into law in January 1970, was used by environmental lawyers to challenge many federally funded projects on procedural grounds for not preparing acceptable environmental impact statements. In all, thousands of lawsuits have been filed by the environmental groups, and the volume of litigation continues unabated.[67]

According to Lettie Wenner's study of the outcome of 1,900 environmental cases in federal courts during the 1970s, the environmental point of view prevailed about half the time.[68] This is a respectable record, particularly since the loss of a case was not always fatal to the environmentalists' goals. Sometimes the delay occasioned by an ultimately unsuccessful legal challenge was enough to cause the cancellation of the project. On other occasions, such as the Alaska Pipeline, projects were modified as a result of the litigation in a way that made them more environmentally acceptable. Above all, the awareness that an environmental group might litigate its disagreement with proposed regulations enhanced its influence in shaping these regulations.

None of these outcomes would have been possible without the court decisions in the 1960s that liberalized the rule of standing. Before this time, the courts allowed federal agencies a great deal of discretion in how they chose to conduct their affairs as long as their decisions were in keeping with the authority given them by Congress. Even if an agency's legal mandate was questionable in this narrow sense, prevailing legal doctrines required an environmental group to prove an agency's action injured it or some of its members in a direct economic or physical way before the group's standing to raise the issue in court would be recognized. This test effectively ruled out most environmental litigation, because many environmental issues involve intangibles such as threats to scenic beauty or to wildlife or to the environment generally. The perceived hopelessness of litigation under these circumstances is well expressed in a lawyer's memorandum to the Wilderness Society in 1969 advising it not to waste its money joining a proposed suit against the Forest Service: "It is the law that the United States cannot be sued without its consent. . . . The principles listed above are elementary. I do not think the plaintiffs have a chance."[69]

These barriers were suddenly removed by landmark legal decisions that (*a*) broadened the class of interests entitled to seek judicial review and (*b*) recognized injuries other than economic or direct physical damage as grounds for standing.[70] These decisions are part of what Richard Stewart has called a "fundamental transformation" of administrative law.[71] Instead of assuming that the administrative agencies' decisions necessarily represented the public interest, the courts increasingly were willing to review challenges to these decisions brought by other interested parties, such as public interest groups. Because the lawyer consulted by the Wilderness Society had not comprehended the revolution or assimilated the Scenic Hudson decision, the society lost the opportunity to join the suit based on the Wilderness Act it had done so much to get passed. Other environmentalists pressed forward and won a landmark case that inaugurated a great deal of litigation on wilderness issues in the years to come.[72]

The second obstacle to effective advocacy that the environmental movement overcame was the threatened loss of the groups' nonprofit tax status if they engaged in litigation or lobbying. Without the money to conduct litigation, standing would have been of little avail to the infant environmental law firms. Litigation is very expensive. In the late 1960s, the Ford Foundation stood ready to foot the bill, but under the tax laws it was prohibited from doing so unless the law groups received a ruling from the Internal Revenue Service (IRS) that qualified them as "public foundations" under section 501(c)(3) of the tax code.[73] NRDC applied for this status in February 1969. In October the IRS abruptly announced that it was temporarily suspending all ruling on applications for tax exemptions from "public interest law firms" and other similar organizations pending a review of their eligibility for this status, which the IRS clearly questioned. The Ford Foundation immediately deferred a $310,000 grant to NRDC, and its contemplated support for EDF and the Sierra Club Legal Defense Fund, among others, was put in question. The fate of the young environmental law movement hung in the balance.[74]

A little more than a month later, the IRS backed off. The favorable outcome was a tribute to the environmentalists' political clout and the respectability of the environmental law groups. NRDC and the environmental community waged an intense campaign to influence the decision of the IRS. NRDC's board members, themselves establishment figures, were able to enlist the help of well-placed individuals in the legal profession, the government (especially Russell Train, chairman of the Council on Environmental Quality), and the Republican party, as well as editorial writers, to petition the IRS on behalf of the environmental law groups.[75] On November 12, 1969, the IRS commissioner threw in the towel and released a set of interim guidelines that cleared the way for the environmental law groups to receive their coveted nonprofit tax status.

Lobbying was another matter. The conservation groups were all 501(c)(3) organizations, qualifying as educational public foundations. Not only did this status relieve them from paying income taxes, but also it gave them highly preferential postal rates, as well as the right to receive tax deductible contributions and grants from foundations. According to the letter of the law, however, their educational activities were to be confined to presenting full and fair expositions of both sides of pending legislation.[76] The 1954 Tax Act specified that no "substantial" part of the activities of a 501(c)(3) organization could be devoted to influencing legislation in a particular direction, either by encouraging letter-writing campaigns to legislators or by lobbying them directly. Because most of the groups were dedicated to protecting the environment as well as educating the public, this limitation was very constricting indeed.

One early response to this situation had been the creation in 1954 of an umbrella organization, funded by non-tax deductible money, called the Citizens Committee on Natural Resources. Its purpose was to support a full-time Washington lobbyist, W. Spencer Smith, who worked on behalf of the several groups that helped pay his salary. For the next fifteen years, Smith was the only registered environmental lobbyist on Capitol Hill. Since he served many masters, several of whom were "wise use" conservation organizations such as the Wildlife Institute, this arrangement did not suit the needs of many of the environmental organizations. The Wilderness Society and the Sierra Club, for their part, pushed ahead on their own during the 1950s and the 1960s with strong campaigns to influence legislation and hoped for the best. The Wilderness Society's long campaign for the Wilderness Act, which was passed in 1964, went unchallenged by the IRS. Two years later, however, the Sierra Club lost its tax-exempt status after a highly publicized campaign against the proposed Grand Canyon dams.

Congress revised the relevant portion of the tax law in the Tax Reform Act of 1969. On the face of it the new law did not improve the situation. According to an interpretation of this law by some environmentally sympathetic tax attorneys, any legislation truly important to an environmental organization "cannot be openly opposed or supported" without running the real risk of losing its 501(c)(3) status.[77] However, the 1969 act did offer some additional leeway. One approach, adopted by FOE, Environmental Action, the Environmental Policy Institute, and the Sierra Club (involuntarily), was the sister organization strategy. These groups elected the less advantageous 501(c)(4) tax category that allowed them unfettered advocacy, but at the same time they each set up a sister 501(c)(3) organization (such as the Environmental Action Foundation) to receive tax-exempt contributions for research and education work.[78] Other groups followed a "5 percent strategy" by restricting their lobbying expenses to no more than 5 percent of their budgets on lobbying, a figure believed to be a "safe" level, although no one really knew for sure.

The outcome of these strategies was a large increase in the number of Washington environmental lobbyists. Figures are not available, but it seems likely that the groups employed (including Spencer Smith) the equivalent of two full-time lobbyists in 1969.[79] By 1975, the number was about forty. Some groups, most notably the National Audubon Society, were reluctant to risk the 5 percent strategy and unable to implement the sister organization strategy. Audubon took the lead in pressing for a revision in the tax law that would define a permissible (and reasonable) level of lobbying for the 501(c)(3) public interest groups. A coalition of these organizations won a significant victory in the Tax Reform Act of 1976, which, among other things, allowed them to spend up to 20 percent of their budget on lobbying.[80] Once this law was passed, the already substantial number of full-time environmental lobbyists was increased, and some important new players entered the lobbying lists. Audubon quickly opened a Washington lobbying office, and the environmental law groups hired lobbyists for the first time. By June 1985, the twelve organizations counted a total of eighty-eight Washington lobbyists.[81]

The last of the three obstacles to effective political action on the part of the environmental groups was the amateur character of their advocacy efforts. Until the late 1960s, the number of paid staff concerned with policy issues was small, and many groups were dominated by leaders who possessed charismatic qualities and exerted strong personal leadership.[82] When advocacy campaigns were undertaken, the groups often relied on volunteers for legal advice, lobbying, and congressional testimony. Fund raising, personnel and accounting procedures, and planning—both substantive and financial—were often ad hoc.

The pressures for professionalization converged from three directions. First, as the pace of legislation increased along with the complexity of the issues and the need to track their implementation through the bureaucratic labyrinth, the need for full-time professional advocacy staff became urgent. Much of this need was in Washington, working with Congress and the various agencies. The groups headquartered in other parts of the country either established Washington offices or moved their headquarters to the capital.[83] Second, the groups' rapid increase in membership stretched their management capacity. When membership growth skidded to a halt in the early 1970s and expected revenues failed to materialize, the ensuing financial crunch exposed the weakness of some of the groups' financial planning and controls. Since then, the need for professional management and expertise has become more and more urgent in order to keep the environmental ship from floundering. Third, to keep their nonprofit tax status and satisfy the reporting requirements of nonprofit regulators, the groups were forced to adopt complex organizational arrangements and maintain detailed records of their expenditures by purpose.

The result was a shift from an essentially amateur management to a professional form of advocacy characterized by paid advocacy staffs, planning exercises, budgets, and financial controls. The dramatic increase in the number of Washington environmental lobbyists has already been noted. In addition, legal and scientific specialists were hired in increasing numbers.[84] The groups formalized their financial operations and hired professional fund raisers. In 1984–85 more than half of the groups hired new executive heads,[85] and many boards placed management ability near the top of the list of the characteristics they sought.[86] By the end of the 1970s, many of the lobbyists who had been hired a decade before were still at their posts or, another sign of professionalism, had taken better-paying jobs with the Carter administration or with other environment organizations. Their accumulated experience was an important resource for the groups.[87]

The transition from amateur to professional lobbies was aided by the continued societal consensus about the need for environmental protection. This consensus reduced the movement's need to rely on charismatic figures to engender public support and minimized the organizational inefficiency and instability that often accompanies this type of leadership. It is striking that, with the exception of Friends of the Earth founder David Brower, none of the groups' leaders is a nationally known spokesperson for the environmental cause. The groups' authority does not rest on personal leadership in the way that Public Citizen is identified with Ralph Nader or the Southern Christian Leadership Conference with Ralph Abernathy. Their boards of directors—some elected by the membership in competitive elections, some not—have played a crucial role in ensuring organizational continuity when leadership problems occurred. Over the past decade, for example, the boards of the Wilderness Society, Environmental

Defense Fund, and Defenders of Wildlife, among others, have fired executive directors.

The Wilderness Society, which celebrated its fiftieth anniversary in 1985, offers a particularly dramatic example of the movement's professionalization. In the 1960s, its activism and effectiveness were largely due to the individual efforts of Howard Zahniser and his successor as executive secretary, Stewart Brandborg. As its membership, budget, and staff expanded, the society fell into such organizational disarray that the board was forced to fire Brandborg in 1976.[88] For the next three years, under two executive secretaries, the organization floundered. In 1978, another executive director, William Turnage, hired an all-new staff whose hallmark, according to the society's historian, was "professionalization."[89] The new chief lobbyist was hired away from the Sierra Club, former senator Gaylord Nelson accepted a newly created post as the society's full-time chairman, and a well-known economist came to head the society's new Economic Policy Department, the result of a major foundation grant. A Forest Management Department was staffed with professional foresters, ecologists, and resource policy specialists. By the end of 1984, the society had twenty full-time Washington lobbyists, eight regional offices, and a team of more than a dozen specialists assembled to analyze, propose responsible alternatives to, and, if necessary, to appeal the more than one hundred forest fifty-year plans that were scheduled to be issued by the federal government in 1985 and 1986.

CONCLUSION

The environmental movement is exceptional among American social movements in the large number of its separate organizations. In this chapter I have focused on the evolution of twelve major environmental lobbies which among them have several million members and almost a hundred Washington lobbyists.

The emergence of the environmental movement and of its ability to continue as a potent force in environmental politics is due to a combination of factors. Broad changes such as higher levels of affluence, the discontent of the 1960s, the decline of political parties, and increasing levels of pollution set the stage for widespread concern about environmental issues. The existence of the older groups meant that a core constituency was already mobilized. Financial support was available at crucial points from Ford and other environmentally sympathetic foundations, labor unions, and well-heeled individuals. The new technology of direct mail was a necessary condition for recruiting mass memberships, another important source of funds as well as of legitimacy for the groups. Of even greater importance is the environmental issue itself. Its universal character, diversity, evocative symbolism, and importance as a meaningful critique of modern society, even now, when concern about material issues is resurgent in America, all make the environmental issue itself the movement's most important resource. Public opinion has consistently been sensitive to environmental problems and supportive of the movement's actions.

The movement's dominant strategies to achieve its goals are to lobby and litigate at the national level. Its ability to do so with reasonable effectiveness in the face of strong opposition from traditionally powerful business and industrial interests was greatly assisted by several important institutional developments, which include changes in the rules of standing and the tax laws, and the groups' ability to professionalize their advocacy.

In the next decade the environmental movement will be hard-pressed to maintain its institutional and political momentum. Foundation support is harder to come by than ever. Reliance on the contributions of isolated individuals through direct mail is vulnerable to long-term shifts in public concern. Two of the movement's strengths—its diversity and its professionalism—may ultimately prove troublesome. The large number of environmental groups and the financial weakness of several of the smaller organizations raise questions about institutional survival. Increased professionalization carries with it the dangers of routinization in advocacy, careerism on the part of staff members, and passivity on the part of the volunteers. These are only straws in the wind, however. Warnings that the movement is facing imminent hard times have been issued periodically over the past decade. For the present and the foreseeable future, environmental problems are likely to worsen rather than improve, and to the extent that the public believes this is the case there is every reason to believe the environmental movement will remain a significant political force.

NOTES

The author wishes to thank Susan J. Pharr, John McCarthy, Christopher Leman, and Michael J. Lacey for their comments on earlier versions of this paper, and Kathy D. Wagner-Johnson for her research and editorial assistance.

1. Owing to space limitations I will not describe the movement's numerous and significant local, state, and regional organizations in any detail, despite the fact that some of them play important roles at the state level and even, on occasion, in national politics. Several of these groups are large. The California-based Save the Redwoods League has sixty thousand members, and the Massachusetts Audubon Society, which is independent of the National Audubon Society, has seventy

thousand. The subnational groups are often an important resource for the national groups and vice versa. For example, national groups seeking a wilderness designation for a particular area often rely on local wilderness activists to mobilize local support and for testimony about the area's wilderness values. The local groups need the help of the national wilderness groups to lobby for their cause in Congress. Grassroots environmental activism is one of the movement's great strengths, but also, predictably, a source of friction, as the locals sometimes criticize the nationals as being arrogant and manipulative, and the nationals criticize the locals for being parochial and unrealistic in their goals.

2. Doug McAdam, John D. McCarthy, and Mayer Zald, "Social Movements: Building Macro-Micro Bridges," in *Handbook of Sociology*, ed. Neil Smelser (Beverly Hills: Sage Publications, 1988).
3. Grant McConnell, "The Conservation Movement—Past and Present," *Western Political Quarterly* 7 (1954): 463.
4. Ibid., 467.
5. Samuel P. Hays, *Conservation and the Gospel of Efficiency* (Cambridge, Mass.: Harvard University Press, 1959), 185–98.
6. Stewart Udall, *The Quiet Crisis* (New York: Holt, Rinehart and Winston, 1963).
7. Rachel Carson, *Silent Spring* (Greenwich, Conn.: Fawcett Crest Books, 1962).
8. Ibid., 218. She did not use the word "ecological" to describe this worldview in her book, although her approach is clearly ecological in the sense in which the word came to be used in the late 1960s to characterize the environmentalist approach to environmental problems.
9. Carl W. Buchheister and Frank Graham, Jr., "From the Swamps and Back: A Concise and Candid History of the Audubon Society," *Audubon* 75, no. 1 (1973): 4–45.
10. Anthony W. Smith, "The Environmental Challenge," in *Toward an Environmental Policy: The Policy and Editorial Comment of the National Parks and Conservation Association over the Years of the Expanded Program, 1958–1971* (Washington, D.C.: National Parks and Conservation Association, 1971), 176. (First published in *National Parks and Conservation Magazine: The Environmental Journal*, Apr. 1970.)
11. William B. Devall, "The Governing of a Voluntary Organization: Oligarchy and Democracy in the Sierra Club" (Ph.D. diss., University of Oregon, 1970).
12. Ibid., 120.
13. Roderick Nash, "Conservation and The Colorado," in *The Grand Colorado*, ed. T. H. Watkins (San Francisco: American West Publishing Co., 1969); Stephen Fox, "We Want No Straddlers," *Wilderness* 48, no. 167 (1985): 13.
14. Devall, "The Governing of a Voluntary Organization," 217.
15. The club's support of the campaign to establish a Bay Conservation and Development Commission to "save" San Francisco Bay in the late 1960s was its first major battle involving the protection of an urban environment.
16. Edgar Wayburn, "Survival Is Not Enough," *Sierra Club Bulletin* 55, no. 2 (1970): 2.
17. Ibid.
18. Jack L. Walker argues more generally that outside sources of funding such as this are a major factor in the rise and fall of social movements, "The Origins and Maintenance of Interest Groups in America" (paper presented Oct. 29, 1981, at the Woodrow Wilson International Center for Scholars, Washington, D.C). For the environmental movement, at least, this generalization is truer for the emergence of new groups than it is for their persistence. Foundation support has greatly aided the groups over the years; see Jeffrey M. Berry, *Lobbying for the People: The Political Behavior of Public Interest Groups* (Princeton, N.J.: Princeton University Press, 1977), and Heritage Foundation, "The Environmental Complex," Institution Analysis no. 4 (Washington, D.C.: The Heritage Foundation, 1977), for details. Nevertheless, the groups' persistence and, especially, their strength would be far less without their ability to mobilize gifts from individuals via direct mail.
19. In the late 1950s, a group of Long Island citizens, led by a well-known ornitholo-

gist, tried to obtain a court injunction against local DDT spraying. Despite appeals that went all the way to the Supreme Court, no relief was granted. The best short account of the court battles on DDT in the 1960s is Joel Primack and Frank Von Hippel, *Advice and Dissent: Scientists in the Political Arena* (New York: New American Library, 1974).

20. This fund was provided with $100,000 by the Ford Foundation, with the understanding that it would be used for the EDF effort in Wisconsin (interview with Arthur Cooley). Ford chose this funding vehicle in order to avoid identifying itself too closely with what was a new type of venture for the foundation.
21. Interview with Gordon Harrison, Feb. 15, 1978.
22. EDF received a total of $994,000 from Ford from 1971 to 1977. Ford also funded the Sierra Club Legal Defense Fund, which served largely as the Sierra Club's legal arm ($603,000) and the Center for Law in the Public Interest, a Los Angeles group that conducted environmental litigation in California ($1,584,000). Ford Foundation, "Offices of Resources and the Environment: Grants 1965–1977" (New York: The Ford Foundation, 1977).
23. Gordon Harrison, *Serving the Earth* (New York: The Ford Foundation, n.d.).
24. The NRDC founders' organizational model was the NAACP Legal Defense Fund.
25. Amory Lovins, "Energy Strategy: The Road Not Taken?" *Foreign Affairs* 55 (1976): 65–96; and *Soft Energy Paths* (New York: Harper, 1977).
26. Another lasting consequence of Earth Day was the influx of new ideas and a new generation of activists into the environmental movement generally. A number of the student ecoactivists now occupy high positions in groups such as the Sierra Club, the Wilderness Society, and Friends of the Earth.
27. Interview with Denis Hayes, 14 October 1976.
28. All but Environmental Action and Defenders of Wildlife belong to an informal coalition of leaders called the Group of 10 that meets periodically to discuss common strategies and problems. Rochelle L. Stanfield, "Environmental Lobby's Changing of the Guard Is Part of Movement's Evolution," *National Journal* 17, no. 23 (1985): 1350–53. When their interests coincided, they were joined in lobbying on certain issues by various nonenvironmental lobbies, such as the League of Women Voters, Common Cause, the Consumer Federation of America, the UAW, Ducks Unlimited, and the National Rifle Association. There are also a number of environmentally related organizations, such as the now defunct Solar Lobby and Zero Population Growth, which pursue issues of direct concern to environmentalists but do not lobby on either the land-wildlife or pollution issues so central to the environmental agenda.
29. David Vogel, "The Public-Interest Movement and the American Reform Tradition," *Political Science Quarterly* 95, no. 4 (1980–81): 607–27.
30. Brock Evans, "Lobbying: A Question of Resources," *Sierra* 63 (Oct.-Dec. 1978): 54–55. Andrew S. McFarland calls this set of beliefs the "civic balance" perspective, *Public Interest Lobbies: Decision Making on Energy* (Washington, D.C.: American Enterprise Institute, 1976), 89. It is most clearly articulated by Ralph Nader and the leaders of Common Cause, but well defines the implicit beliefs of most Washington environmental leaders and lobbyists. Andrew S. McFarland, *Common Cause: Lobbying in the Public Interest* (Chatham, N.J.: Chatham House Publishers, 1984).
31. See joint policy statements issued by the groups such as *The Unfinished Agenda*, ed. Gerald O. Barney (New York: Thomas Y. Crowell 1977), and *An Environmental Agenda for the Future* (Washington, D.C.: Island Press, 1985). The latter represents "a consensus reached by the leaders of 10 major conservation groups." Of the groups considered in this chapter only Defenders of Wildlife and Environmental Action were not represented.
32. The roots of their radicalism are quite different. FOE's comes from a view of the moral primacy of nature at this stage in human economic development, whereas EA's is based on the belief that capitalism is inherently destructive to the environ-

ment. Thus FOE would be much more sympathetic to deinstitutionalization, a tenet of "deep ecology" that calls for economic change away from large-scale industries, than EA, which would be concerned with the program's effect on workers. William B. Devall, "The Deep Ecology Movement," *Natural Resources Journal* 20, no. 2 (1980): 299–322.

33. Allan Schnaiberg, *The Environment: From Surplus to Scarcity* (New York: Oxford University Press, 1980), 371.
34. Robert Cameron Mitchell, "How 'Soft,' 'Deep,' or Left? Present Constituencies in the Environmental Movement for Certain World Views," *Natural Resources Journal* 20, no. 3 (1980): 345–58.
35. Robert Cameron Mitchell, "From Elite Quarrel to Mass Movement (The Anti-Nuclear Power Movement)," *Society* 18, no.5 (1981): 76–84.
36. Denton E. Morrison, "How and Why Environmental Consciousness Has Trickled Down" (paper presented at a conference, Distributional Conflicts in Environmental Resource Policy, International Institute for Environment and Society, Mar. 26–27, 1984, at Science Center, West Berlin), 227–33.
37. Mitchell, "How 'Soft,' 'Deep,' or Left?" 354.
38. Mitchell, "How 'Soft,' 'Deep,' or Left?"; Riley E. Dunlap, "Public Opinion and Environmental Policy," in *Environmental Politics and Policy,* ed. James P. Lester (Durham, N.C.: Duke University Press, 1989).
39. Riley E. Dunlap, "Paradigmatic Change in Social Science: From Human Exemptionalism to an Ecological Paradigm," *American Behavioral Scientist* 24 (1980): 5–14; Riley E. Dunlap and Kent D. Van Liere, "Commitment to the Dominant Social Paradigm and Concern for Environmental Quality," *Social Science Quarterly* 65, no. 4 (1984): 1013–28; Lester W. Milbrath, "Environmental Beliefs and Values," in *Political Psychology*, ed. Margaret G. Hermann (San Francisco: Jossey-Bass Publishers, 1986), 97–138; Morrison, "How and Why Environmental Consciousness Has Trickled Down."
40. Robert Cameron Mitchell, "Public Opinion and Environmental Politics in the 1970s and 1980s," in *Environmental Policy in the 1980s: Reagan's New Agenda,* ed. Norman J. Vig and Michael E. Kraft (Washington, D.C.: CQ Press, 1984).
41. The Environmental Policy Institute (then named Center) has relied solely on large contributions from well-heeled patrons and foundations since its founding, and this source of contributions has, apparently, been stable over the years. They experimented with recruiting a mass membership at one point in the late 1970s, but gave it up within a year as more trouble than it was worth.
42. EPI overcomes the problem of legitimacy by working with local environmental organizations, such as river preservation groups, on congressional legislation of mutual interest to both parties.
43. Mancur Olson, *The Logic of Collective Action: Public Goods and the Theory of Groups* (Cambridge, Mass.: Harvard University Press, 1971); Harriet Tillock and Denton Morrison, "Group Size and Contributions to Collective Action: An Examination of Mancur Olson's Theory Using Data from Zero Population Growth, Inc.," in *Research on Social Movements, Conflict, and Change*, ed. Louis Kreisberg (New York: JAI Press, 1979).
44. Kenneth R. Godwin and Robert Cameron Mitchell, "The Implications of Direct Mail for Political Organizations," *Social Science Quarterly* 65, no. 3 (1984): 829–39.
45. Larry J. Sabato, *The Rise of Political Consultants: New Ways of Winning Elections* (New York: Basic Books, 1981).
46. McFarland, *Common Cause*, 79–81. Environmental groups send prospect mailings that often amount to millions of pieces to lists of people identified as potential members. These lists are typically the membership lists of other, like-minded groups, which are obtained in exchange, but subscription lists to certain magazines or lists of contributors to certain political candidates may also be used. Thanks to the low postal rate available to nonprofit groups for bulk mailings, a return rate as low as one percent may suffice to enable a mailing to eventually pay for itself, once

the new recruits renew their memberships. The technology is heavily dependent on the use of computers to address the prospect mailings, manage the members' accounts, and direct the sequence of renewal requests (some groups send as many as seven) to delinquent members.

47. Direct mail is not without its difficulties and drawbacks. Soliciting by direct mail is subject to a saturation effect whereby above a certain level for a given group, ever larger numbers of mailings fail to give a viable return. Moreover, direct mail campaigns require a large capital investment in prospect mailings that do not pay for themselves until a year or two later, when the new members renew their memberships. Sabato, *The Rise of Political Consultants*, 226–33. Contrary to John D. McCarthy and Mayer N. Zald's hypothesis that the greater the resources a group received from isolated constituents, the less stable its income ("Resource Mobilization and Social Movements: A Partial Theory," *American Journal of Sociology* 88, no. 6 [1977]: 1228), income from direct mail is unlikely to fluctuate rapidly since it is based on thousands of individual decisions whose aggregate outcome is relatively stable, especially when compared to the alternatives of relying on a relatively few foundation grants or big gifts from individuals.
48. McCarthy and Zald, "Resource Mobilization and Social Movements," 1215. See Edward J. Walsh and Rex H. Warland, "Social Movement Involvement in the Wake of a Nuclear Accident: Activities and Free Riders in the TMI Area," *American Sociological Review* 48, no. 6 (1983): 764–80, for a review of this debate and a test based on a local instance of social movement mobilization.
49. Excluding EPI, which does not rely on direct mail.
50. This is for the regular individual membership. More expensive categories of membership are offered to families. Some groups offer less expensive memberships for senior citizens and students. Although the groups offer some individual benefits, the most notable of which is the group's publication, only a small portion of the membership fee goes for this purpose.
51. For a theory of public interest lobby membership mobilization that emphasizes the importance of losses versus gains, see Robert Cameron Mitchell, "Silent Spring/Solid Majorities," *Public Opinion* 2, no. 3 (Aug.-Sept. 1979): 16–20, 55.
52. Defined in terms of direct members—the federation also had a million or more affiliate members at this time.
53. Mitchell, "Public Opinion and Environmental Politics in the 1970s and 1980s."
54. Since many people support more than one group, this figure probably represents about 1.4 million individual environmentalists.
55. These data come from a survey of a thousand-person random sample of the groups' membership and from a national telephone survey of the general public, both conducted by the author in 1978.
56. Paul H. Weaver, "Regulation, Social Policy, and Class Conflict," *The Public Interest* 13, no. 50 (1978): 45–63.
57. The "new class" was thought basically to consist of upper middle class professionals. Bernard Frieden, *The Environmental Protection Hustle* (Cambridge, Mass.: M.I.T. Press, 1979); William Tucker, "Environmentalists and the Leisure Class," *Harper's*, Dec. 1977, 49–80, and *Progress and Privilege: America in the Age of Environmentalism* (Garden City, N.Y.: Doubleday, Anchor Press, 1982); Tim Peckinpaugh, "The Specter of Environmentalism: The Threat of Environmental Groups" (Republican Study Committee, Special Study, Washington, D.C., Feb. 12, 1982).
58. Vogel, "The Public-Interest Movement and the American Reform Tradition."
59. Joanne Omang, "Watt Finds Time to Hear Audubon Society," *Washington Post*, May 15, 1981.
60. Everett Carl Ladd, "Clearing the Air: Public Opinion and Public Policy on the Environment," *Public Opinion* 5 (Feb.–Mar. 1982): 19; William Schneider, "The Environment: The Public Wants More Protection, Not Less," *National Journal* 15 (Mar. 26, 1983): 676–77; Riley E. Dunlap, "Public Opinion on the Environment in the Reagan Era," *Environment* 29 (1987): 6–11, 32–37.

61. Mitchell, "Silent Spring/Solid Majorities"; Ladd, "Clearing the Air"; Morrison, "How and Why Environmental Consciousness Has Trickled Down."
62. Jay D. Hair and Patrick A. Parenteau, "Special Report: The Results of the National Wildlife Federation's Associate Membership Survey and Affiliate Leadership Poll on Major Conservation Policies" (Washington, D.C.: The National Wildlife Federation, 1981). The federation, with its high proportion of working class members, was a crucial test case for Watt's challenge.
63. It viewed with favor the efforts of an elusive activist, known as "the Fox," who for a time embarrassed corporations by such gestures as pouring noxious wastes from their discharge pipes on the rugs in their headquarters' reception areas. More recently, groups like Greenpeace and a small wilderness group called Earth First! have used direct action to try to block objectionable private or government activities.
64. Evans, "Lobbying: A Question of Resources," 54.
65. McFarland, *Public Interest Lobbies*; Jeffrey M. Berry, *Lobbying for the People*.
66. Efforts of this sort were previously restricted to the nonpartisan League of Conservation Voters.
67. Suits were filed by EDF and NRDC as well as by some of the other groups who acquired litigation capabilities of their own, most notably the Sierra Club and the National Wildlife Federation. Groups without a legal staff also file cases by engaging outside counsel.
68. Lettie M. Wenner, *The Environmental Decade in Court* (Bloomington, Ind.: Indiana University Press, 1982).
69. Dennis M. Roth, *The Wilderness Movement and the National Forests: 1964–1980*, prepared for U.S. Department of Agriculture, Forest Service (Washington, D.C.: Government Printing Office, 1984), 21.
70. Court decisions in cases such as *Scenic Hudson Preservation Conference v. FPC* (354 F.2d 508, 2d Cir. 1965, *cert. denied*, 384 U.S. 941, 1966) and *Office of Communications of the United Church of Christ v. FCC* (359 F.2d 994, D.C. Cir. 1966) enabled environmental groups to obtain standing to argue against agency decisions that injured their members' values for scenic beauty, environmental protection, and recreation. Even though the Supreme Court rebuffed the Sierra Club's attempt to extend further the new doctrine of standing when it claimed standing as a surrogate plaintiff for all present and future users of the Mineral King Valley without specific reference to its members' use of the valley (*Sierra Club v. Morton*, 405 U.S. at 737, 1972), this decision had no practical effect on the ability of environmental groups to litigate, because the groups could still act as a representative plaintiff for those of its members who suffered an individualized injury from a given agency action. Karen Orren, "Standing to Sue: Interest Group Conflict in the Federal Courts," *American Political Science Review* 70 (1976): 739; Joseph F. DiMento, "Citizen Environmental Litigation and the Administrative Process: Empirical Findings, Remaining Issues and a Direction for Future Research," *Duke Law Journal* 22, no. 2 (1977): 409–48. Thus, in an important case decided a year later (*U.S. v. SCRAP*, 412 U.S. at 669, 1973), the members of an ad hoc environmental group that asked that railroads be restrained from collecting a surcharge that would make recyclable materials more expensive to transport were given standing on the grounds that some of them, such as campers and sightseers, used the forests in the Washington, D.C., area, which would ultimately be affected by this surcharge.
71. Richard B. Stewart, "The Reformation of American Administrative Law," *Harvard Law Review* 88, no. 8 (1975): 1669–1813.
72. *Baker vs. Forest Service* as cited by Roth, *The Wilderness Movement and the National Forests: 1964–1980*, 23.
73. Groups receiving this status enjoy other benefits described below. The loss of this status does not necessarily mean that a group loses all the benefits of a nonprofit designation, as they will qualify for designation as a "social welfare organization"

under section 501(c)(4). Organizations of this type are permitted to engage in one-sided advocacy and be tax-exempt as long as the primary purpose of the advocacy is to promote the common good and general welfare. They cannot receive tax deductible contributions, however, nor are they eligible for the low postal rates accorded the public foundation.

74. Richard Corrigan, "Tax Report/Public Interest Law Firms Win Battle With IRS Over Exemptions, Deductions," *National Journal* 2, no. 47 (1970): 2541–49.
75. Ibid.
76. *Law and Taxation: A Guide for Conservation and Other Nonprofit Organizations*, prepared for the Conservation Foundation by Berlin, Roisman, and Kessler (Washington, D.C.: The Conservation Foundation, 1970).
77. Ibid., 25.
78. The tax law permitted an arrangement whereby the (c)(4) organization could contract with its (c)(3) partner for the research needed for its lobbying efforts. See Berry, *Lobbying for the People*, for a discussion of the tax status of nonprofit lobbying groups.
79. In the discussion that follows, "lobbying" includes attempts to influence both the passage of legislation and its implementation by administrative agencies. For firsthand accounts of how the Sierra Club developed its Washington lobbying efforts, see Ann Lage, *Building the Sierra Club's National Lobbying Program, 1967–1981; Brock Evans: Environmental Campaigner; From the Northwest Forests to the Halls of Congress; and W. Lloyd Tupling: Sierra Club Washington Representative, 1967–1973* (Berkeley: University of California Regional Oral History Office, 1985).
80. John Walsh, "Lobbying Rules for Nonprofits: New Option Sets Specific Limits," *Science* 196 (1977): 40–42.
81. This figure is for full-time equivalents. Some staff members devote only part of their work to lobbying.
82. Michael McCloskey, "Wilderness Movement at the Crossroads, 1945–1970," *Pacific Historical Review* 41 (1972): 346–61.
83. Staff were also needed outside of Washington. Groups working on wilderness issues, for example, found they needed to hire field representatives to mobilize local grassroots support for proposed wilderness areas. Other groups followed suit on issues such as toxic waste dumps, which were site specific and required grassroots pressure in Washington.
84. The need to develop a scientific basis for their cases led the law groups to hire scientists who were both professionally competent and able to engage in advocacy rather than bench science. Robert Cameron Mitchell, "Since Silent Spring: Science, Technology and the Environmental Movement in the United States," report no. 5 in *Proceedings of Institute for Studies in Research and Higher Education Conference on Scientific Expertise and the Public* (Oslo, Norway: The Norwegian Research Council for Science and the Humanities, 1979), 171–207. Increasingly, as the 1970s wore on, other environmental groups saw the need to acquire scientific staff, including economists, capable of addressing technical issues either by pulling together the available knowledge or, in some cases, by conducting basic research aimed at key issues.
85. Stanfield, "Environmental Lobby's Changing of the Guard."
86. Grant P. Thompson, "New Faces, New Opportunities: The Environmental Movement Goes to Business School," *Environment* 27, no. 4 (1985): 6–30.
87. In 1985, the Environmental Policy Institute boasted that its staff had an average of seven years of experience in their fields, "while one-third of the Senate and 45 percent of the House have only been in Washington since 1980." Bureau of National Affairs, "Environmental Groups Plan to Use Greater Strength, Sophistication in Challenging Reagan during Second Term," *Environment Reporter* 75, no. 2 (1985): 1671.
88. Fox, "We Want No Straddlers."
89. Ibid.

3

PARKS, WILDERNESS, AND RECREATION

Joseph L. Sax

Parks and wilderness areas are not the product of the modern growth of regulatory government or of the expansion of the federal role in American life. They are, instead, elements of a peculiar byway of American government, an experiment in public ownership and management that has no significant counterpart in our national experience.

ROLE OF THE FEDERAL GOVERNMENT BEFORE WORLD WAR II

When the Constitution was adopted, no one anticipated that the federal government would become the permanent proprietor of nearly a third of all American land. Indeed, for more than a century, the government assiduously sold or gave away its lands to settlers, to the railroads, and to the states with no thought of maintaining a public domain. But by the time the era of disposition more or less came to an end with the Taylor Grazing Act of 1934,[1] tens of millions of acres were already reserved in national parks and national forests—lands that would become the focus of our contemporary national recreation, wilderness, and preserve policies.

Although there was some controversy over public ownership of these lands (especially the national forests), and although that controversy reemerges periodically,[2] as a practical matter the propriety of large-scale federal land ownership and management was a settled issue long before the era of "big national government" or federal regulation even began.

Perhaps the most useful way to conceive of the traditional federal recreational land system is by reference to what it was not: it was not essentially a regulatory regime, with government and regulatee in an adversary posture struggling over the content of rules and regulations. In fact, the governance of the national parks and national forests was so open-ended that it cannot be usefully compared with any example drawn from the contemporary structure of administrative law.

For the National Park Service, from the day of its creation in 1916, the governing statute was the organic act (a law that sets out the basic mission and responsibility of a bureau), which simply stated that the purpose of the parks was "to conserve the scenery and the natural and historic objects and the wild life therein and to provide for the enjoyment of the same in such manner and by such means as will leave them unimpaired for the enjoyment of future generations."[3]

Similarly, the Forest Organic Act of 1897, which was not significantly supplemented for more than half a century, provided that "no national forest shall be established, except to improve and protect the forest within the boundaries, or for the purpose of securing favorable conditions of water flows, and to furnish a continuous supply of timber for the use and necessities of citizens of the United States."[4] For decades before the enactment of the 1964 Wilderness Act, the Forest Service administratively designated "primitive" areas that were managed much as congressionally designated wilderness is managed today.[5] Precisely because such classifications were not legislatively mandated, the Forest Service had the flexibility to respond by modifying or withdrawing administrative designations.

The Bureau of Land Management, which had charge of almost two-thirds of the federally owned land (much of which is valuable for recreation), operated under no organic act at all until 1976.[6]

Why was such an open-ended and administratively free-wheeling system both possible and workable? One factor was the relative abundance of the resource. The national parks were little used in the period before the automobile became the key element in vacation travel, and in the period around World War I a major task of the Park Service was to encourage visitation and build a constituency.[7] It was only after World War II ended that the park system began to face major problems of crowding and insufficient facilities.[8]

Another factor was the almost uniform style of recreation that existed in the early decades of this century. Reports and photographs of the era show clearly that outdoor scenes were dominated by upper middle-class families vacationing in genteel style.[9] Nothing suggested that a time would come, as it did in the 1960s, when there would be a veritable riot at Yosemite, spawned by a confrontation between "traditional" visitors and the large numbers of young people who had come to the park with lifestyles and expectations imported from the streets of San Francisco and Berkeley.[10]

The third factor that made the pre-World War II system so manageable was that the Park Service had always limited national parks to relatively few places of superlative natural endowment, leaving to state, regional, and local authorities the job of providing ordinary recreation in (more or less) ordinary places. Although the Park Service aided those local officials, it never assumed overall responsibility for fulfilling the nation's recreational needs.[11]

After World War II, states and localities became less able to meet growing needs, and Congress stepped in to fill the void. The result was a startling increase in the size and scope of the national park system. Congress established national seashores and lakeshores along the oceans and Great Lakes, national recreation areas, and even national urban parks near New York, San Francisco, and Cleveland. These places, close to both populations and industry, generated a kind and level of controversy that was almost entirely new to the Park Service.[12] Towns were found in the middle of national park areas, as at Cape Cod and the Fire Island National Seashores. Local people felt that they needed these places for conventional recreation, activities of a kind that seemed very much at odds with the older preservation mandate.[13] The newer parks even had residential subdivisions and power plants as neighbors.[14] The consequences, inevitably, were growth in the amount and change in the nature of conflict, and much more adversarial activity than had ever been experienced before.

THE PARK SERVICE AND PLANNING

Despite this growth and change, the administrative structure of the national park system has changed very little over the past four decades. The parks still operate under the authority of the Organic Act of 1916,[15] which is treated with a Constitution-like reverence; new units of the park system are still created under acts of Congress that are similar to the establishing laws of the pre-World War II period. However, the management of the parks has changed dramatically.

A revealing index of the transformation that has taken place is the change in planning procedures. The Park Service has been committed to a planning process for many decades, reaching back to its earliest days under Steven Mather and Horace Albright. Master plans existed for many parks in the early 1930s,[16] and by the time of World War II a master plan existed for every unit of the national park system.[17] It was not until the early 1970s, however, that the nature of the planning process underwent a fundamental reorientation.

The place was Yosemite National Park, and the time, 1973. A new concessioner, the recreational conglomerate Music Corporation of America (MCA), acquired the old Yosemite Park and Curry Company. The Park Service, working with (some would say under the undue influence of) MCA, published a preliminary master plan for the park that called for considerable development, including modernized and increased hotel accommodations in Yosemite Valley and enlarged winter usage.[18] It was not surprising that the plan generated intense controversy, nor would it have been surprising if the Park Service and MCA had backed down in the face of strong opposition.

What was surprising, and new, was how the Park Service responded. It "went public," opening the planning process to a series of public meetings, preparing a workbook that showed in detail all the alternatives that might be adopted, and inviting a sort of referendum on its plans from the many thousands of individuals and organizations to whom it distributed the workbook.[19] Nothing could have been less characteristic of an agency with strong views about how parks should be managed and a self-image of elite leadership. The Yosemite planning process set a model for park planning that has been widely followed since.

The change in style of federal park management occurred without much formal change in the legal structure of the national parks. The other major federal recreation management agencies, the Forest Service and the Bureau of Land Management, have undergone similar changes in the way they perform, but in the context of more formal statutory changes enacted in their governing legislation.

THE FOREST SERVICE AND MUSY

The first of the major post-World War II statutory changes was the enactment in 1960 of the Multiple-Use, Sustained-Yield Act (MUSY) as a new organic law for the Forest Service.[20] Recreation was the central issue leading to the new law. The Forest Service had been operating for many decades under the Forest Organic Act of 1897, which described its functions in terms of only two uses, timber and watershed protection.[21] In fact the Forest Service had been active in the recreation business for a long time and had been doing multiple-use planning since the 1930s,[22] although it had no express legislative authority to do so.

The MUSY Act was not forced upon an unwilling Forest Service by the Congress. On the contrary, the Forest Service itself was the initiator and proponent of the MUSY Act.[23] One incentive for the new law was simply that demand for recreation areas was growing during the postwar period, increasing the possibility for conflict with the forests' traditional primary constituency, the wood products industry. The Forest Service believed that legislation would legitimate the multiple-use policy it had practiced for years and free it to continue managing the forests as it believed best.[24]

Other forces were at work as well. The National Park Service, also responding to growing demands for recreation, was in an expansionist mood, and the Forest Service hoped a statutory mandate to manage the forests for recreation would ward off the threat of losing valuable recreational land to the ambitions of the Park Service. There was nothing new in this. The Park Service and the Forest Service had battled for decades over control of prime recreational federal land, and the Forest Service's interest in providing recreation was partly a defensive tactic to thwart loss of such lands to the national park system.

A third incentive was the growing strength of the preservationist, or wilderness, element of the recreational constituency. Led by the Sierra Club, this constituency not only opposed the logging of high-quality recreational land but also insisted that it be left as wilderness, a position that the Forest Service felt undermined its own commitment to multiple use. Support of the MUSY legislation, including recreation as one equal (not preeminent) use of the forests, could strengthen the hand of the Forest Service in dealing with wilderness advocates and also, it seemed then, cut off the growing pressures for wilderness legislation.[25]

In 1960 the MUSY Act was passed, providing that "the national forests are established and shall be administered for outdoor recreation, range, timber, watershed and wildlife and fish purposes."[26] Although Forest Service officials believed that the law would preserve the managerial discretion it had long enjoyed, at least some observers saw the future more clearly. The National Lumber Manufacturers Association (which opposed the bill) insisted that "by making all uses equal in priority the forest manager will probably have to act on the basis of public pressure."[27] In fact, as with the national parks, this is a generally accurate description of what happened. Growing demand; constituencies enlivened by recognition of their "rights," their political power, and their visibility; and increased distrust of the Forest Service by recreationists who saw little change in primary commitment of the service to timber harvesting set the stage for a variety of public pressures.

Any hope the Forest Service may have had in 1960 that it would remain as free as it had been was quickly dashed. Its planning was now exposed to the equal "rights" created by the MUSY Act, a development that was intensified with the enactment of the National Environmental Policy Act (NEPA) at the end of 1969.[28] The NEPA required the agencies to incorporate an environmental impact statement into the planning process, formalizing planning procedures and opening the planning process to public scrutiny and to citizen-initiated lawsuits.

THE WILDERNESS ACT: A NEW ERA

Enactment of the MUSY Act did not alleviate pressures for wilderness legislation. Only a few years later, in 1964, Congress passed the Wilderness Act, which imposed on the Forest Service a duty to study areas within the forests that still retained wilderness qualities and turned over to Congress the right to designate official wilderness (rather than leaving it to the Forest Service). Congressional designation not only deprived the Forest Service of the right to reallocate such lands to other uses, but also imposed a set of stringent, congressionally mandated standards (such as prohibitions on structures and the use of motorized vehicles) that determined the management of wilderness.[29]

The Wilderness Act may well be viewed as the single most significant congressional enactment separating the old era of administrative discretion and expertise from the modern period of legislative skepticism toward the federal land-management bureaucracy. In the calendar of contemporary environmental law, the 1964 act is an early event, and it arose out of a series of specific concerns about the way the Forest Service used its discretion. As noted earlier, the Forest Service was the first federal agency to use a wilderness designation, which it did without any congressional mandate. Whether it deserves great credit for this far-seeing innovation, promoted by such extraordinary employees as Aldo Leopold and Robert Marshall, or whether wilderness designation was merely a ploy to hold on to high-quality lands in the face of Park Service ambitions is a matter about which there will never be total agreement.

What is indisputable is that in the years following World War II, some of the most prized administratively designated wilderness areas, such as the French Pete Valley in Oregon (returning from wilderness to general forest designation to permit logging) and the Gila Wilderness in New Mexico (partially opened to commercial logging) were being jeopardized by Forest Service reconsideration.[30] Events such as these generated pressures for the Wilderness Act. The approach taken in that act set the tone for subsequent federal land legislation by diminishing agency discretion and setting out explicit legislative standards to guide and control administrative action. The Wilderness Act may be thought of as the beginning of the legislative version of what has come to be known in the courts as the "strict scrutiny, hard look" approach.[31] In addition to requiring elaborate studies and legislatively establishing permissible uses, Congress also set out its own definition of wilderness rather than leave that question to the Forest Service, resulting in one of the rare flights of poetic language to be found in the United States Code: "A wilderness, in contrast with those areas where man and his own works dominate the landscape, is hereby recognized as an area where the earth and its community of life are untrammeled by man, where man himself is a visitor who does not remain. . . ."[32]

The developments just described fundamentally modified the posture of the Forest Service vis-à-vis its constituents, and modified it most unfavorably as to the preservationist constituency that traditionally had been among the weakest of its clients. From being among the most independent and self-governing of federal bureaucracies, the Forest Service now found itself among the most challenged. An agency which virtually had never had to defend its decisions in court became a familiar defendant.[33] Its timber management plan in Alaska's Tongass National Forest was challenged on the ground that it violated the MUSY Act by committing essentially the entire forest to timber harvesting.[34] In the White River National Forest in Colorado, it had to defend against the charge that it had undercut its wilderness study obligations by permitting timber harvesting in a "de facto" wilderness area that had not been either rejected or designated by the Congress.[35] A number of NEPA suits called into question the environmental dimension of the agency's forest planning procedures.[36] Even its professional expertise in clearcutting was condemned in a widely circulated external study on the Bitterroot National Forest in Montana, and then enjoined in a celebrated lawsuit at the Monongahela National Forest in West Virginia.[37] The *Monongahela* case, in which the court prescribed in detail methods of permissible timber harvesting based on the wording of an 1897 statute, threw Forest Service management into chaos and in effect demanded a congressional response modernizing the mandate for management of the national forests.[38] The result was the National Forest Management Act of 1976, which began as a law about clearcutting techniques but evolved into a more general organic act.[39]

For the purposes of this paper, only a few larger elements of this complex and elaborate law are significant. Overall, the statute is a predictable example of the congressional style of the 1970s: more planning, more public participation, more restrictions on traditional discretion, and more diversification of the mission of the agency. The statute requires a set of planning studies (which it explicitly requires to be made on the basis of interdisciplinary work), including an inventory; an assessment of uses, demands, and supplies; a long-term program for management of the forests; and plans for each unit of the system. At some levels the planning requirements are quite abstract, demanding that the Forest Service "state national goals that recognize the interrelationships between and interdependence within the renewable resources,"[40] and "provide for diversity of plant and animal communities based on the suitability and capability of the specific land area in order to meet overall multiple-use objectives."[41] In other respects, the statute is quite detailed and specific, setting out requirements for timber harvesting that ensure "that such lands can be adequately restocked within five years after harvest,"[42] controlling the method of clearcutting,[43] and setting out explicit standards for timber deemed suitable for harvesting.[44]

While the statute is complicated and abstract enough to ensure that the Forest Service will retain a good deal of discretion, there can be no doubt that Congress sought to control discretion to a much greater degree than ever before, in significant part to ensure greater attention to those environmental factors necessary for outdoor recreation, including wilderness preservation. One amusingly specific example of the legislators' determination to avoid "business as usual" is a requirement that the size of tracts designated for clearcutting be approved "by the responsible Forest Service officer one level above the . . . officer who normally would approve the harvest proposal."[45]

The law is also calculated to ensure active public participation, requiring that land management plans be available to the public at convenient locations, that there be public hearings and comments, that advisory boards "representative of a cross section of groups interested in the planning . . . and management . . . and the various types of use and enjoyment of the lands" be formed,[46] and that a number of specific decisions of a controversial type be made with public participation.[47]

THE BUREAU OF LAND MANAGEMENT

The Bureau of Land Management (BLM), the largest (in terms of acreage) and least known of the major federal recreation and wilderness agencies, also was subjected to a comprehensive organic act by the Congress in 1976, the Federal Land Policy and Management Act (FLPMA).[48] The BLM took the position that it, like the Forest Service, was a multiple-use agency, and certainly in some respects it was. But until the 1976 law was enacted, it had no statutory authority to manage its lands for multiple use and sustained yield,[49] nor did it have an express environmental mandate. BLM was created from the grazing service established by the Taylor Grazing Act of 1934, and certainly the range industry was its primary constituency, just as the forest products industry had been the primary constituency of the Forest Service.[50]

In enacting FLPMA, Congress followed much the same path it had taken with the Forest Service, diversifying the mission of the agency, restricting its discretion, and inserting a major role for public participation in its planning and management decisions. The act mandates an elaborate and comprehensive planning effort, using "a systematic interdisciplinary approach to achieve integrated consideration of physical, biological, economic, and other sciences."[51] It demands explicit attention to environmental issues, requiring that priority be given to identifying and protecting "areas of critical environmental concern";[52] it sets out a process for the study of potential wilderness areas and a scheme for their management, pending designation or rejection by the Congress.[53] It also adopts multiple use as the management principle for the BLM lands, but it gives a broader and more environmentally focused definition to that standard than do the MUSY Act and the National Forest Management Act: "recreation, range, timber, minerals, watershed, wildlife and fish, and natural scenic, scientific and historical values; and harmonious and coordinated management of the various resources without permanent impairment of the productivity of the land and the quality of the environment . . . and not necessarily to the combination of uses that will give the greatest economic return. . . ."[54]

Recreation areas are provided through several other federal programs, among the most important of which are the reservoir management of the Bureau of Reclamation and the Corps of Engineers, the Wild and Scenic Rivers System (which in effect protects the shorelines of designated rivers from federally permitted development), and the National Wildlife Refuge System, managed by the Fish and Wildlife Service in the Department of the Interior.[55] The refuges have been the subject of the most interesting statutory changes in recent years. Originally they were not conceived as sites of recreation at all, and the Migratory Bird Conservation Act of 1929 authorized acquisition of reserves "for use as inviolate sanctuaries for birds"[56] (though activities such as mining and grazing have long been permitted on some refuge lands).

By 1962, however, "in recognition of mounting public demands for recreational opportunities on areas within the National Wildlife Refuge System," Congress enacted the Refuge Recreation Act, which authorized administration of the refuges for recreation "as an appropriate incidental or secondary use . . . only to the extent that is . . . not inconsistent with . . . the primary objectives for which each particular area is established."[57] Like the other modern statutes noted earlier, the Refuge Recreation Act is a response to burgeoning public demand and results in a diversification (and, inevitably, complication) of the mission of an agency whose task previously was singular and much more focused. Although the statute provides for neither elaborate planning nor citizen participation (to say nothing of citizen-initiated lawsuits to enforce the law), it does incorporate an early indication of the growing congressional skepticism of bureaucracies and of the effort to constrain administrative discretion. After stating that recreation is a legitimate use, although incidental or secondary, Congress obliged the secretary of the interior to refrain from opening refuges to recreational use "until [he] shall have determined that such . . . use will not interfere with the primary purposes for which the areas were established."[58] A similar provision appears in the National Wildlife Refuge System Administration Act of 1966.[59]

CITIZENS BECOME INVOLVED

It is not surprising that, despite these provisions, there was no challenge in the courts to wildlife refuge management until the 1970s. Citizen participation, as a legal matter, did not really begin until the late 1960s and did not fully develop until state laws such as the Michigan Environmental Protection Act of 1969[60] and the environmental impact statement provision of the National Environmental

Policy Act of 1969, signed in January 1970,[61] gave impetus to citizen litigation. After that time, it was inevitable that laws imposing constraints on federal land managers, such as those governing the refuges, would begin to be scrutinized.

Typical of the new citizen involvement cases was *Defenders of Wildlife v. Andrus*, in which a secretarial order opening Nevada's Ruby Lake Wildlife Refuge to recreational motorboating was successfully challenged on the ground that the use was not incidental or secondary and on the finding, first, that no "determination" had been made that the boating was consistent with the primary purpose of the refuge and, second, that such a finding, later made, was not supported by evidence.[62]

Decisions such as this characterized the transformation that was occurring for all the federal land management agencies: great legislative control, in the form of constraints that were to become judicially enforceable rights;[63] greater scrutiny by both the Congress and the courts, with an accompanying sharp decline in agency discretion; a legalization of the old tradition of scientific management; a broader management mandate; a wider range of constituents and great conflict among them; and greater readiness on the part of constituent users to assert their demands as rights. Why and how all this came about, and what it portends for the future, are the subjects of the following sections.

PARK AND FOREST MANAGEMENT CHANGES WITH THE TIMES

The management of parks, recreation, and wilderness seems to present nothing so dramatically new as the Clean Air Act,[64] the complex and expensive edifice of the 1972 Federal Water Pollution Control Act,[65] or the discoveries of toxic substances that so dominate today's environmental news.[66] National park and forest management seems to be among the most stable and steady of federal enterprises, despite the relatively recent enactment of the Multiple Use-Sustained Yield Act and the National Forest Management Act. Wilderness designation as a management policy dates back to 1921,[67] and even the first legislatively established wilderness was created in 1930.[68] For all this apparent continuity, however, these areas of public land management have been no less shaken by the drama of the environmental movement beginning in the 1960s than have the problems of pollution, contamination, and toxicity.

The story is nowhere better revealed than in the evolution of federal wilderness policy. How did wilderness change from being an obscure corner of Forest Service administrative policy to a major issue on the congressional agenda and a central symbol of the ideological, or "preservationist," element of the environmental movement? To answer this question is to expose the very core of modern environmental politics.

Wilderness areas had long existed, but to Forest Service officials wilderness areas were but one element in the utilitarian calculus of supply and demand for forest products and but one feature in the political calculus of perpetual jousting between the Department of Agriculture and the Department of the Interior. That wilderness areas could become a symbolic manifestation of the reemerging environmental movement of the post-World War II period, a centerpiece for a showdown on values rather than a commodity issue, was a realization that came to Forest Service officials far too late for them to retain control of the question.

The decade following the end of World War II brought not only a vast increase in demand for raw materials and for automobile-based recreational areas, but also a release from the restrictiveness of the Great Depression and the war years. The era of conspicuous consumption, idealized in the fin-bearing "petrohogs" of Detroit, so familiar a feature of the mid–1950s, was also the time when the old conservation movement began to awaken from its torpor. The early signs were not much noticed by the public at large, but in retrospect they were there all the same. The Congress—urged on by a few committed members such as John Dingell, Hubert Humphrey, and John Saylor—led the way with laws like the Fish and Wildlife Coordination Act of 1948.[69] At the same time, the first of the bills that would later become the Wilderness Act was introduced by Senator Humphrey.[70] Even the modern style of environmental lawsuit began in this period, although it was not publicly noticed for nearly a dozen years: in the early 1950s, a sportsmen's council in Wisconsin, opposing construction of a dam on the Namekason River because it threatened fishing and boating, successfully demanded that the Federal Power Commission take account of nature protection in granting licenses on navigable rivers.[71]

As early as 1951, the obscure leader of an obscure organization, Howard Zahniser of the Wilderness Society, gave a talk at a Sierra Club conference in San Francisco titled "How Much Wilderness Can We Afford to Lose?" and made the novel suggestion that Congress establish a national wilderness preservation system.[72] A few years later, at the 1954 meeting of the Society of American Foresters, John

P. Gilligan, who had just completed a Ph.D. dissertation on Forest Service wilderness management, gave a talk during which he noted a number of instances in which wilderness designations had been deleted to make room for mining, grazing, roads, and other developments. Gilligan's speech, reprinted in the Wilderness Society magazine, attracted attention not only for its data but perhaps even more for its observation of the "steady trend favoring mass use over high quality benefits" and "the irresistible emphasis on dollar rather than social values of our free enterprise economy."[73]

Gilligan's point was hardly new; it echoed back at least to the mid-nineteenth century.[74] But its familiarity was its strength. The oldest and most stable theme in the American conservation movement is founded on a reaction against the perceived excesses of materialism, set against the disappearance of the frontier and the wilderness. Writings of Henry David Thoreau and Mark Twain, establishment of the first national parks in the period following the Civil War, the early campaign to save the redwoods, and the bitter battle over Yosemite's Hetch Hetchy reservoir were all outpourings of public feeling in the same cause.[75] Like a seed awaiting the right conditions for germination, the country's symbolic attachment to its frontier—at once real and mythical—awaited circumstances that would bring it into flower. The period of the mid–1950s was such a propitious moment, denoted by renewed prosperity and its related excesses, along with the real pressures that recreation and industrialism were imposing upon the land.[76]

What Zahniser, Gilligan, and others saw early in the wilderness area was only a version of what Rachel Carson and others dramatically brought to public attention in the ensuing decade.[77] For such efforts to succeed in mobilizing a constituency, however, not only the right timing and the right conditions were needed, but also adversaries who, by resisting the current of the times, permitted a small movement to grow and flourish. In this, environmental advocates could not have been better served. The unrestrained attacks on Rachel Carson by the chemical industry are well known and well documented.[78] The wilderness effort experienced no such clamor, but its opponents did, unwittingly, advance the cause.

By the 1950s, even the National Park Service helped fuel the movement, following the replacement of the preservation-oriented Newton Drury as director by the more recreation-minded Conrad Wirth. For example, it responded to post-World War II pressures on the parks with an extensive road building and development program known as Mission 66. In response to criticism that the program would only intensify use and impair the parks' natural features, Director Wirth responded (speaking of construction of a road into Wonder Lake in Mt. McKinley National Park): "That does not mean that the park is no longer a wilderness. The road is a wilderness road, to bring people into the wilderness."[79] The Forest Service was even more blunt. It first came out in direct opposition to the enactment of wilderness legislation, arguing that it would "strike at the heart of the multiple-use policy," and industry representatives took a stance calculated to raise the temperature of the debate, excoriating the "colossal selfishness . . . of those professional forest lovers . . . [who] want these regions treated as parks for their private delectation at the expense of the homeless. . . ."[80]

AGENCIES FAIL TO PERCEIVE THE PUBLIC MOOD

None of these responses was surprising, considering the traditional missions and perspectives of those who enunciated them. The point is rather that the officials failed to sense the changing public mood; failed to appreciate that periods of unremitting development routinely generate a preservationist reaction; and, most important, did not see that the reaction was symbolic and ideological in large part, so that invocations of phrases such as "multiple use," "meeting demand," and the like were not perceived as responsive at all. Indeed, such replies revealed a near-total failure to see the problem Gilligan had addressed in remarking on "the irresistible emphasis on dollar rather than social values."[81]

The question of why both public land management agencies and commodity users were so (self-defeatingly) unresponsive cannot be definitively answered; yet their response is central to understanding why the environmental movement, even as early as the 1950s, when it consisted largely of "professional forest lovers," was able to set in motion such far-reaching changes and to mobilize so vast a constituency in the ensuing decade.

Despite indications of fundamental shifts in public demands, the federal land agencies showed little inclination to change. They kept doing what they had been doing, avoiding new initiatives, and following the path of least resistance. For the Park Service, a common response to increased visitation was to provide new facilities like the old ones, only in increased number and scope. The operative assumption was that old constituencies continued to have the same importance as they had previously. Thus, the Forest Service, whose principal clientele had been commodity users, continued to respond as it always had (as by deleting areas from wilderness designation at the urging of those users) and was slow to recognize the burgeoning importance of recreation users (especially the new backpacking generation) as a clientele.[82]

Moreover, the bureaucracies found it difficult to respond to symbolic or ideological issues, or perhaps it would be more accurate to say that they were reluctant to see their own activities and operating assumptions in symbolic terms. The Forest Service response to the wilderness issue is illustrative. Among the points of opposition to proposed wilderness legislation that seemed most telling to Forest Service officials was the need for "flexibility" in management, which meant freedom to shift from one use to another as conditions dictated. This, of course, was entirely antithetical to what proponents of wilderness legislation wanted, which was permanence, a secure heritage of wilderness to pass on to future generations. One proponent, David Brower of the Sierra Club, described this concern as a need for "a very important right that belongs to future generations—the right to have wilderness in their civilization, even as we have it in ours."[83]

Forest Service officials simply could not hear this sort of rhetoric. To them, flexibility was a necessary concomitant of multiple-use management, and multiple use was simply good scientific, rational forestry. It did not occur to them that multiple-use management was an ideology too, a set of values just like wilderness preservation, and that they were engaged in a controversy about values, not about science. That is why one could witness, over and over again, Forest Service personnel shaking their heads in frustration over the inability of conservationists to realize that the trees they were trying to save were "overmature."[84] Forest Service officials seemed unaware that to those interested in wilderness preservation the maturity of a tree, which has meaning only in terms of timber harvesting, was beside the point.

A DECISIVE VICTORY FOR THE MOVEMENT

Wilderness advocates had chosen the right time and the right circumstances. Only one other element was needed to bring the environmental movement back to life: a decisive victory in battle. This they had with the struggle over Echo Park Dam in Dinosaur National Monument in 1954. The Echo Park story has been well and often told, and need not be repeated at length here.[85] In brief, it was a question of a long-standing plan for hydropower dams to be built in the upper Colorado River Basin, one of which, the Echo Park Dam, would have flooded a part of Dinosaur National Monument, then a little-known and infrequently visited element of the western wilderness.

Why the conservation movement decided to take its stand there may never be completely clear, but, as it turned out, the movement's leaders chose well. One element in the decision was doubtless the parallel to the titanic battle over the flooding of Hetch Hetchy valley in Yosemite a half-century earlier. Another may have been the support and leadership of the National Park Service in opposition to the dam, for the Park Service saw in the Echo Park proposal a precedent for hydropower incursions on its territories, an issue that it had fought in Yellowstone and elsewhere over the years. The environmental movement, considering its weakness at that time, probably could not have succeeded without an influential ally such as the Park Service. The Echo Park controversy was complicated by other features that greatly helped what was then a very modest coalition of conservation organizations: economic analysis had focused on the upper Colorado projects as examples of poor economic planning, with large costs and dubious benefits.

California, as the beneficiary of the water that came down the Colorado, was nervous about development in the upper basin. Indisputably the cause was aided by the arrogance and inflexibility of the Bureau of Reclamation, whose head, Michael Straus, at one point spoke out against "the self-constituted long-distance protectors of Dinosaur National Monument. . . . From their air-conditioned caves overlooking the undeveloped wilderness areas of Central Park . . . these self-appointed guardians have taken it upon themselves to safeguard the canyons of Dinosaur . . . for the handful of brave souls who dare to explore the area by boat."[86]

The open conflict between two bureaus in the Interior Department—the Park Service and the Bureau of Reclamation—ensured that the matter would become a cause célèbre. The split between Californians and people in the upper basin (especially Utah) gave the case high political visibility. And the timing, which offered a national platform to such well-known writers as Bernard DeVoto and encouraged editorial opposition in papers like the *New York Times*, generated a flood of mail in congressional offices. One Ohio representative was quoted as saying, "I have never been so besieged about anything as I have about Echo Park. . . . I have just got through a primary campaign and almost from every platform they asked, 'How do you stand on Echo Park?' Then the fur begins to fly."[87]

By the end of 1955 the battle was over, and the Echo Park Dam proposal had been eliminated. The conservation movement had taken its stand against what must have seemed the most invulnerable sort of pork-barrel project, and it had won. The victory was certainly a catalyzing event, setting the stage for the first wilderness bills to obtain serious consideration in Congress and for the postwar environmental movement to take its place as a serious force in national politics. The wilderness battles of the following years—the proposed Grand Canyon dams that brought the Sierra Club to national attention with its ads in the national press; the fight over the Forest Service's 1961 plan to open a corridor in the Selway-Bitterroot wilderness on the Montana-Idaho border to multiple-use development and the ensuing controversy over clearcutting; and the disappointing response of the Forest Service to conservationist efforts to preserve the North Cascades wilderness in 1959—strengthened the visibility of the environmental movement and ensured that wilderness preservation would remain a national issue.[88]

THE PARK SERVICE AND THE FOREST SERVICE LOSE GROUND

For the Forest Service, which became the principal bureaucratic casualty of the wilderness movement, it was a question of too little, too late. The Forest Service had unintentionally sent the message that protection of the wilderness depended on private citizen organizations taking their case to Congress and the courts rather than on the actions of public land management agencies. The Forest Service, the "inventor" of wilderness areas, became the bête noire of wilderness politics.

The Park Service had not been the target of early wilderness controversies, and holding large areas of land as wilderness was highly compatible with its traditional mission. However, it too was concerned about the new style of wilderness law and administration. From its perspective, as one observer noted, "large areas of Park Service wilderness would severely restrict management discretion . . . the commandments of the Wilderness Act might cramp [its] style."[89]

The result was that both the Forest Service and the Park Service were reluctant to move on to the second stage of the Wilderness Act program—evaluating other untrammeled lands within their jurisdictions and recommending to Congress and to the president which areas should be designated as statutory wilderness. The bureaucrats failed to see that the very skepticism about administrative discretion that had led to laws like the Wilderness Act was also strongly insinuating itself into the judicial system. The environmental movement, with its newfound strength, was about to try its power in a new forum—the courts. The conventional bureaucratic tactic of shaping legislative directions at the administrative stage ran head-on into the determination of environmental groups to cement their legislative victories in the courts.

THE JUDICIARY INTERVENES

As the 1960s ended and administration of the Wilderness Act was getting under way, the judiciary was gearing up to intervene in administrative decision making to an extent never known before. The first great modern case involved a challenge to Federal Power Commission licensing of a Consolidated Edison pumped storage plant on the Hudson River in New York.[90] There the court of appeals recognized citizen standing to sue (the right of concerned citizens to go to court to enforce the laws). The court acknowledged the right of a private citizen organization to participate in the licensing and also recognized its right to demand that regulatory agencies affirmatively implement environmental protection elements in the law rather than standing by as "an umpire, blandly calling balls and strikes."[91] This case, called the Storm King litigation, opened the door to environmental organizations as litigants in the courts, rather than as mere supplicants.

Storm King was followed shortly by a series of cases involving pesticide regulation (brought by the Environmental Defense Fund, which was determined to bring the work of Rachel Carson into the courtroom);[92] by the first of a series of challenges to the federal highway program;[93] and, after 1970, by cases concerning the implementation of the National Environmental Policy Act, which, by requiring federally prepared environmental impact statements, gave the environmental movement a means to put traditional administrative discretion under the judicial microscope.[94]

Environmentalists now had a weapon far more potent and easier to invoke than occasional and protracted approaches to Congress of the kind used in the Echo Park Dam controversy. They used the courts to ensure implementation of the Wilderness Act, as well as the other laws that Congress was increasingly willing to enact in the environmental field. Congressional and judicial actions fed on each other. The legislature passed increasingly detailed laws, with ever more specific mandates to the administrators. The more specific Congress was, the more the courts were willing to intervene, on the ground that there was "law to apply," rather than just general discretionary directions.[95] Once the problem of standing to sue was decided favorably for environmentalists, the administrative agencies were converted from experts above the fray to defendants forced to explain themselves to federal judges.[96]

For a long time, the agencies did not know what had hit them, and their responses generated increasing judicial involvement. The first important case involving judicial review of implementation of the Wilderness Act concerned the White River National Forest in Colorado.[97] The question in this 1969 case was the status of an area adjacent to an administratively designated primitive area. The Forest Service permitted timber harvesting in the adjacent area, which was of wilderness quality, on the ground that it was only required to study and present to Congress its report on the primitive area itself. But since the wilderness statute permitted the president to recommend enlargements of primitive areas, environmentalists argued that the adjacent area must be held in its wilderness status pending the decision of the president and the Congress. Otherwise, they said, the Forest Service would be able to preempt presidential and congressional decisions about land that could be designated statutory wilderness by depriving it of its wilderness quality before the president and Congress could consider them.

This was the earliest version of an issue that has since become the major question in wilderness management: what to do about "de facto" wilderness land before Congress decides whether to establish a wilderness there. For our purposes, detailed consideration of the statutory requirements is less important than the fact that the Forest Service could not imagine that the answer to that question would be taken in hand by a federal judge, at the behest of a group of environmentalists, and would be removed from its managerial judgment. That such questions would be resolved in the courts as matters of statutory interpretation seemed to undermine everything in which the Forest Service believed and on which it had been founded: expertise, scientific forestry, discretion, and professional independence.

When the Forest Service got to court in the White River National Forest case, it did exactly the wrong thing. It argued that timber management was relegated by law to Forest Service discretion and was unreviewable in court; that the plaintiff conservation organizations had no standing to sue; and that the government was not subject to suit because of the doctrine of sovereign immunity. If the plaintiffs were permitted to prevail, they said, "projects planned by the executive branch and funded by the legislature . . . [would] be frustrated by the judicial branch at the whim of any citizen who disagrees with the justification of such projects." "Congress," they said, "had wisely left these technical matters to the technicians."[98]

A TIDAL WAVE OF LITIGATION

As the case went forward, more and more information came out. It turned out that the plan for harvesting the area in question had actually been made in 1962 and had remained unchanged despite the subsequent enactment of the Wilderness Act in 1964. As the judge observed, the Forest Service assumed that because they had already built a road in the area to get out the timber, "the existing road would more or less go to waste if they didn't use it for industry purposes. That's their philosophy."[99] Having failed to get the case dismissed on technical grounds, the Forest Service turned to a second strategy, one that had rarely failed them in the past. They regaled the judge with the prospect of closed mills, lost jobs, and a national shortage of timber. The judge was unmoved. "This parade of horribles doesn't impress me at all," he said. "I don't think the sky will fall, and I don't accept this 'Chicken Little' approach to the law. One little judicial decision isn't going to disturb the mazes of bureaucracy described to me today—it won't even cause a ripple."[100]

The judge was too modest. *Parker v. United States* caused more than a ripple. One might say it generated a tidal wave of litigation, reinforcing all the suspicions about bureaucratic arrogance and inertia that had come to the surface in the Echo Park controversy and putting administrators in the spotlight in a way that periodic legislative battles could never do. Courts operate at the most detailed level of inquiry, with attention focused on documents and the testimony of witnesses. Every detail about response to particular constituent pressures, every datum that suggests whether planning has been done in response to a statutory mandate, all the documentation on costs and benefits, the interest rates used, and the presence or absence of studies on impacts on wildlife and the local economy are brought into the courtroom.

The *Parker* decision was only the first in a long series of suits generated in the 1970s with the Forest Service and other land management agencies as defendants.[101] The ease of access to courts ensured that the newfound power of the environmental movement would not be limited to a rhetorical gain in the statute books or to those few issues on which a congressional constituency could be mobilized. The agencies themselves ensured—as by their behavior in the *Parker* case—that environmentalists would be a permanent and powerful constituency.

SCIENCE, VALUES, AND LAND MANAGEMENT

The story recounted in the preceding sections makes clear that the American system is capable of producing genuine change. Reform is possible, but it frequently occurs in ways that the reformers do not expect. Both Congress and the courts acted as if they were bringing a more "scientific" perspective to federal land management—demanding systematic, interdisciplinary studies; calling for comprehensive environmental impact statements; and scrutinizing administrative decisions to ensure that the formal record supported the decision made.

The scientific form often conceals quite a different substance. That has been the case here in two respects: first, although one may demand scientific study of proposed wilderness, the decision as to which land should be designated as wilderness, or how much land should be designated, is not really a scientific, or technical, question at all. One can have vast amounts of data and still be entirely unsure whether a given tract should be recommended as wilderness, held for motorized recreation, or turned over to mineral development. The demand for extensive data has constrained agencies, but it has not resolved value or priority questions except to a very limited extent. It is no longer possible to commit a forest entirely to timber harvesting or to manage a park for developed recreation in a designated wilderness, but important choices are still being made by the land management bureaucracies.[102] What has changed most significantly is that something must go to all the relevant constituencies. The agencies have become mediators among a wide range of clienteles.

Second, no matter how much information is produced, the responsible agency still must concern itself with constituency pressures. However valuable a tract may be as wilderness and however little value its timber may produce, a local mill and a local community may still depend on timber harvesting in the area.[103] The need of the decision-making bureaucracy to survive, to avoid controversy that imperils it, remains a powerful (perhaps the most powerful) influence on its recommendations.

This is not to say that the procedural reform that has taken place thus far (demanding more technical data and forcing agencies to be responsive to more diverse constituencies) is without effect. The process does permit us to know something about the relative values of land for various purposes, and it no doubt encourages the wilderness designation of areas that have very high values for wilderness and very little value for commercial development. It also forces agencies to respond to changing values. It is no longer possible for the Forest Service to subordinate wilderness to logging whenever those interests are in conflict, as it was wont to do in the 1950s and 1960s. Perhaps most importantly, the various contending interest groups must defend their demands in light of increased knowledge of the costs to other constituencies.

For all its seeming incoherence and fragmentation, one need not lament the decision-making process described here. Indeed, it can be seen as a reflection of the strong inclination in American life to hold extremes at bay, opting for a middle course, recognizing diversity as legitimate, not seeking some single all-embracing solution, and going at problems incrementally. The nature of the process, broken down into small pieces, gives relative ease of access to a wide variety of interest groups, which means that there are no permanent all-or-nothing winners or losers. The fact that today's loser can come back tomorrow gives legitimacy to many interests and many clienteles.

SHARING POWER SERVES DIVERSE INTERESTS

The history of the wilderness movement is instructive. The ability of the system to put constraints on the Forest Service, and to formalize skepticism of bureaucratic discretion, reveals openness in the governmental process. The sharing of power among the administration, the courts, and the legislature, rather than having one branch of government decide wilderness questions alone, also opens the process to a range of interest groups with different opportunities for access. As values change—as the excesses of development give way to a yearning for more wilderness, for example—the system makes transitions manageable without extreme disruption. The ability of interest groups to introduce into the decision-making system their own studies and planning documents to challenge proposed policy permits adjustments to be made and creates a process that allows unsatisfied interests to be heard and to bring some pressure to bear. The power of various groups creates incentives to search out alternatives that will be less harmful to each affected group.

Slow and muddled as it sometimes seems, the process is rich in tolerance for differences, open to adjustment, decentralized in operation, and skeptical of authority and of authoritative solutions. It assumes that interest groups—given sufficient leverage over decision making—will come out with acceptable results. In short, it reflects the American style of governance and the American character.

NOTES

1. 48 Stat. 1270, 43 U.S.C. sec. 315.
2. See, for example, J. D. Leshy, "Unraveling the Sagebrush Rebellion: Law, Politics and Federal Lands," *University of California, Davis Law Review* 14 (1980): 317.
3. 39 Stat. 535, 16 U.S.C. sec. 1. Congress did sometimes specify policies in establishing individual parks. The law creating Bryce Canyon National Park, for example, recognized the property rights of existing miners and homestead entrants. 43 Stat. 593, 16 U.S.C. sec. 401. The Everglades National Park provided for management as a wilderness, and prohibited development that would interfere with "primitive natural conditions." 48 Stat. 817, 16 U.S.C. sec. 401c.
4. 30 Stat. 34, 16 U.S.C. sec. 475.
5. 78 Stat. 890, 16 U.S.C. sec. 1131; Harold K. Steen, *The U.S. Forest Service, A History* (Seattle: University of Washington Press, 1976), 152–62.
6. George C. Coggins and Charles F. Wilkinson, *Federal Public Land and Resources Law*, 2d ed. (Mineola, N.Y.: The Foundation Press, 1987), 141.
7. Robert S. Shankland, *Steve Mather of the National Parks* (New York: Alfred A. Knopf, 1951), chap. 12.
8. Conrad L. Wirth, *Parks, Politics and the People* (Norman, Okla.: University of Oklahoma Press, 1980), chap. 9.
9. Earl Pomeroy, *In Search of the Golden West: The Tourist in Western American* (New York: Alfred A. Knopf, 1957).
10. Conservation Foundation, *National Parks for the Future* (Washington, D.C.: The Conservation Foundation, 1972), 197.
11. John Ise, *Our National Park Policy: A Critical History* (Baltimore: Johns Hopkins University Press, 1961), 296, 363.
12. See, for example, James N. Smith, "The Gateways: Parks for Whom?" in Conservation Foundation, *National Parks for the Future*, 213; J. William Futrell, "Parks to the People: New Directions for the National Park System," *Emory Law Journal* 25 (1976): 255.
13. See, for example, Joseph L. Sax, *Mountains Without Handrails* (Ann Arbor: University of Michigan Press, 1981).
14. For example, Kay Franklin and Norma Schaeffer, *Duel for the Dunes: Land Use Conflict on the Shores of Lake Michigan* (Urbana: University of Illinois Press, 1983).
15. 39 Stat. 535, 16 U.S.C. sec. 1.
16. Ise, *National Park Policy*, 361.
17. Ibid., 449.
18. U.S. Department of the Interior, National Park Service, *Yosemite National Park Preliminary Draft Master Plan* (12 Aug. 1974); U.S. Congress, House, *National Park Service Planning and Concession Operations: Joint Hearing Before Certain Subcommittees of the Committee on Government Operations and the Permanent Select Committee on Small Business*, 93d Cong., 2d sess., 20 Dec. 1984; see also Sax, *Mountains Without Handrails*, 73.
19. Yosemite Planning Team, *The Workbook, Yosemite Master Plan: Guidelines for the Design of Alternatives* (Washington, D.C.: National Park Service, 1975); see also Michael Mantell, "Preservation and Use: Concessions in the National Parks," *Ecology Law Quarterly* 8 (1979): 33.
20. 74 Stat. 215, 16 U.S.C. sec. 528.
21. 30 Stat. 34, 16 U.S.C. sec. 475.
22. Steen, *U.S. Forest Service*, 298.
23. Glen O. Robinson, *The Forest Service: A Study in Public Land Management*, prepared

for Resources for the Future (Baltimore: Johns Hopkins University Press, 1975), 41.

24. Steen, *U.S. Forest Service*, 297–307.
25. The Sierra Club was not at all enthusiastic about the multiple-use bill. See Grant McConnell, "The Multiple Use Concept in Forest Service Policy," *Sierra Club Bulletin* 44 (October 1959): 14.
26. 74 Stat. 215, 16 U.S.C. sec. 528.
27. Steen, *U.S. Forest Service*, 306.
28. 83 Stat. 852, 42 U.S.C. sec. 4321.
29. 78 Stat. 893, 16 U.S.C. sec. 1133. Congressional designation was sought by industrial users, who thought it would favor them, and opposed by environmental groups. Both misconceived the political popularity of wilderness.
30. Michael Frome, *Battle for the Wilderness* (New York: Praeger, 1974), 129–38; Craig W. Allin, *The Politics of Wilderness Preservation* (Westport, Conn.: Greenwood Press, 1982), 111; Steen, *U.S. Forest Service*, 303.
31. See, for example, William H. Rodgers, Jr., "A Hard Look . . . Environmental Law Under Close Scrutiny," *Georgetown Law Journal* 67 (1979): 699.
32. 78 Stat. 890, 16 U.S.C., sec. 1131(c).
33. Administrative controversy, however, was far from unknown to the Forest Service. See Herbert Kaufman, *The Forest Ranger: A Study in Administrative Behavior* (Baltimore: Johns Hopkins University Press, 1960), 78–80.
34. Sierra Club v. Hardin, 325 F.Supp. 99 (D. Alaska, 1971); see also Sierra Club v. Butz, 3 E.L.R. 20292 (9th Cir., 1973).
35. Parker v. United States, 309 F.Supp. 593 (D. Colo., 1970), *aff'd*, 448 F.2d 793 (1971), *cert. denied*, 405 U.S. 989 (1972).
36. American Timber Co. v. Bergland, 473 F.Supp. 310 (D. Mont., 1979); Citizens Against Toxic Sprays v. Bergland, 428 F.Supp. 908 (D. Ore., 1977).
37. U.S. Congress, Senate, Arnold Bolle, *A University View of the Forest Service*, 91st Cong., 2d sess., 1970, S. Doc. 91–115; West Virginia Division of the Izaak Walton League of America, Inc. v. Butz, 522 F.2d 945 (4th Cir., 1975).
38. Ibid.
39. 90 Stat. 2949, 16 U.S.C. sec. 1600.
40. 90 Stat. 2949, 16 U.S.C. sec. 1602(5)(D).
41. 90 Stat. 2949, 2952, 2958, 16 U.S.C. sec. 1604(g)(3)(B).
42. Ibid., (E)(ii).
43. Ibid., (F).
44. Ibid., (k).
45. Ibid., sec. 1604(g)(3)(F)(iv).
46. Ibid., sec. 1612(a).
47. See Thomas B. Stoel, Jr., "The National Forest Management Act," *Environmental Law* 8 (1978): 550, 565–66.
48. 90 Stat. 2743, 43 U.S.C. sec. 1701.
49. The BLM had temporary multiple-use authority under the Classification and Multiple Use Act of 1964, 78 Stat. 968, which expired in 1970.
50. See generally, Marion Clawson, *The Bureau of Land Management* (New York: Praeger, 1971).
51. 90 Stat. 2747, 43 U.S.C. sec. 1712.
52. Ibid.
53. 90 Stat. 2785, 43 U.S.C. sec. 1782.
54. 90 Stat. 2745, 43 U.S.C. sec. 1702(c).
55. See generally George S. Coggins and Charles F. Wilkinson, *Federal Public Land and Resources Law*, 815 et seq.; U.S. Department of the Interior, National Park Service, *Preserving our Natural Heritage*, vol. 1, *Federal Activities* (Washington, D.C., 1975).
56. 45 Stat. 1223, 16 U.S.C. sec. 715d.
57. 76 Stat. 653, 16 U.S.C. sec. 460(k).

58. Ibid.
59. 80 Stat. 927, 16 U.S.C., sec. 668dd(d) (1)
60. Mich. Comp. Laws Ann. secs. 691.1201–.1207, Mich. Stat. Ann. secs. 14.528(201)-(207).
61. 83 Stat. 852, 42 U.S.C. secs. 4321–61.
62. U.S. Dist. Ct., D.C., 11 ERC 2098 (1978). Perhaps the most significant litigation over the refuge system involves potential oil and gas development in the Arctic National Wildlife Refuge in Alaska. See Trustees for Alaska v. Hodel, 806 F.2d 1378 (9th Cir., 1986).
63. See William Rodgers, *Energy and Natural Resources Law Cases and Materials* (St. Paul: West, 1983), 190 et seq.
64. 84 Stat. 1713, 42. U.S.C. sec. 7401.
65. 86 Stat. 816, *rev'd* in 1977, 91 Stat. 1567, 33 U.S.C. sec. 1251; see the Act of June 30, 1948, c. 758.
66. For example, the so-called Superfund Law, 94 Stat. 2767, 42 U.S.C. sec. 9601.
67. Allin, *Politics of Wilderness Preservation*, 70.
68. Ibid., 79.
69. 72 Stat. 563, 16 U.S.C. sec. 661.
70. For a background account, see Michael McCloskey, "The Wilderness Act of 1964: Its Background and Meaning," *Oregon Law Review* 45 (1966): 288; Delbert V. Mercure and William M. Ross, "The Wilderness Act: A Product of Congressional Compromise," in *Congress and the Environment*, ed. Richard A. Cooley and Geoffrey Wandesforde-Smith (Seattle: University of Washington Press, 1970).
71. Namekagon Hydro Co. v. Federal Power Commission, 216 F.2d 509 (7th Cir., 1953). Contrast State of Washington Department of Game v. Federal Power Commission, 207 F.2d 391, 398 (9th Cir., 1953) ("If the dams will destroy the fish industry of the river, we are powerless to prevent it"), *cert. denied*, 347 U.S. 936 (1954).
72. Frome, *Battle for the Wilderness*, 138.
73. Allin, *Politics of Wilderness Preservation*, 102–3.
74. The leading history is Roderick Nash, *Wilderness and the American Mind* (New Haven: Yale University Press, 1973).
75. See, for example, Holway R. Jones, *John Muir and the Sierra Club: The Battle for Yosemite* (San Francisco: Sierra Club, 1965); S. R. Schrepfer, *The Fight to Save the Redwoods: A History of Environmental Reform 1917–1978* (Madison: University of Wisconsin Press, 1983).
76. Frome, *Battle for the Wilderness*, 133, noting the rise of clearcutting on national forests in the 1950s.
77. Rachel Carson, *Silent Spring* (Boston: Houghton Mifflin, 1962).
78. Frank Graham, Jr., *Since Silent Spring* (Boston: Houghton Mifflin, 1970).
79. Allin, *Politics of Wilderness Preservation*, 111.
80. Ibid., 111–12.
81. Ibid., 103.
82. A revealing example is found in the memoir of Conrad L. Wirth (see n. 8), who was director of the National Park Service for twelve years during the 1950s and 1960s. For a case study and theoretical exploration of this observation, see Ben Twight, *Organizational Values and Political Power: The Forest Service Versus the Olympic National Park* (University Park, Pa.: Pennsylvania State University Press, 1983).
83. Allin, *Politics of Wilderness Preservation*, 116.
84. See the discussion of the Parker case in Joseph L. Sax, *Defending the Environment* (New York: Alfred A. Knopf, 1971), 195.
85. For example, O. Stratton and P. Sirotkin, *The Echo Park Controversy*, Inter-University Case Program, no. 46 (University, Ala.: published for the program by University of Alabama Press, 1959); Allin, *Politics of Wilderness Preservation*.
86. Stratton and Sirotkin, *Echo Park Controversy*, 49.
87. Ibid., 93.

88. Frome, *Battle for the Wilderness*, chap. 9.
89. Allin, *Politics of Wilderness Preservation*, 146–47.
90. Scenic Hudson Preservation Conference v. Federal Power Commission, 354 F.2d 608 (2d Cir., 1965), *cert. denied*, 384 U.S. 941 (1966).
91. Ibid., 620.
92. Environmental Defense Fund, Inc. v. Hardin, 428 F.2d 1093 (D.C. Cir., 1970).
93. Road Review League v. Boyd, 270 F.Supp. 650 (S.D.N.Y., 1967).
94. William H. Rodgers, Jr., *Environmental Law* (St. Paul: West Publishing Co., 1977), 701: "Much of NEPA is a sleeper exceeding the boldest expectations of its creators. . . ."
95. Citizens to Preserve Overton Park, Inc. v. Volpe, 401 U.S. 402 (1971).
96. The U.S. Supreme Court had backed off somewhat from this position by the late 1970s. See, for example, Vermont Yankee Nuclear Power Corp. v. NRDC, 435 U.S. 519 (1978).
97. Parker v. United States (see n. 39); the discussion in Sax, *Defending the Environment*, 195.
98. Sax, *Defending the Environment*, 197–98.
99. Ibid., 200.
100. Ibid.; see also Earl S. Wolcott, Jr., "Parker v. United States: The Forest Service Role in Wilderness Preservation," *Ecology Law Quarterly* 3 (1973): 153.
101. See, for example, nn. 34, 36, and 37. Among the most recent cases involving the use of the environmental impact statement for the Forest Service study of wilderness areas in California is California v. Block, 690 F.2d 753 (9th Cir., 1982).
102. For example, Wilderness Public Rights Fund v. Kleppe, 608 F.2d 1250 (9th Cir., 1979), *cert. denied* 446 U.S. 982 (1980) (boating in Grand Canyon); Conservation Law Foundation v. Clark, 590 F.Supp. 1467 (D. Mass., 1984) (off-road vehicle use at Cape Cod National Seashore).
103. See, for example, *In re Appeal of Land Management Plan and Environmental Impact Statement for the Grand Mesa, Uncompahgre and Gunnison National Forests*, before the Chief, U.S. Forest Service (1983) (a challenge to harvesting of "uneconomic" timber to sustain the local economy).

4

PUBLIC LAND POLICY: CONTROVERSIAL BEGINNINGS FOR THE THIRD CENTURY

Frank Gregg

The postwar history of the two largest land systems owned by the United States is one of rapid change and intensifying controversy. The roots of controversy are found in profound philosophical differences about appropriate relationships of man and nature, differences centering on the relative emphasis to be given to the use of land resources for economic purposes as distinct from the recognition, protection, and appreciation of natural values.

The stakes are substantial. The national forests administered by the Forest Service in the U.S. Department of Agriculture (USDA) include over 185 million acres of lands and inland waters. The Bureau of Land Management (BLM) will be responsible for about 240 million acres after completion of transfers of federal lands to the state of Alaska and to Native American institutions under the Alaskan Native Claims Settlement Act of 1971, a process that may go on for many years. The combined total of more than 425 million acres is roughly 19 percent of the total land area of the fifty states.[1] Almost 400 million acres of the total are in the eleven western states and Alaska, in concentrations ranging from 76 percent (Nevada) to 21 percent (Washington) of the land areas of individual states.

The lands contribute natural resource commodities to the nation's economy. Water is the most precious resource in the semiarid West, and the high-elevation forested watersheds—largely under Forest Service jurisdiction—yield much of the runoff for the western river systems. The federal lands offer major supplies of timber, especially in the great Douglas fir regions of the Pacific Northwest. Energy resources of coal, oil and gas, uranium, oil shale, tar sands, and geothermal sites on the federal lands are key elements in national strategies to reduce dependence on foreign oil. The federal lands are major sources of metallic minerals, including precious metals and copper, and modest sources of a wide range of a group of minerals identified as having strategic importance.

The economies of a number of the more rural western states and of hundreds of rural areas throughout the region are supported in major part by a range livestock industry that has grazed the lands since they were settled.

Federal lands are of primary significance to millions of Americans, not as a storehouse of resource commodities, but as a common heritage of open spaces, mountain and canyon vistas, wildlife, clean and free-flowing streams, unpolluted and undeveloped lakes—a heritage valued by those who use the lands directly for recreation and renewal, and by millions of others whose appreciation is derived from popular literature and film.

The intensity of controversy over use of these lands since World War II is a function of increasing demands for commodity production, for recreation access and use, and for preservation of scenic and wildland values. A brief history of the public lands and of the two agencies that administer them will provide background for a discussion of the postwar period.[2]

The government has been the nation's largest landowner since the earliest years of our history. Following the Declaration of Independence, several of the thirteen states claimed title to large areas of land that had belonged to the crown, extending far to the west of the original thirteen states.

States that had no such holdings argued that the "Western lands" should be ceded to the federal government in the interest of equality among the states. The Continental Congress concurred; the cessions were made between 1791 and 1802. These original "public lands" exceeded 235 million acres, including all or parts of Ohio, Indiana, Illinois, Michigan, Wisconsin, Minnesota, Alabama, and Mississippi.

This beginning was followed by acquisition of 1.6 *billion* additional acres of land and inland waters over a period of sixty-four years, beginning with the Louisiana Purchase of 1803 and ending with the purchase of Alaska from Russia in 1867. In short, substantially all the lands outside the boundaries of the original thirteen states and Texas* came under the jurisdiction of the United States as "public domain," that is, were originally the property of the government of the United States and were subject to retention, disposal, and use and management under laws enacted by Congress.

For most of the nation's history, disposal of the public domain was the overriding policy. Lands were given or sold to war veterans as rewards for public service; sold as townsites to individuals and governments; granted or sold to homesteaders to encourage settlement and improvement of agricultural lands; granted to states to provide revenues for public schools, universities, and other purposes; granted to states and railroad corporations to encourage development of transport systems; and claimed and patented for mining under various laws crafted especially to encourage mineral discovery and production. This list is far from complete: the point is that by the end of the nineteenth century most of these lands had passed into nonfederal ownership in what must be the largest land distribution undertaking of all time.

Throughout this period, the principal agency of the government responsible for the public domain was the General Land Office (GLO), after 1849 as part of the U.S. Department of the Interior. Disposal, not management, was GLO's mission.

Early Agency History

What is now the national forest system was created initially by a series of presidential actions withdrawing lands from public domain status, and therefore from homesteading and other disposal statutes, under an 1891 act revising a number of public land laws. The act, which has come to be known as the Forest Reserve Act of 1891, contained a section added in conference between the House and the Senate authorizing the president to establish public reservations of public lands "wholly or in part covered with timber or undergrowth . . . ." President Benjamin Harrison reserved a total of more than thirteen million acres in the first two years following enactment, apparently without controversy. President Grover Cleveland added twenty-five million acres, including more than twenty-one million acres in 1897

---

*Lands now within the State of Texas were acquired through Texas's war for independence against Mexico and were never a part of the federal public domain. Portions of Oklahoma, Colorado, and New Mexico were acquired by the United States from Texas in 1850 and became public domain lands.

only a few days before leaving office—and without consulting with the governors or congressional delegations involved, generating an outcry against the exercise of federal authority to retain and manage land that has become a standard feature of political discourse in the public land states.

By the end of Theodore Roosevelt's presidency in 1913, over 150 million acres had been reserved for national forests, substantially all from public domain lands in the West. The balance (the total under Forest Service jurisdiction is now about 187 million acres) was added over the next seventy years. Most of the land area added since 1913 has been through acquisition of lands, often cutover and farmed-out lands in the Lake States and the mountainous areas of the East and South, and transfers of lands in Alaska from BLM in the Alaska National Interest Lands Act of 1980.

The initial purpose of the 1891 "forest reserves" was unclear. Earlier reservations, such as Yellowstone in 1872, had been made in order to preserve unique natural values. The secretary of the interior at the time assumed preservation of special natural qualities to be a priority objective. Protection of watershed values through protection of forest cover was a compelling reason for support in the water-short West. But much of the sustained political support for forest reservations had come from a small band of people who had been exposed to or trained in professional forestry concepts in Europe, largely in Germany. Management of forest ecosystems for sustained production of timber under the direction of professional foresters was the unifying objective of this group.

The forest reserves were under the jurisdiction of the agency of disposition, the General Land Office of the Department of the Interior. Those managers dedicated to forest management as a profession and a cause, however, were in the Division of Forestry of the Department of Agriculture, which had been established two decades earlier to study timber supplies and ways of meeting needs. The forestry management group pressed for a utilitarian management policy.

In 1897, the Congress addressed the use versus preservation question by specifying three basic purposes for the forest reserves: to "preserve and protect the forest . . ."; "for the purpose of securing favorable conditions of water flows"; and "to furnish a continuous supply of timber . . . ." Utilization of timber as a stated objective was a critical factor to emerging forestry professionals and interests, as watershed protection was to western water users. Utilitarian values, on balance, were emphasized.

By 1905, Gifford Pinchot, the competent and messianic leader of the professional forest management utilitarians, had succeeded in persuading Theodore Roosevelt to transfer the forest preserves to the Department of Agriculture, with Pinchot as "forester." The agency was officially titled the U.S. Forest Service.

By that time, the use versus preservation philosophical cleavage was sharply drawn, with John Muir, an early supporter of the forest preserves who became the first president of the Sierra Club, as the most effective opponent of Pinchot's utilitarianism. Pinchot had received his forest training in Germany and had been the first to apply forest management to a substantial land area in the United States, the privately owned Biltmore Forest. His dedication to professional decision making and to economic utilization of natural resources was profound. He developed and exploited with astonishing success a special relationship with the president. By the end of the Roosevelt administration, the Forest Service administered 150 million acres as the result of huge additional reservations prompted by Pinchot. Roosevelt and Pinchot had defined conservation as a cause directed at efficient economic utilization of natural resources under strong governmental leadership, based on scientific concepts applied with an extraordinary degree of power and discretion by trained natural resource professionals. Success in securing administrative control of a national system of forests reinforced the sense of mission of forestry professionals and institutionalized their philosophical convictions in the Forest Service.[3]

The Pinchot-dominated era also left a lasting philosophical and political chasm between the utilitarian, economic efficiency, and professional dominance philosophy, on one hand, and the belief in preservation of natural values as important quite apart from any economic use, on the other. The wildland, scenic resource, and natural area preservation groups were evident in the late 1800s. These forces grew to such effective power as to establish the National Park Service over Forest Service objection in 1916, and to secure millions of acres of Forest Service land over a period of decades for placement in the growing national park system. The aura of profound distrust between the two broad constellations of institutions and individuals in these camps remains palpable today.

Pinchot was ultimately fired by President Taft in 1911 for, perhaps predictably, his tactics in a feud with the secretary of the interior. His legacy as forester and conservation administrator has endured in the Forest Service tradition of professional leadership relatively free of political constraint. Its management system has become a model of decentralized public administration.[4] The key features include an extraordinary degree of authority delegated to local managers, who are in direct contact with resource users. The field manager is guided by a manual (originally a small handbook, now a veritable library), which establishes criteria and procedures for the exercise of his discretion.

The field manager is aided and constrained by a hierarchy above him providing research and technical support, as well as guidance—and periodic inspection. A pattern of voluntary conformity to objectives and methods in this huge agency has been reported by researchers, citing common professional backgrounds (forestry), formal and informal training, peer pressure, and promotion policies as supporting factors.

By the end of World War II, the Forest Service had reason for confidence. It had enjoyed forty years of steady growth in competence and esteem, had survived repeated efforts to place the agency back in the Department of the Interior, and continued to enjoy its position of relative autonomy in the Department of Agriculture.

Periodic rumblings of discontent had failed to move the Congress to consider disposal of the national forests seriously or to impose significant constraints on management authorities. The service had survived a series of battles with the organized western livestock industry, and during World War II it managed to avoid a repetition of the overgrazing in the name of war emergency that had taken place during World War I.[5] The Forest Service had powerful allies in Congress, in many politically active natural resource groups, and in professional societies.

The history of what is now the second great land system managed for multiple purposes—the public lands under jurisdiction of the Bureau of Land Management—differs strikingly. The agency was created in 1946 by a merger of two agencies of the Department of the Interior: the General Land Office (GLO), the historic instrument of public land disposal, and the Grazing Service, an agency established only a dozen years before with the primary mission of regulating livestock grazing on public domain lands.

Public lands currently administered by BLM were not selected for retention by the Congress or the executive branch. The BLM-administered lands are, in a reasonably precise sense, remnants; they are those portions of the public domain that were not placed in other categories of federal administration (national forests, parks, monuments, wildlife refuges, or other categories) or disposed of to nonfederal owners. The lands tend to be semiarid and arid rangelands, in mountainous regions lying below and adjacent to the higher-elevation national forest lands. They were judged to be unsuitable for agriculture by homesteaders, and they are often at the lower end of biological productivity scales. Management is complicated by scattered and intermingled ownership patterns, arising from the checkerboard pattern of land disposal often used by the federal government and from selection of the better sites (such as waterholes, stream corridors, and adjacent lands) by private individuals within larger blocks of public lands.

The General Land Office had existed since 1812, but it had little else to offer as a bureaucratic asset. As the instrument of public land disposal for over a century, GLO had been irreversibly—and often unfairly—stained by abuse and scandal in the processes of disposition. The agency had statutory responsibility for leasing of minerals (oil, gas, and coal were those most in demand) on federal lands. The laws provided nominal discretion in granting or denying leases, but GLO acted more as a processor of documents and occasional referee among claimants than as a decision maker in minerals management. Additional responsibilities in land record keeping, cadastral surveys, and continued land disposal did not add up to a sense of institutional purpose. Only its responsibility for two million acres of prime "O&C" timberland in western Oregon (revested in the federal government when certain original grantees failed to live up to the terms of their grants) constituted an affirmative management mission.*

GLO's responsibilities on the remaining public domain lands were shared with its sister agency in the Department of Interior, the Grazing Service. The Grazing Service was only a decade old. It had been established by the Taylor Grazing Act of 1934, which provided for administration of livestock grazing on the public domain. Livestock had grazed the public lands as "open range" for generations. The tragedy of the commons applied with a vengeance: serious overgrazing was endemic; violence, the threat of violence, and fraud were familiar weapons in establishing who was to graze where.[6]

The act was significant. It asserted for the first time congressional intent to secure purposeful management of the remaining public domain for the limited purposes of grazing administration and rangeland rehabilitation. It also signaled, dimly, the end of the disposal era.

The two messages were mixed. Initiative for the bill came from a

---

*The Congress in 1917 approved an act reasserting ownership of lands that had been granted to the Oregon and California Railroad Company in 1866—hence, "O&C" lands.

livestock industry far more concerned about establishing rights to use the range against claims of other grazers than about rangeland management. Establishment of an authoritative, professional management system on the Forest Service model was distinctly not envisaged.

Managerial intent was eroded by a qualification that all authorities granted under the act were "pending final disposal" of the lands, a phrase that constrained congressional interest in appropriations for management and improvement and left open, at least for tactical political purposes, the prospect of divestiture.

The government's role as manager under the act was weak.[7] The first administrator of the Grazing Service, Farrington Carpenter of Colorado, chose from the beginning to use associations of stockmen as the preferred mechanism for allocation of grazing use privileges, and did so with such effectiveness as to inspire admiration among students of public administration.[8]

The objectives of the livestock users in governmental administration were clear:[9] enough structure to sustain the allocation system, but not enough to impose any conditions on use—either fees for grazing use, grazing intensities, or grazing patterns. The successful efforts of the stockmen to sustain operator dominance over Grazing Service employees is documented in some of the best of the immense mass of public land literature, ranging from amending the act to make district advisory boards of grazing permittees mandatory, to requiring proof of state residence of grazing service staffs and forcing the transfer of field staff members who challenged operator preferences.

The impression of universal venality and obstruction in much of the literature of the period is undoubtedly exaggerated. Levels of overgrazing were reduced substantially, and competence and courage were to be found in both headquarters and field offices of the service.[10] But the combination of aggressive industry political action, championed by eager voices in the western congressional delegations and opposed only fitfully by interior secretaries and rarely by anyone else, combined to create an image of incompetence and lack of professionalism that plagued the Grazing Service throughout its life, and followed its successor agency, the Bureau of Land Management, into the 1970s.[11]

If anyone doubted the determination of the organized public land livestock industry to avoid any significant measure of control by the Grazing Service, the doubt was clearly resolved in 1946.[12] The service had been the subject of a series of hearings beginning in 1941, led by Senator Patrick A. McCarran of Nevada. The hearings were carefully staged to evoke operator protest over the efforts of the Grazing Service to manage use of the land (the grazing fee issue, as always, was a recurring theme). The House refused to appropriate funds for the service unless fees were raised; the Senate preferred to forego any appropriations rather than raise the fee; appropriations were cut 85 percent from the previous year; and the Grazing Service as a separate agency died a merciful death.

As Sally Fairfax observes darkly:

> The remnants of the Grazing Service were joined with the General Land Office by an executive reorganization. Thus, without any money or any mission, the Bureau of Land Management was born in the wreckage of the Grazing Service.[13]

There remained, however, a core of competent staff members from the Grazing Service who were determined to achieve their vision of effective rangeland management. In philosophical orientation, they resembled their forestry counterparts in the Forest Service.

The Unseen Agenda

Neither the confident Forest Service nor the newly organized BLM was aware of the broad social forces that would soon emerge to place new pressures and expectations on the public lands and management agencies. The postwar economic boom that reached full flower in the 1950s and continued with minor interruptions until the late 1970s was not anticipated immediately after World War II.

Demand for forest products had risen dramatically during the war, but before the war the timber industry in the Northwest (where the bulk of the Forest Service commercial timber is found) had encouraged the Forest Service to limit sales in order to protect private timber markets in a depressed economy. It was not certain that this pattern would not reemerge.

The range wars were familiar turf; both agencies were well aware that controversy would continue, although they had different expectations about their capacity to do battle.

Pressure for more effective consideration of wildlife as a resource desired by hunters and fishermen on the public lands was palpable during the 1930s and had begun to intensify at the beginning of the 1940s, as federal funds under the Pittman-Robertson Act strengthened the professional insights and resolve of state game and fish agencies. Organized sportsmen's groups provided a modest support base.[14]

Wilderness preservation was a recognized management objective of the national forests and was a focal point for determined efforts by several organized conservation groups. However, wilderness was not widely considered to be a class of use equal to commodity production. Wilderness preservation was not a major management issue on the BLM lands.

Outdoor recreation was unconstrained on both land systems. The Forest Service encouraged and accommodated it within its concept of multiple-use management. Recreation was not a priority management issue to BLM. The coming recreation explosion was a function of a lively postwar economy that had not yet emerged. "Environmentalism" as it was later defined was largely unknown in a nation anxious about the postwar economy.

Mining and mining location continued to take place, largely independent of agency discretion, under the General Mining Act of 1872, known (reverently, to miners) as the *1872 Mining Law*. BLM (taking over from GLO) handled mineral leasing on both land systems; issuance of oil, gas, and coal leases continued to be demanded and offered with little constraint, especially on BLM lands.

In short, the Forest Service entered the postwar period confidently. BLM enjoyed the services of many competent people, but they had no reason to expect support for their ambitions to develop a management system similar to that of the Forest Service.

EVOLUTION OF ENVIRONMENTAL FORCES

The first intense postwar controversies were generated by the perennial grazing issue. By the mid–1950s, the most aggressive pressure for change in public land policy was prompted by public concern for a mixed aggregation of values and environmental experiences.

Interest in wilderness preservation had produced an administrative system for wildland protection within the national forests dating back to 1920.[15] Wilderness interests, relatively small in number but generally sophisticated and persistent, continued to press for prompt and generous designations under a second review process initiated in 1938. Timber, mining, and grazing interests organized themselves in opposition, and the battle over the use of wildlands was joined.

As the postwar economy accelerated, an astonishing outpouring of demand for many forms of outdoor recreation activity on the public lands emerged. Recreation activities such as camping, picnicking, hiking, and boating generated demand for developed facilities. Both the activities and the facilities themselves can damage wilderness, and might have led to a split between wilderness and other recreational interests. The attractiveness of relatively natural landscapes, however, was central to the outdoor experience for many, perhaps most, of the public land recreation users. As a general rule, the experienced wildland-oriented conservation leaders had little trouble in forging alliances with organizations responsive to the broad public interest in outdoor recreation on public land.[16]

In the early 1960s, Rachel Carson's *Silent Spring* helped stimulate concern for environmental contamination, attracting new participants to conservation activism. The public initially feared the effects of chemical pesticides on wildlife and ecosystems, but the scope of fear was steadily broadened over the decades by news of new contaminants and by the growing evidence of human health effects and other dangers.

Ultimately, in response to profound disenchantment with value systems attributed to government and corporate institutions, significant numbers of Americans sought and advocated alternative lifestyles.[17] These styles tend to deemphasize wealth accumulation and other totems of "moving up" and to emphasize instead family, personal worth and health, self-realization, and a preference for the "natural" in food, appearance, and landscapes. These preferences include varying degrees of rejection of public land policies and practices that alter or degrade land, water, and air from health, ecological, recreational, and aesthetic perspectives.

By the mid–1970s, a complex system of private organizations had emerged, functioning at national, state, and local levels, with allies in the media and strong links to public land agency staffs.[18] An environmental movement of considerable depth had come of age, had secured persistent public support of much of its agenda as measured by public opinion polls and public policy responses, and was to survive its first direct challenge in the early days of the Reagan administration.

Environmental Policy Successes

Most of the body of legislation affecting management and protection of the public land systems and many changes in administrative policies and practices were adopted at the prodding of environmental quality interests.

Demand for outdoor recreation facilities in the 1950s led to increased appropriations for the Mission 66 and Operation Outdoors programs of the National Park Service and the Forest Service, respectively. Interest in preservation of wild and scenic lands led to some early transfers of land from the Forest Service to the National Park Service, and from BLM to the Park Service and the Fish and Wildlife Service. The Forest Service response to demands for large-scale wilderness designation was considered by wilderness advocates to be slow; a legislative solution in the form of the Wilderness Bill was introduced (1956) and eventually enacted (1964).[19]

The Kennedy-Johnson years produced, besides the Wilderness Bill, a series of mandates for wildland, scenic, and recreation values in public land administration, stimulated to a substantial degree by recommendations of the Outdoor Recreation Resources Review Commission (ORRRC), a study commission established by Congress and composed of a mix of congressional and presidentially appointed members under the chairmanship of Laurance S. Rockefeller. ORRRC reported in January 1962, blessing the basic concepts of wilderness legislation,[20] and recommending a series of additional actions. Stewart L. Udall, secretary of the interior throughout the Kennedy and Johnson presidencies, capitalized on the ORRRC report and made his "New Conservation" a major theme of both administrations.

The Land and Water Conservation Fund Act of 1964 was enacted in direct response to ORRRC recommendations, providing an infusion of funds for acquisition of lands within the national forests as well as for a dramatic expansion of the national park system. The Wild and Scenic Rivers Act and the National Trail System Act followed in 1968, with both BLM and the Forest Service as administering agencies for specific areas established by Congress. Other new categories of federal land administration (national seashores, lakeshores, recreation areas) were invented. National recreation areas and scenic rivers and trails on Forest Service and BLM lands constituted (with wilderness) overlays of environmentally inspired special management objectives and practices on the multiple-use land systems.

The 1970s were no less prolific. The National Environmental Policy Act (NEPA) required public disclosure of effects of proposed and alternative activities, both site-specific and programmatic, on the public lands and elsewhere. The Endangered Species Acts of 1966, 1969, and 1973 eventually made protection of endangered species and habitats high-priority considerations in specific land use decisions.

Federal water pollution control laws were strengthened to include, among other things, programs to control non-point source pollution and to regulate dredging and filling activities. Both affect forest management and, to a lesser extent, rangeland management.

Health concerns led to strengthening of air quality and pesticide laws; the latter have had immediate effects on public land management, primarily in constraining pesticide use for forest and rangeland vegetation management purposes.

Amendments to the Administrative Procedures Act in the 1960s and 1970s required agencies to open more fully their rule-making processes to the public, and to establish and follow procedures designed to assure evenhanded administration.

Since the early 1970s, using adequacy of environmental impact statements pursuant to NEPA, procedural requirements under the Administrative Procedures Act, and relaxed rules on standing as prime tools, the environmental community has found allies in the courts. The opening was pressed vigorously to influence administration of major program areas (coal leasing, livestock grazing, timber management, and wilderness review procedures, for example) as well as site-specific issues.[21]

LAND AS COMMODITY: A POWER BASE

The industries that use the public lands are far from homogeneous. Mining, timber, livestock, and oil and gas interests are composed of large and small operations with mixes of individual and corporate ownerships. Policy differences often emerge within industries, but they tend to close ranks in specific cases of conflict with other interests. On fundamental issues of conflict with environmental interests over allocation and regulation of use, the commodity groups coalesced to protect their interests.

Hardrock mining (including precious and base metals and uranium); coal, oil and gas, and other energy mineral production; timber production; and livestock grazing have long been established as major commodity uses of the public lands.[22] These industries are organized to protect their interests at all points of decision and leverage, and they have decades of experience in doing so.

The public lands are also used as rights-of-way for electric transmission lines, pipelines, and highways. Particularly in recent years, siting of energy generation and conversion facilities, from electric generating stations to coal gasification plants, has become a major use of land. And the public lands have been seen by the federal government as sparsely populated sites available for a variety of noncommodity uses (often unwelcome), such as nuclear testing and waste disposal sites and missile testing sites. BLM in particular has felt the pressure of the search for sites for noxious facilities.

For decades, elected officials in the western states were quite dependably aligned on the side of economic users in controversies over either agency control or environmental constraints. As the West became an urban region in the postwar years and, with the rest of the country, participated in the outdoor recreation boom and in the growing public concern for environmental quality and quality of life, political leadership responded, changing to meet the new demands being made by the public. This shift, a historic one for public land management, is reflected in the mixed pattern of policy responses in the postwar years. The following pages describe a few major policy issues in which commodity production and environmental interests clashed sharply.

The Range Wars

Fresh from its decimation of the Grazing Service, the western livestock industry launched an effort to achieve ownership of public land grazing allotments immediately after World War II. In 1947, a Joint Livestock Committee on Public Lands advanced a proposal that would allow grazing permittees on BLM lands to acquire their allotments under preferential conditions; in the national forests, acquisition of grazing allotments was to require an intervening step: grazing lands in the national forest would be transferred to the Interior Department and then offered for purchase to grazing permittees, again under preferential terms.[23]

The "Great Land Grab," as it was called by *Harper's* editor Bernard DeVoto, attracted unusual attention from national media, fanned by the writings and urgings of a small group of officials and staff members of western and national conservation organizations. The divestiture proposal did not arouse serious congressional interest.

The late 1950s, the 1960s, and the early 1970s were periods of relative peace on the range. Livestock grazing use on the BLM lands was reduced over the twenty years from 1950 to 1970 by about a quarter, with controversy generally localized.

In 1974, environmental interests renewed the range wars with fresh vigor, bringing a successful NEPA lawsuit that required BLM to prepare site-specific environmental impact statements assessing the effects of alternative levels and patterns of livestock grazing on more than 170 million acres of BLM lands in the West.[24] The burden of preparing environmental impact statements (EIS) fell to the Carter administration. It accepted the requirement imposed by the court as an instrument for making decisions about the use of public lands by livestock that would have been required in any event by the multiple-use mandate of the Federal Land Policy and Management Act (FLPMA) in 1976.

A strategy was developed (and published under the title "Managing the Public Rangelands")[25] that involved conducting intensive range inventories, setting livestock use carrying capacities, and achieving desired levels of use within three years of completing each EIS. Early EISs (a total of 245 were to be prepared over a decade) dictated reductions in about half of the grazing allotments, averaging about 10 percent from recent actual use and considerably more from legally authorized use levels.[26]

The political atmosphere was charged by the perception of a "War on the West" attributed to the Carter administration in the form of attempts to end the tradition of unrestricted entry to public lands for hardrock mining; the federal water project "hit list" (many of the projects were in the West); the "RARE II" reanalysis of Forest Service lands for wilderness potential; the pending wilderness review of BLM lands for wilderness (mandated by FLPMA); and other grievances.

The politically astute public land livestock industry had little difficulty in landing a series of counterpunches. A federal court order sharply constricted the three-year carrying capacity goal.[27] Range scientists refused to support a method of establishing livestock carrying capacity based on soil and vegetation inventories, arguing for monitoring as a more reliable indicator.[28] Direct political appeals to Congress led to a 10 percent limit on reductions in any one year unless consented to by the operator.[29]

Most significantly, a movement was initiated by three BLM livestock permittees in Nevada (who were also state legislators) to stimulate political opposition to BLM's aggressiveness in livestock and other commodity use controls. They found a warm response in segments of the West, and the movement served as a rallying point for traditional western economic interests against the Carter administration and two decades of environmental initiatives. The Sagebrush Rebellion, as the movement came to be called, proposed that the public lands be transferred to the states as a matter of constitutional interpretation. The purpose, however, was political support for relief from anticipated restraints by the Carter administration on interests associated with commodity use on the public lands, notably livestock grazing.[30] Eventually, Sagebrush, which is discussed in more detail later, became a coalescing forum for western expression of a wide range of grievances about a distant and obtrusive federal government on issues ranging from mandatory school busing to Occupational Safety and Health Administration (OSHA) regulations. By any political standard, Sagebrush was a rousing success.

In the midst of Sagebrush, the director of the Bureau of Land Management came close to achieving consensus on a modified approach to making and implementing rangeland decisions through the EIS process. A coalition was in fact developed, which the western governors both joined and helped lead. The prospects and incentives, however, changed with the election of Ronald Reagan and Interior Secretary James Watt's early signals that the livestock industry would not need to bargain with the environmental community about its place on the range. Curiously, by the midpoint of the Reagan administration the livestock industry was exploring with the environmental community the prospect of reviving the rangelands coalition. There is increasing frustration within the environmental community, however, about what they believe is continued severe environmental damage caused by livestock grazing. A significant minority of environmental leaders are calling for an end to livestock grazing as a lawful use of the public lands.[31] The coalition has again dispersed.

Timberland Management

As the backlogged demand for housing accelerated in the early postwar period, the timber industry sought and received sharp increases in access to timber on the national forests and on BLM's valuable O&C timberlands in western Oregon.

The cut on the national forests rose from less than 4 billion board feet in 1950 to more than 9 billion in 1960; sales on the O&C lands also more than doubled, from about 400 million to 1 billion board feet. After 1960, however, while the postwar economy was generating increased demands for timber and forest products generally, the availability of federal timber leveled off. The peak year for timber cut on the national forests was 1966 at 12.1 billion board feet, with a median of slightly over 10 billion over the 1970–1980 decade. Sales from the O&C lands, after a peak of 1.66 billion board feet in 1970, averaged about 1.16 billion board feet over the following ten years.[32]

This period of large-scale timber harvest also included sharp increases in appropriations for reforestation and other forest management activities. The period was welcome indeed to professional foresters in the Forest Service and BLM who were committed to the concept of the managed forest as the social justification for the profession and for the forest systems. As a general proposition, the Forest Service (and BLM on the O&C lands) agreed with industry on the wisdom of expanded timber harvest. Even in the heat of controversy over wilderness, with the Forest Service moving steadily, if reluctantly, to accommodate wilderness interests, the industry did not see the agencies as enemies.

A principal reason may have been a sense of kinship among the professionals in industrial forestry, BLM's forestry-dominated staff in the O&C region, and the Forest Service generally. Unlike the situation in the livestock industry, the timber industry and agency professionals shared some fundamental assumptions, including (although differing in details that became important as industry demands accelerated) a commitment to sustained-yield management under professional control.

As the impact of increased timber cutting was seen—most dramatically in the Northwest in the early postwar years—environmental organizations in the Northwest and nationally (most vehemently the Sierra Club) objected strenuously to the sharp acceleration of road construction by the Forest Service,[33] accusing the agency of deliberately driving roads into wildlands to preempt potential wilderness designation.

In any event, the industry, even with the sympathy of much of the Forest Service professional staff, was unable to stem the loss of commercial timberlands to timber harvest. Losses came in early administrative designations, as in the Three Sisters Wilderness Area in Oregon; in subsequent congressional establishment of new national parks in the California redwoods and the North Cascades area of Washington; in Wild and Scenic Rivers and other special designations; and, critically, in the ongoing process of wilderness review and congressional wilderness designation. Nevertheless, much timber was sold and harvested, major investments were made in roads needed for timber management, and substantial investments in reforestation were made. The national forests and BLM's O&C lands became fully integrated into the timber economy of the Northwest.

The wilderness phenomenon, of course, challenged the longstanding assumption of utilitarian values—of timber production—as an overriding objective in forestland management. The ethos of Pinchot and the profession was also being challenged on the wisdom of professional practice itself as applied on those lands that remained available for timber harvest.

The rejection by major segments of the public of economic utility as the dominant objective in natural resources management, and of professional expertise as the determinant of how natural resources are to be managed, is strikingly demonstrated in the recent history of national forest management.

Citizens and public officials challenged, over a period of years, the practice of clearcutting on the Monongahela National Forest in West Virginia.[34] The Forest Service responded to concerns about damage to scenery, streams, wildlife habitats, and recreation sites with certain changes in on-the-ground practice, but too slowly to satisfy critics. A similar public assault was mounted against clearcutting and terracing of steep slopes on the Bitterroot National Forest in Montana, a case that was most significant for challenging sustained-yield management on sites where economic return from timber sales would not cover the costs of sale administration and subsequent reforestation and management.

Ultimately, Monongahela anticlearcut advocates exploited the language of the 1897 Forest Service Organic Act, which limited the agency to harvesting "mature" timber that had been "marked and designated," a far cry from the all-age harvest practices of clearcutting. A lower court found in favor of the citizen groups in 1973;[35] the federal appeals court upheld the ruling in 1976, enjoining any further commercial harvesting on the three small tracts chosen to test the issue.

Assuming that the decision could well lead to a nationwide moratorium on timber harvest, the Forest Service moved to seek amendments to the Organic Act. The bargaining processes that followed led to the National Forest Management Act of 1976 (NFMA). The NFMA reserved a substantial degree of ultimate discretion of forest management practices to the agency professionals, but imposed a land use planning requirement on the agency that bristled with procedural detail. It required interdisciplinary teams to do the planning, specified access points for public participation and imposed serious constraints (if not prohibitions) on the exercise of professional judgment in the on-the-ground application of professional concepts.

In short, the era of Forest Service discretion over major land use allocations ended with the Wilderness Act of 1964, which established the Congress as a direct decision maker on the uses to be permitted on millions of acres of national forests. Professional discretion in management practice on lands available for multiple-use management was eroded by the Monongahela suit and the Bitterroot controversy, as well as by the series of environmental laws, such as those on water quality, endangered species, and surface mining regulation.

Controls over pesticide use (another tool for accelerating timber production widely supported by forest professionals) were first imposed in 1947, sharply strengthened in 1972, and modified in 1975. Constraints on use of pesticides is another and important limitation on professional practice.

Timber production is also at issue in a third controversy, this time involving an alliance between the Forest Service and environmental and conservation interests. Historic practice in the Forest Service calls for annual increments of timber harvest that will assure a continuous, nondeclining, even flow of timber. Industry and many economists and professional foresters object. Because the critical Pacific Northwest public land timber supplies are largely mature "old-growth" forests, no longer growing and susceptible to mortality, critics argue that the best policy is to accelerate old-growth harvests, producing a bulge during the intensified harvest period and a subsequent but temporary decline in yields as reforested areas become ready for harvest.

John Crowell, the first assistant secretary in the Reagan era to directly supervise the Forest Service, ordered substantial increases in timber offerings. President Jimmy Carter had tried to do so out of concern for inflated housing costs but was met with an "all deliberate speed" response by the agencies.[36] Accelerated timber harvest schedules in many national forest plans prepared under NFMA in the 1970s and 1980s have been challenged through administrative appeals and lawsuits.

Finally, management of timber, including logging and reforestation, on lands where economic returns are extremely questionable has drawn increasingly severe criticism.[37] In this case, even segments of the timber industry have joined in the criticism, preferring to see scarce dollars invested in more productive sites. Timber sales and harvest in national forest land use plans have been scaled back in response to political pressure and fear of successful litigation. The below-cost timber sale issue has attracted a novel coalition of environmental advocates, fiscal conservatives, and free market economists in opposition to Forest Service timber policies.

Timber continues to be available for harvest in the Northwest and other areas. But the cumulative challenges to timber production raise serious questions as to whether the fundamental assumptions of the forestry profession will prevail over time.

Minerals

For most of the years of the postwar period, the mining industry has believed itself to be in a defensive position in public land management. Creation of new national parks out of national forest land, wilderness designations, and restraints on mining while federal lands are studied for wilderness potential have generated alarm in the mining industry about "lockup" of potentially valuable mineral resources.

Beyond the wilderness issue, the hardrock mining industry has been beset by persistent demands for amendment or repeal of the Mining Law of 1872, the fiercely guarded authority under which minerals may be sought, claims filed, and minerals mined with relatively little control by the land management agencies. A bill to change the location and patent system provided by the 1872 law to a lease arrangement, in which discretion would be vested in the secretary of the interior, was offered in the 1970s, but was abandoned by its sponsor, Congressman Morris K. Udall, when political retaliation by the industry almost succeeded in defeating his bid for reelection. The agencies are currently silent, but they have long wished for authority to regulate mining uses with the same degree of authority available for other uses.

Surface effects of hardrock mining on both Forest Service and BLM lands continued to be unregulated in any formal sense until the 1970s. The Forest Service finally adopted regulations requiring the preparation and approval of mine plans, reclamation strategies, posting of bonds for compliance, supervision, and the like in 1974. BLM offered a set of regulations in 1976; they were withdrawn, but ultimately were issued in a modified form in the last days of the Carter administration.

Energy minerals and "strategic minerals" have been subjects of special attention and controversy. The western public lands contain over 30 percent of the nation's coal reserves. Relatively unconstrained leasing in the 1960s and early 1970s was severely criticized as speculative in character, and as depriving the federal government of appropriate revenues from the leases. A voluntary moratorium on further leasing was instituted by the secretary of the interior in 1971.

The Arab oil embargo brought a dramatic increase in interest in western coal development, at a pace genuinely frightening to citizens and political leaders in Montana, Wyoming, Utah, and Colorado—the so-called OPEC states. Environmental litigants succeeded in getting a federal court to enjoin further coal leasing in the mid-1970s under an NEPA suit.[38] Leasing was resumed in 1980 in a program developed by the Carter administration to relate leasing levels to foreseeable demands, avoid environmentally sensitive areas, strengthen state influence on leasing decisions, and coordinate leasing with regulation of mining activities under the Surface Mining Control and Reclamation Act of 1977 (SMCRA).

Responding to both Appalachian and western concerns for environmental damages caused by strip mining, Congress provided authority in SMCRA for stringent regulation of surface mine operation and reclamation and for designating areas of special vulnerability as "unsuitable" for coal mining. SMCRA, in combination with the Carter administration's coal leasing policies, was perceived by some in the industry and by many politicians as a tool for delaying western coal production. Coal leasing, resumed in 1980 under Carter, was enjoined again when new procedures initiated by the Reagan administration were successfully challenged by environmental litigants.

The oil and gas industry was relatively unconstrained by the Forest Service and BLM for much of the postwar era. Wilderness protection became a serious constraint in the 1970s, when discoveries were made along the Overthrust Belt in the northern Rockies. The Overthrust Belt, a geological phenomenon stretching from Canada to Mexico, traverses national forest lands in Montana, Wyoming, Idaho, and Utah, lands that contain both already-designated and potential wilderness areas.

Access for exploration and production to attractive sites, especially those in Montana and Wyoming, was constrained. Exploration methods in potential wilderness areas were strictly regulated to assure no impairment of surface values. The Reagan administration, through Interior Secretary James Watt, tried—unsuccessfully—to open the areas to exploration and production despite requirements of wilderness laws and regulations.

With enactment of the Federal Land Policy and Management Act in 1976, BLM was required to inventory the public lands under its jurisdiction for potential wilderness designation by Congress, and to protect wilderness values in the potential area until Congress decided on designation. The oil and gas industry felt these constraints keenly, and joined other commodity interests in working to limit wilderness expansion, to ease management restrictions on both designated and potential wilderness, and to soften the impact of regulation under the environmental laws of the 1970s.

The oil and gas industry was forced by circumstances to form coalitions with other interests, especially mineral interests and livestock grazers. Many individuals and small-to-medium-sized concerns became active supporters of the Sagebrush Rebellion, both as a divestiture possibility and as a medium for forcing policy change. The major oil companies, however, chose to stay aloof from the more strident proposals for reshaping public land policies.

On balance, even in government response to the energy and other commodity crises of the 1970s and 1980s, environmental concerns in public land management competed effectively in the policy arena.

Beyond the coal and oil and gas controversies, efforts by Presidents Nixon and Carter to establish authoritarian agencies to ease siting and construction of energy facilities were defeated. Both offered variations of the energy siting czar; both would have superseded authority held by federal land agencies as well as by states and localities; and both were defeated.

An attempt to use scarcity to give minerals a priority in public land use—built around alleged national security problems posed by dependence on foreign sources of strategic minerals—was launched in 1981 by James Watt. This effort foundered quickly when distinguished minerals economists challenged its basic assumptions, the environmental community launched a well-organized counterattack, and the White House was ambivalent, apparently out of concern over the possible inflationary effects of higher mineral prices.

The New Baseline: The Reagan Era

The Reagan administration candidly attempted to establish the primacy of commodity production in public land policy. The period may be described as the renaissance of the public land politics of the Old West—of economic dependence on direct use of resources in mining, timber, and livestock grazing, and of taming the land and the rivers for settlement and for economic opportunity.

The Sagebrush Rebellion was led by interests—livestock grazing, hardrock mining—quite self-consciously attempting to evoke images of the trail boss (Marlboro man and all) and the lonely prospector in battle with unfeeling bureaucrats, easterners, and environmentalists. Sagebrush became much more—it triggered a chorus of complaints about the Carter administration and the federal government generally, but its origins and imagery were in the frontier tradition of the natural resource industries.[39]

The Reagan administration rhetorically endorsed Sagebrush, but moved promptly under Secretary Watt to respond to commodity interests with changes in policy and administration rather than ownership. In fact, the secretary claimed credit—accurately—for defusing Sagebrush as a public land disposal idea.

It seems reasonable to assume that livestock permittees, small miners, coal companies, oil and gas operators, geothermal steam developers, public land timber customers, and others were cheered by the administration's championing of their legitimacy and their contributions to society, and by tangible benefits resulting from delays in livestock grazing cuts, increased mineral and timber offerings, and countless specific field-level administrative actions.

The apparent confrontational compulsion of James Watt, however, helped broaden public and media support for determined efforts by the environmental community and its political allies to hold ground against reversing environmental goals and reducing the protections fashioned in the 1960s and 1970s. Watt was forced out of office. His successors have not challenged the basic legal framework of environmental laws, although they candidly pursue commodity-oriented policies in the land use and budget allocations.

The policy and program tilts toward commodity interests took place within the familiar arena of utilitarian values versus environmental quality values. The Reagan era also spawned a series of proposals outside the mainstream of the natural resources controversy.

One such proposal, labeled "asset management" was a broad effort to sell "unneeded" government-owned real property to produce liquid assets. The program placed primary rhetorical emphasis on the case for disposal of urban lands and buildings that would be better sold for large sums of money than held for presumably marginal federal uses. The asset managers, institutionalized in a Federal Property Review Board operated out of the White House, persisted in including both BLM and Forest Service lands as potential surplus, and even set quotas for dollars to be generated by public land disposal.

The Congress, environmental forces, and the western governors saw asset management as Sagebrush in disguise.

Secretary Watt kept his distance from asset management after some early confusion, and made it clear eventually that the federal multiple-use land systems would not, in any significant sense, become part of the Property Review Board's disposable inventory. In both political and revenue-generating terms, asset management proved as dismal a failure as Sagebrush was a political success.

Some administration officials also pressed for policy reforms of various kinds that were generally characterized as "privatization" proposals. "Privatization" came to mean many things. One proposal, associated with Nevada Senator Paul Laxalt and then-presidential counselor Edwin Meese, was a throwback to the Great Land Grab. Lands in BLM grazing allotments were to be sold to permittees at fire-sale prices. The prospect of large-scale sale of national forest lands was casually discussed. No enduring support emerged.

Privatization also came to include a policy requiring federal agencies to give priority to contracting out all manner of governmental services rather than executing them through federal employees. Eventually, the Congress constrained this initiative as it applied to the federal natural resource agencies, although the Forest Service announced in 1988 a program to encourage private operation of certain recreation facilities and services.[40]

Various other schemes for encouraging private individuals and organizations to undertake management functions on the federal lands were advanced. A BLM proposal for "cooperative management agreements" would have granted to selected livestock permittees an unusual degree of autonomy in managing allotments without ongoing BLM supervision. The proposal, offered twice by BLM director Robert Burford, was vigorously opposed by environmental interests, constrained as the result of a lawsuit, and ultimately confined to a few test cases. The most promising idea in the "cooperative privatization" category involved arrangements for nonprofit organizations to manage areas or activities of special meaning to them, such as natural areas, wildlands, and recreation sites. Sensible cooperative arrangements have also been made with state fish and game departments and with state and local park and recreation agencies.

The word "privatization" has also been used—in part by critics—to describe the vastly increased offerings of coal, oil and gas, geothermal, timber, and other resources by the Interior and Agriculture Departments.

The administration worked diligently to achieve, in precise terms, the privatization of public resources of economic value by leasing them to private interests while holding the land nominally in federal ownership.

On some federal lands, private rights to development of specific resources constrain potential for other uses. These lands can be said to be "privatized." But for the most part, "asset management" and privatization are dead issues as large-scale disposal methods.

AGENCY ROLES IN INNOVATION

The public land management agencies have rarely been innovators in postwar policymaking. The primary reason is this: their leaders were simply too busy coping with the staccato series of new policy proposals from others to assert across-the-board leadership. It is also unclear that anyone would have followed.

The agencies (or rather, the Forest Service) also had something already in hand that they wanted badly to protect. The tradition of professional agency staff's exercising an extraordinary degree of discretion was secure. Authority to allocate land to various uses was not yet threatened. Freedom to select among technologies in professional practice was unchallenged except in the normal give-and-take of field-level management.

The tightly organized Forest Service bureaucracy was also flourishing, with new professionals added to the ranks as public demands for new services in such areas as outdoor recreation and fish and wildlife habitat enhancement required new expertise.

BLM did not begin the period with these advantages, but it set about assembling them. The decentralized administrative structure of the old Grazing Service was put to use by the new agency. Budgets increased; new staff was added. By the 1960s, without a clear legal charter, BLM had developed a management system and philosophy—multiple use and all—that was remarkably similar to that of the Forest Service.

Demands and conflicts intensified geometrically during the postwar period. The institutional styles that developed encouraged conflicting interests to seek their interests by participating in administrative processes, led by the agencies and leading to resource allocation decisions made by the agency to be implemented through on-the-ground management practices specified by agency professionals.

Both agencies are multiprofessional, with foresters remaining dominant in leadership positions in the Forest Service. Range specialists and land and realty specialists (from the Grazing Service and General Land Office heritages respectively) and foresters (from the O&C lands) shared dominance in BLM in the postwar period, despite growing minerals management responsibilities. The Conservation Division of the U.S. Geological Survey was made a part of BLM in 1983, a massive infusion of minerals professionals that may have substantial long-term implications.

While conventional wisdom distinguishes between a professional Forest Service and a far less professional BLM, Paul Culhane finally documented in his 1981 *Public Land Politics* what should have been obvious: that both agencies are staffed from the same civil service rosters of professionals required to meet identical qualifications.[41] The aura of multiple use around a core concern for renewable resource utilization and management is palpable in the professional milieu of both agencies, as is the inevitable bias toward professional dominance in decision making. At the top, however, in the contrast between the custom of choosing the chief of the Forest Service from among career professionals and using a more political process to select the BLM director from a widely varied pool of candidates, the distinction is compelling.

Much has been made of what environentalists and scholars see as sharp and enduring biases toward utilitarian management objectives. Despite apparent responsiveness by the agencies to commodity production policy tilts during the Reagan administration, there is reason to question the assumption that the Pinchot philosophy still permeates the agency atmospheres. The Reagan era followed a long period of environmental initiatives; some swing of the pendulum was inevitable. Concern for the economy generally as well as for energy and strategic mineral independence in the late 1970s and early 1980s gave special impetus to commodity production. Even the Carter administration worked diligently to enhance timber and mineral production from the public lands. Political leaders, not agency professionals, were pressing for these objectives.

Utilitarianism lives in the core of the forestry, range, and minerals professionals in the agencies, but younger professionals in forestry and rangelands, at least, tend to bring a new order of environmental sensitivity to their professional commitments. Compulsion toward relatively intensive use and management will continue, probably most intently among foresters, but it should be offset by economics (intensive management is uneconomic in marginal timber and forage production areas) and by internal and external pressures from environmentally oriented professionals and interest groups. Restoring staff reductions in wildlife, wildland, recreation, and archaeological professions would do much to restore agency responsiveness to environmental quality concerns.

The Forest Service Pattern

The 1960 Multiple Use and Sustained Yield Act (MUSY) was an early effort to contain diverging objectives in public land management within the managerial framework of the national forest system.

Most scholars agree that MUSY was a Forest Service initiative. The act defines the concepts of multiple use and sustained yield. Most significantly, wilderness is recognized as consistent with the multiple-use concept. Outdoor recreation is first in an alphabetical list of statutorily recognized multiple uses. The act was fashioned, it seems clear, to disarm pressures for transferring prime recreation, scenic, and wildland areas to the National Park Service by signaling Forest Service capacity to meet the need for wildland protection and recreation opportunity in a multiple-use framework. Accommodation and containment in the context of agency-led administrative processes were the objectives.

RPA and NFMA

Similar objectives influenced Forest Service roles in two more purposeful systemic enactments, the Forest and Rangeland Renewable Resources Planning Act of 1974 (RPA) and the National Forest Management Act of 1976 (NFMA), but on these occasions the hand of outside influences was heavy.[42] The Resources Planning Act was led to enactment by congressional friends of the Forest Service, notably the late Senator Hubert Humphrey, out of frustration (shared by the service) over the perceived blindness of the federal Office of Management and Budget to "appropriate" levels of investment in national forest management. The Forest Service consented to and helped shape, but did not initiate, the RPA legislation.

The National Forest Management Act was formulated hastily when the Forest Service was effectively enjoined from carrying out an orderly program of timber sales by a lawsuit exploiting limitations on timber harvest authority under the 1897 Organic Act (the lawsuit and its implications were discussed earlier in this chapter).

Briefly, the requirements of the RPA and the NFMA are as follows: The RPA requires the Forest Service to develop and apply a data base that makes it possible to consider potential demands on and outputs from the national forests in the context of total demands for such goods and services. The RPA also assesses alternative management strategies, investment levels, and outputs, and it produces a recommended program for consideration in appropriation processes and in providing policy direction for management of individual national forests. The NFMA requires the Forest Service to prepare land use plans for individual national forests, and spells out rules for doing so. Heavy emphasis is placed in the NFMA on public participation, interdisciplinary teams, inventories of natural resource conditions and trends, demand assessments, and alternative management strategies, all to be analyzed against public preference and environmental and economic criteria.

The output of NFMA is a comprehensive plan for each of the national forests, to be prepared for fifteen-year periods with opportunities for amendment. Although the bills were enacted separately, the Forest Service has attempted to coordinate them, that is, to tie national goals and priorities to local and regional plans, and to use national demand data and goals for various resource outputs to influence the choice of resource products to be produced in each region and in each national forest. All these are to be coordinated with recommended national budget allocations for various programs and administrative units.

An astonishing amount of money and manpower has been and is being applied by the Forest Service to establishing workable inventory and demand assessment systems, information management systems, public participation processes, and budgeting systems for the integrated operation of RPA and NFMA.[43]

In addition to apparent faith in its ability to accomplish this herculean task, some other articles of faith may be divined in the nature of the Forest Service response to the enactment of RPA and NFMA:

- faith in the utility of projection of goods and services likely to be demanded by the national and local economies;
- faith in the ability to predict consequences of management actions applied over time, and in the stability of desired outcomes relative to ecosystem response times (as in reforestation);
- faith in the continuing availability of money to sustain the costs of the system (assumed to decline over time);
- faith in people's willingness to participate in the system in the face of human history suggesting the overweening importance of outcomes, not process.

In any event, the Forest Service has deeply committed itself from the beginning to making the RPA-NFMA system work. It is tempting to suggest that the agency's response can be traced not only to its feelings of competence, but also (somewhat cynically) to the perception that the system forces parties at interest to play by a complex set of rules fully understood only by Forest Service staff.

BLM: The Federal Land Policy and Management Act

The Bureau of Land Management took advantage of the environmental era to seek a firm charter legitimating and strengthening its role as manager of public land resources. The agency had developed an administrative structure, a multiple-use management philosophy, and a management system based on application of professional concepts and practices without the explicit blessings of Congress over a period of decades since its birth in 1946.

A congressionally chartered Public Land Law Review Commission, which reported in 1970,[44] had been briefly looked to by BLM staff as a vehicle for developing a firm policy base for management of the public lands; however, the commission failed to meet these expectations.

With little encouragement from outside the agency, except for the tenacious support of Michael D. Harvey of the staff of the Senate Committee on Interior and Insular Affairs, BLM staff assisted in development of several versions of an organic statute. The Federal Land Policy and Management Act, known as FLPMA, was enacted in October 1976, thirty years after the shotgun marriage of the General Land Office and the Grazing Service.

FLPMA formally recognized the Bureau of Land Management and directed the secretary of the interior to manage the public lands for multiple uses under the principle of sustained yield. The act also declared retention of the public lands as policy, with disposal required to meet national interest tests, and brought BLM into the mainstream of a pivotal public land controversy by requiring a review, to be completed by 1991, of BLM lands for potential placement in the National Wilderness Preservation System.

A key provision is the requirement that decisions on management of specific sites be based upon land use plans that consider the "best mix" of uses, are supported by a continuous program of natural resource inventories, and are formulated with public participation and in consultation with other federal agencies, Indian tribes, and state and local governments.

The secretary is also instructed to prepare every four years an estimate of what he believes to be the "efficient and effective" level of appropriations for BLM management, known as the four-year authorization process. This estimating process, a modest BLM counterpart to RPA, has not proved significant.[45]

The Forest Service response has led to an astonishingly ambitious effort to rationalize (and presumably to retain control of) forest decision making from the national policy level to the ranger district by linking RPA and NFMA requirements in an integrated system. That system is now under attack as being costly, as centralizing decision making and undercutting the agency's long tradition of delegation of authority to field managers, as being insensitive to local preferences and ecosystem conditions, and, in practice, as being biased toward quantifiable commodity outputs without appropriate regard for difficult-to-measure noncommodity values.[46] The outcome of RPA-NFMA strategies will have powerful effects on the future of national forest policy and administration. In significant ways, the future of the Forest Service as manager is at issue.

BLM Management Methods

The BLM planning system is arguably more decentralized, and more sensitive to the interplay of local demands, local preferences, and the condition and capacity of affected ecosystems. It is conceivable that the BLM style may prove more enduringly useful as a way of inviting conflicting interests to address conflicts in substantive rather than procedural and litigious terms. If so, the perception of western communities about the relative merits of the two agencies and their management styles may change.

BLM has been shaken by painful reminders of the lack of respect and support from secretaries of the interior that have plagued the agency and its predecessor. But the agency and the lands have become visible through confrontation; another generation of divestiture proposals has been beaten back; and it has been recognized for its professionalism.

POLICY STALEMATE

Since the first wave of Reagan era initiatives to reverse the environmental trend of the postwar era, a condition approaching stalemate has developed in public land policy dynamics. Control of the House of Representatives, and later the Senate, by the Democratic party has been a critical factor. So has the enduring preference of the American people for high standards in environmental protection, including a powerful commitment to wilderness as a major land use.[47]

The basic statutes developed in the 1960s and 1970s remain intact, including MUSY, RPA, NFMA, FLPMA (the management systems), and the complex network of laws that constrain agency flexibility by placing overlying requirements on the public lands. NEPA, the Endangered Species Act, the Administrative Procedures Act, the Wilderness Act, the Federal Water Pollution Control Act, and pesticide regulation programs are examples.

The more dramatic of the administration's initiatives—wilderness leasing of oil and gas, vast increases in timber sales—have been stopped, delayed, or modified by combinations of public outrage, congressional action, and litigation. Even the genuine Reagan mandate of 1984 did not immediately change the balance of power.

A popular president, backed by the political power and economic resources of the American business community, with the support of the Senate, failed to reestablish commodity production in a clear position of dominance in public land policy.

In the broad arena of competition for effective public support, the environmental community has done well indeed. In the absence of war or economic disaster, the stalemate could be broken either by the resumption of environmental enhancement and protection as the highest public policy goal or by further shifts to commodity production.

The "Real West": A Divided Region

With respect to the public lands, one factor that will continue to express itself is the profound social and economic change taking place in the West. The Old West of primary economic dependence on natural resource-based industries is fast disappearing. The West is, of course, largely an urban region in population distribution. The growth areas in the region's economy (high tech, services) roughly parallel those of other regions. Its natural resource-based industries will continue to decline relative to the economy as a whole.

The image of the West as political captive of the resource-based industries is nostalgia, not reality. The West shared and in some cases led in the growing national concern of the 1960s and 1970s for environmental amenities, ecological sensitivity, and environmental contamination, as illustrated by California's initiatives in air pollution control and coastal resource protection, Oregon's land use planning and (limited) control system, and the pre-SMCRA development of surface mining controls by Montana and Wyoming.

There is little reason to doubt that Secretary Watt, from his long association with business and corporate interests nationally, as an employee of the U.S. Chamber of Commerce, and with traditional western natural resource user interests as leader of the Mountain States Legal Foundation, felt he spoke for the West on the wilderness issue.

At the height of the wilderness leasing controversy, members of the Montana, Wyoming, and New Mexico congressional delegations (including, in particular, Republicans Malcolm Wallop of Wyoming and Manuel Lujan of New Mexico) returned from a congressional recess shaken by unmistakable opposition to an action assumed to be "western." They promptly withdrew their support for the leasing proposals, and Watt backed off. The political prospects of substantial additions to the wilderness system even in a hostile administration were suddenly improved.

Western support for wilderness illustrated the lack of any compelling base of support in the region for extreme changes in public land policy. Sagebrush is useful as an illustration. Although Sagebrush as a rallying point for antifederal and anti-Carter grievances was a smashing success, only one western governor, Robert F. List of Nevada, explicitly endorsed disposal of the federal lands. A few governors stood aside; most were publicly or privately opposed. Two who opposed disposal most vigorously and visibly, Richard D. Lamm of Colorado and Bruce Babbitt of Arizona, were reelected in 1982 with overwhelming majorities despite clear accountability on the issue.

The West's response to privatization and asset management was unequivocal. The western governors ridiculed the proposals as Sagebrush in disguise, as ludicrous in claiming to offer hope of reducing the national debt, and as ignorant in generating great increases in economic activity on lands largely already available to commodity development. Faced with buying grazing allotments, ranchers demurred. Miners feared problems in the sanctity of claims, as well as loss of the extraordinary advantages they enjoy on the public lands under the 1872 mining law.

The Western Governors as Stabilizing Force

Responding initially to concerns about social and environmental impacts of rapid energy development following the 1973 oil embargo, the western governors have become major players in public land policy and have functioned as a powerful moderating force.

Images of the West as an area to be sacrificed to energy independence through crash programs for development of coal, oil shale, oil and gas, electric generating stations, transmission lines, synthetic fuel plants, and other energy-driven developments brought new structures (such as the Western Governors' Policy Office, WESTPO) with supporting professional staffs into prominence as regional policy forums.

The governors successfully opposed efforts to override state authority on energy facility siting in the Nixon-Ford and Carter administrations; bargained with mixed success for energy impact assistance money both from energy developers and the federal government; helped stymie (temporarily) restrictive water project authorization and funding policies under Carter; helped stimulate and politically legitimize opposition to Sagebrush; secured a powerful voice in coal leasing (and potential oil shale and tar sand leasing) during the Carter administration; helped build a temporary coalition of conservation and livestock interests on rangeland management during the Carter era through a Rangelands Subcommittee; and helped force the Reagan administration to abandon wilderness leasing proposals and initiatives that proposed sharply reducing the voice of the states in coal leasing.[48]

The policy effect of activism by the governors has been one of moderating drastic shifts in policy and in patterns and intensities of land use. The reason, it may reasonably be assumed, rests in electoral politics: in the elevation of public land policy on the scale of public interest to a level inviting or forcing gubernatorial intervention, and in the governors' visibility in responding to the mix of urban, environmental, and commodity production interests that contend for priority in public land values.

Building on concern for the future of the West in the context of rapid energy development, the governors have expanded the regional agenda to include efforts by the federal government to use the public lands (especially BLM lands) as "dumps" for such unpopular land uses as MX missile basing sites, and radioactive and other toxic waste sites. In these cases, the land use planning and environmental impact statement requirements of FLPMA and NEPA became important tools of western interests in delaying action and exposing impacts.

The net effect of the emergence of governors as policy participants has been strongly positive to a BLM still weak in congressional political support. A strong case can be made for strengthening incentives and opportunities for state government involvement in public land decisions as a means of moderating the wild swings of federal policy in an era of confrontational politics.

PRIVATIZATION AS THEOLOGY

Even after its gradual disappearance from Reagan administration oratory, privatization has been kept alive—if faintly—by a small group of advocates. The basic argument reflects belief that private ownership of public land resources will produce efficiency in management and, through the invisible hand of the market, will produce the optimum mix of goods and services demanded by society, including noncommodity values (wildlands, wildlife, recreation activities).[49] Protection of ecosystem productivity is widely assumed to flow automatically from the self-interest of private owners in future economic uses. This relatively pure dogma is often accompanied by colorful and sometimes accurate criticisms of public management.

The arguments for public benefit are based on faith in a market of informed consumers with an unconstrained choice in selection of goods and services available from an unconstrained number of competitive suppliers—circumstances not discernible in the "market" that would emerge as an allocator of public land goods and services. Many rights to commodity uses are already held under permits and leases; capital requirements limit entry of new firms in several fields; the fact that recreation and wildland users are scattered all over the country makes organization of these interests as a competitive factor in resource allocation difficult. The assertion as an article of faith that the private market will automatically protect ecosystem productivity is particularly willful in the light of the long history of soil and water abuse, erosion, and environmental contamination by the private economy.[50]

Economic Efficiency: A Mainstream Argument

Marion Clawson, one of the more prolific and innovative of public land scholars, is one of a number of mainstream economists who see economic efficiency as an adequate reason for drastic changes in public land policy. Clawson has proposed a leasing system for public land surface resources that would invite competition among commodity users, as well as competition between commodity interests and environmental and recreational interests.[51]

On commercial timberlands, successful bidders would acquire private rights to utilization of timber for one hundred years, would pay up front for standing merchantable timber plus an annual rental, and would bear all costs of management. The lessee would be free of specific management requirements (federal and state environmental laws would presumably apply, as they would on the "privatized" public lands under the outright disposal proposals); performance bonds and compensation for lessee improvements would "provide strong financial incentive for maintenance of the forest stand."

Grazing permittees on the more productive rangelands would be invited to lease lands on which they now hold permits for a fee based on beef prices. The term would be fifty years, and lessees would be subject to performance bonds and penalties for resource degradation. Grazing leaseholders would not secure rights to any merchantable timber.

As much as sixty million acres of public lands offering any combination of resource values (including established wilderness areas) would be made available for lease by groups, associations (with particular reference to environmental and recreation interests), and private persons. If commercial timberlands or productive grazing lands were sought under this provision, nonprofit lessees would pay the same price as commodity interests. On other public lands, no down payment would be required; the annual rental would be one dollar per acre; and no timber could be harvested commercially. Again, all managerial responsibilities would be borne by the lessee. Public access would be controlled by the lessee under all categories of leases.

The most striking aspect of the proposal is a "pull-back" feature. As Clawson describes it, any person or group could apply for a second lease covering a portion of any tract for which a lease had been applied under the program. Clawson assumes that a second applicant may place a high value on part of a tract covered by an original lease application, perhaps an area of high scenic or recreation value. The second applicant could "pull back" part of the tract (but not enough to negate the objective of the original applicant). Pull-back lessees would be subject to the same financial terms as the original applicant.

Clawson argues with some urgency that the lease-pull-back package (which he describes as indivisible) would turn conflicting interests away from confrontational all-or-nothing public land politics toward bargaining about the use of specific sites and groups of sites. His proposal founders immediately on the assumption that environmental, conservation, and recreation groups have any realistic prospect of competing with the private sector on any tracts attractive for leasing for timber or grazing. The constituency for wildland, scenic resource, wildlife, and archaeological resource protection is widely scattered geographically. The task of amassing funds in the amounts that would be required to achieve environmental objectives now supported by tax-financed public management programs would be insurmountable as a widely applicable project.

The efficiency argument is important. There is plenty of opportunity for improving efficiency within the framework of the existing management systems. Insistence on fair market value for commodity use and better-administered fees for recreation uses are high-priority if politically difficult targets. Discipline on the expenditure side, particularly in support of subsidized commodity uses, is a perennial challenge. But Clawson's proposal would leave BLM and the Forest Service with fragmented holdings of those lands *least* likely to generate offsetting revenues to cover costs of administration, making the efficiency case against continued federal ownership and administration of unleased public estate relatively more compelling.

Robert Nelson, an economist and career policy analyst in the Department of the Interior, has proposed a major resorting of ownership and management responsibilities on the basis of land characteristics.[52] Large areas of national forests and BLM-administered public lands valued as recreation lands of less than national significance would be turned over to the states. Recreation lands of national significance but inappropriate for national parks—wilderness, wild and scenic rivers, and so on—would become the charge of a new Federal Recreation Service (perhaps from the remains of the Forest Service). For prime commercial timber lands, administration would be turned over to new "timber management corporations," with stock initially held by federal and local governments (who currently share in timber sale receipts), with the federal shares to be sold over a transitional period. Grazing lands (presumably the more productive) would be leased to ranchers for grazing for fifty to one hundred years with option to renew.

In all the recent wave of proposals, values concerning amenity, recreation, wildland, and wildlife are placed in an inferior position. The aura of conservation as the "gospel of efficiency" permeates the atmosphere, at the initiative of economists rather than the resource professionals of the Pinchot era.

All the proposals would erode the benefits of public ownership: right of access, legislative protection for a wide range of values, psychic satisfaction in ownership and a public legacy for future generations, and avenues of political accountability for management of the public estate. A summary judgment is that the proposals do not meet concepts of equity and preference as these have been and are likely to continue to be defined by American society.

The efficiency argument has had one effect, in the context of reducing government expenditures: a proposal by BLM and the Forest Service to switch jurisdiction on millions of acres of land in order to reduce administrative costs.

Babbitt: Priority for Public Uses

A speech Bruce Babbitt made when he was governor of Arizona anticipated the end of the era of commodity policy initiative and in fact predicted that a new wave of environmental demand was about to crest. Babbitt proposed what both environmental and developmental interests had been wary of: amendments to the basic statutes defining objectives of public land management.

Speaking to the Sierra Club on May 4, 1985, in San Francisco, California, Babbitt argued:

> We also need a new western land ethic for non-wilderness. The old concept of multiple use no longer fits the reality of the New West. It must be replaced by a concept of public use. From this day on, we must recognize the new reality that the highest, best and most productive use of western public land will usually be for public purposes—watershed, wildlife and recreation.

He added:

> Mining entry must be regulated, timber cutting must be honestly subordinated to watershed and wildlife values, and grazing must be subordinated to regeneration and restoration of grasslands. . . . It is time now to replace neutral concepts of multiple use with a statutory mandate that public lands are to be administered primarily for public purposes.

Babbitt's break with multiple use (and agency discretion in determining the ultimate mix of uses) may have been the first by a major statewide political figure in the West. It was, in a political sense, a counterpart to the candid assertion by Reagan and Watt that commodity interest domination is desirable. The tactics differ, however, in that Babbitt proposed to legislate the priority for public uses, while the Reagan administration chose to concentrate on policy tilts under existing law.

Babbitt predicted that a "new wave of national reform" in public land policy was at hand, and that public support for the direction he advocated would emerge:

> We will remember the early eighties, as a time when the environmental movement, driven from the Nation's capital, went home . . . to find its grassroots support, to relate to the people, to sharpen its message and to gain energy and strength for a new wave of reform.

Public land policy will remain a subject of lively debate for some time to come.

NOTES

1. These and other data on public land areas, uses, disposals, etc., are taken either directly or from other sources citing various editions of *Public Land Statistics,*

published annually by the Bureau of Land Management, U.S. Department of the Interior, Washington, D.C.

2. Much of the brief outline of history that follows is summarized with gratitude from Samuel T. Dana and Sally K. Fairfax, *Forest and Range Policy* (New York: McGraw-Hill, 1980), a readable and intelligently interpreted introduction to more than two centuries of public land policy evolution.
3. Samuel P. Hays, *Conservation and the Gospel of Efficiency*, (Cambridge, Mass.: Harvard University Press, 1959), is the primary source on origins and institutionalization of these concepts.
4. Herbert Kaufman, *The Forest Ranger* (Baltimore: Johns Hopkins University Press, 1960), is recognized as a classic in both natural resources and public administration fields.
5. See E. Louise Peffer, *The Closing of the Public Domain: Disposal and Reservation Policies* (Stanford: Stanford University Press, 1951), for a reasonably detailed account of Forest Service-livestock industry relationships.
6. Philip Foss, *Politics and Grass: The Administration of Grazing on the Public Domain* (Seattle: University of Washington Press, 1960); Wesley Calef, *Private Grazing and Public Lands*, (Chicago: University of Chicago Press, 1960); William Voigt, *Public Grazing Lands: Use and Misuse by Industry and Government* (New Brunswick, N.J.: Rutgers University Press, 1976); as well as Peffer, *Closing of the Public Domain*, and Dana and Fairfax, *Forest and Range Policy*, are among those who have produced an unusually rich body of literature on this subject.
7. Foss, Calef, Voigt, and Peffer are prime sources (see n. 6).
8. Carpenter remained proud of his accomplishments until his death in 1980, and entertained visitors with stylized accounts of cattlemen-sheepmen conflicts and extralegal means used to resolve them.
9. See n. 6.
10. Foss's account of Soldier Creek as an example of industry domination was dramatic because of the local grazing service staff resistance.
11. The image persisted in the 1980s. Cecil Andrus, secretary of the interior during the Carter administration, described BLM as the "Bureau of Livestock and Mining" early in his administration. James Watt attacked the bureau as unprofessional and environmentally biased. Andrus had the grace to recant.
12. Voigt's *Public Grazing Lands* is particularly interesting here (if colorfully critical of the industry and its congressional friends) as a first-hand account by a participant in what he (and Bernard DeVoto) called "Barrett's Wild West Show."
13. Dana and Fairfax, *Forest and Range Policy*.
14. The Wildlife Management Institute (a research, advocacy, and educational institution funded by sporting arms and ammunition manufacturers), the new (1936) National Wildlife Federation, the Izaak Walton League of America, a few other private groups, and the association of state fish and game department administrators had developed a reasonably coherent policy advocacy program before World War II, which emerged promptly at war's end. Collaboration between these organizations and sympathetic Forest Service staff members was well developed in 1951, when the author was first professionally employed in the conservation field.
15. Aldo Leopold's early leadership as Forest Service wilderness advocate is richly documented. Donald Baldwin, *The Quiet Revolution* (Boulder, Colorado: Pruett Publishing Co., 1972), makes room for acknowledgment of roughly coincident contributions by Arthur H. Carhart, then a Forest Service landscape architect and later a writer on natural resources policy. Dana and Fairfax, *Forest and Range Policy*, offer a running chronicle of wilderness evolution.
16. Observations based largely on the author's participation in efforts in Colorado and in Washington over two decades to broaden the constituency for conservation causes generally.
17. Samuel P. Hays has written extensively on environmental politics in the post-World

War II era, culminating in his *Beauty, Health, and Permanence: Environmental Politics in the United States, 1955–1985*, (New York: Cambridge University Press, 1987).

18. The Environmental Policy Center, Natural Resources Defense Council, Environmental Law Institute, and Environmental Defense Fund are examples of institutions that rose to prominence in the 1970s.
19. There was considerable controversy within the conservation community about the wisdom of introducing a bill in 1956. Some argued that doing so would galvanize opposition in a political environment not conducive to enactment.
20. David R. Brower, then of the Sierra Club, had first proposed a federally sponsored "Scenic Resources Review" squarely focused on scenic and wildlands. Joseph W. Penfold of the Izaak Walton League argued for a study which would lead to support of federal wildland preservation as one element of a broader response to the outdoor recreation boom. Penfold drafted the ORRRC legislation, and persuaded Colorado Congressman Wayne N. Aspinall, chairman of the House Interior and Insular Affairs Committee, to sponsor it. The author participated in some of the Brower/Penfold interchanges.
21. A rueful attempt at humor in Cecil Andrus's tenure acknowledged judicial oversight of major programs by referring to the judges as *secretary*, as in "Secretary" Flannery, for purposes of range management. No one laughed.
22. For an excellent interpretive summary of postwar commodity, recreation, and other uses of public lands, see Marion Clawson, *The Federal Lands Revisited* (Baltimore: Johns Hopkins University Press, 1983).
23. See Foss, Calef, Voigt, Dana and Fairfax, and Peffer (n. 6).
24. NRDC v. Morton, 388 F. Supp. at 840 (1974).
25. U.S. Department of the Interior, Bureau of Land Management, *Managing the Public Rangelands* (Washington, D.C., 1979).
26. From the author's recollection of data from the first twenty-two EISs completed under the court order. About 16 percent of permittees in these EISs received increases; about 35 percent sustained no change. The cuts from *licensed* as distinct from actual use are significant because western lending institutions historically value private ranch operations holding federal grazing permits at the licensed public land grazing level rather than the apparent carrying capacity.
27. The case was *Valdez v. Applegate*. Valdez was a permittee in the Rio Puerco area of New Mexico. A lower court found in favor of BLM's decisions following EIS completion as a reasonable exercise of administrative discretion. The court of appeals in Denver found that placing decisions in effect (even if phased over a three-year period) during the pendency of any appeal could not be justified without a showing of increased resource damage in the event of a delay.
28. The director at the time anticipated this problem and arranged in 1978 for the National Academy of Sciences to hold a series of symposia on range science issues and methodologies in the hope of revealing a scientific consensus on range inventory, suitability, allocation, and monitoring. No consensus basis for management methods and decisions emerged.
29. Secured by Senator James McClure through Appropriations Act language.
30. Dean Rhodes, a state representative, was particularly candid about his motives and expectations in remarks made to the National Public Lands Advisory Council in 1979 and elsewhere.
31. The author has participated in several sessions involving conservation, livestock, and public land agency staffs discussing rebuilding the coalition in Washington, D.C., and Tucson. A minority of environmental groups argues that history shows no prospect of effective control of livestock grazing and urges the no-grazing option.
32. Clawson, *Federal Lands Revisited.*
33. David Brower used a regional forester headquartered in Portland named Stone as a symbol of road-building, antiwilderness mania in the Forest Service.

34. Dana and Fairfax, *Forest and Range Policy*, provide a condensed account of the Monongahela and Bitterroot controversies.
35. Izaak Walton League v. Butz.
36. The agencies' response was to emphasize the importance of tying old-growth harvests to land use planning for specific public land areas, complete with public participation and EISs. A second tactic was to be outspoken about the budget increases necessary to assure that the promise of future forest productivity was in fact realized.
37. The General Accounting Office (GAO), a watchdog arm of the Congress, was used by critics to legitimize opposition to "below cost" sales. GAO issued a series of reports. See *Need to Concentrate Intensive Timber Management on High Productive Lands* (GAO, 1978), for an analysis emphasizing the need for increased timber production in the face of building material inflation and for return on federal investment. Western governors, national and local environmental organizations, forest community residents, and many others have joined in opposition.
38. Originally NRDC v. Morton.
39. See "Making Sense of the Sagebrush Rebellion: A Long Term Strategy for the Public Lands" (paper prepared for the Third Annual Conference of the Association for Public Policy Analysis and Management, Washington, D.C., October 1981). Nelson has also written an analysis of Sagebrush cum privatization in "Seeking Alternatives to Federal Land Ownership: Assessing the Sagebrush Rebellion and the Privatization Movement" (paper prepared for the National Conference of the American Society for Public Administration, Denver, April 1984).
40. The House Appropriations Subcommittee under Congressman Sidney Yates of Illinois was a major factor here, as the subcommittee was in defense of wilderness and in opposition to sharp acceleration of coal leasing and other Reagan initiatives. The Forest Service initiative was justified in part as necessary because of cutbacks in federal appropriations.
41. Paul Culhane's *Public Land Politics* (Baltimore: Johns Hopkins University Press, 1981) was overdue. It confirms BLM's professional competence and reports accurately on the skill of field managers in both the Forest Service and the BLM in playing interest groups against each other as a way of avoiding "capture" and achieving agency objectives.
42. The literature on these programs is immense. Dana and Fairfax in *Forest and Range Policy* are coherent on evolution. Hanna J. Cortner and Dennis L. Schweitzer (paper prepared for the annual meeting of the Western Political Science Association, San Diego, 1982) offer a sober assessment of constraints faced by the Forest Service, "Local Production and National Budgets: Gluing Together Public Planning for Forestry." Richard W. Behan has proposed repeal of both laws to save the money and revert to a system based on field-level interactions of public preference, information generated from modest resource inventory investments, and professional judgment in the framework of delegated authority: "RPA: NFMA: Time to Punt," *Journal of Forestry*, December 1981.
43. Behan, "The Problems of Planning and the Public Alternative" (paper prepared for symposium, Natural Resource Economics and Policy: Exploration with Journalists, sponsored by the Center for Political Economy and Natural Resources, Montana State University, July 15, 1982), reports an unnamed Forest Service official as estimating annual system costs at $500 million. The figure exceeds the total annual operating budget of BLM.
44. Public Land Law Review Commission, *One Third of the Nation's Land* (Washington, D.C., 1970).
45. U.S. Department of the Interior, Bureau of Land Management, *4–Year Authorization, 1982–1985* (Washington, D.C., 1980).
46. Helen Ingram and Sally Fairfax, political scientists at the University of Arizona and the University of California at Berkeley, have pursued the argument for more intuitive, less costly, and more decentralized management systems with considera-

ble stamina. Examples: Ingram and Dorotha Bradley, "Science vs. the Grassroots: Representation in the Bureau of Land Management" (paper prepared for annual meeting, Western Political Science Association, San Diego, 1982); Ingram and Dan McCool, "Relevance of Management Information Systems to Policy Choices: Lessons for the Bureau of Land Management," in *Developing Strategies for Rangeland Management*, National Research Council and National Academy of Sciences (Boulder, Colo.: Westview Press, 1984); Fairfax in "Beyond the Sagebrush Rebellion: The BLM as Neighbor and Manager in the Western States" (paper prepared for annual meeting, Western Political Science Association, San Diego, 1982).

47. A survey response by 3,034 of 4,872 registered voters in Arizona questioned about wilderness attitudes is revealing. In the survey, conducted by Arizona BLM, 81 percent of respondents said they believed that wilderness is an important use of federal land in the state. See *Arizona Wilderness Public Opinion Survey* (Phoenix: Bureau of Land Management, 1983).
48. Most of these collegial actions were taken through the Western Governors Policy Office (WESTPO), now reorganized as the Western Governors' Association, with offices in Denver.
49. John Baden, alone and with various coauthors, has been the most provocative and often the most creative of the "free market" group. See Baden and Dean Lueck, "A Property Rights Approach to Wilderness Management," in *Public Lands and the U.S. Economy: Balancing Conservation and Development* (Boulder, Colo.: Westview Press, 1984); Baden and Richard Stroup, "Political Economy Perspectives on the Sagebrush Rebellion," *The Public Land Law Review*, Spring 1982; and Baden and Richard Stroup, *Natural Resources: Bureaucratic Myths and Environmental Management* (San Francisco: Pacific Institute for Public Policy Research, 1983). The theology varies little among authors from this segment of the economists' spectrum, but Baden's exploration of mechanisms for encouraging noncommodity interests to assume management responsibilities for (if not ownership of) valued public land environments is worth pursuing. The role of the Appalachian Mountain Club in administration of the Appalachian Trail is the precedent.
50. A number of established scholars in natural resource economics have rendered an important service to environmental interests by attacking the privatization ideologists on economic grounds. See Carlisle Ford Range, "The Fallacy of Privatization," *Journal of Contemporary Studies*, Winter 1984; and Daniel W. Bromley, "Public and Private Interests in the Federal Lands: Toward Conciliation," in *Public Lands and the U.S. Economy*.
51. Clawson, *Federal Lands Revisited*.
52. See Nelson, "Seeking Alternatives to Federal Land Ownership: Assessing the Sagebrush Rebellion and the Privatization Movement" (paper prepared for the National Conference of the American Society for Public Administration, Denver, April 1984).

5

THE FEDERAL GOVERNMENT'S ROLE IN THE MANAGEMENT OF PRIVATE RURAL LAND

Malcolm Forbes Baldwin

Environmental problems of rural land use in America have been observed for centuries, from the degraded tobacco farms of Virginia in the mid-eighteenth century to the soil erosion and overtimbering that plagued the Northeast a century later. Henry Thoreau, George Perkins Marsh, Gifford Pinchot, and other forerunners of the twentieth-century environmental movement voiced concern about land degradation, but until the 1930s federal conservation programs were limited largely to the public lands of the West.

The New Deal enlarged the federal role in private land use in the Midwest and the East through new agricultural programs and natural resource programs of the Tennessee Valley Authority (TVA), but pervasive federal involvement in private rural land management came after World War II. Postwar federal efforts to stimulate industry, commerce, agriculture, and housing significantly changed the use and management of hundreds of millions of acres of private rural land.

The effects of federal programs on the 1.4 billion acres of private rural land in the United States (which constitute two-thirds of the United States) stimulated new legislation in the 1960s and 1970s to avoid or contain land abuses.[1] Congress and the executive branch also established significant federal protection of coastal resources, soils, wetlands, and lands affected by coal surface mining, but these conservation programs have been disjointed, ad hoc responses to perceived environmental problems largely because Congress failed to pass the proposed National Land Use Policy Act in the mid–1970s.

Federal impacts on private rural land span virtually the entire range of federal activities. In addition to the profound indirect effects of federal tax laws on land use, by the end of the 1970s approximately twenty-five federal agencies were directly affecting public and private land use under more than seventy statutes.[2] They have primarily stimulated development, not conservation. The Department of Transportation (DOT), the Department of Housing and Urban Development (HUD), the Corps of Engineers, and the Environmental Protection Agency (EPA) earned reputations as "growth shapers" of urban fringe and rural areas by the late 1970s, as the federal government spent an estimated $12 billion annually for economic development in nonmetropolitan areas.[3] By the mid–1980s, domestic budget cuts and the completion of the forty-three-thousand-mile interstate highway system had reduced or at least contained the impacts of federal programs on private land, but the importance to the use of land of federal public works, economic development, defense, housing, air and water pollution, and many other programs promises to continue.

Juxtaposed to federal private land development stimulants are federal programs of environmental assessment, land planning, conservation, and regulation, many of which are administered under federal statutes through state and local governments. Programs with significant but mostly unquantified conservation effects on private rural land include farmland impact provisions under the Farmland Protection Policy Act of 1981,[4] programs under the Coastal Zone Management Act,[5] federal environmental impact statement requirements under the National Environmental Policy Act,[6] the wetland and navigable water permit programs of the Corps of Engineers,[7] and regulations established under the coal Surface Mine Conservation and Reclamation Act.[8]

This paper examines the shift of the federal role in private rural land management in the 1960s and 1970s; the roles played by environmentalists, scientists, the Congress, and federal agencies; and the extent to which this federal policy shift has withstood the economic and political changes of the late 1970s and early 1980s.

FEDERAL MANAGEMENT OF PRIVATE LAND BEFORE WORLD WAR II

Throughout most of U.S. history, federal land policy consisted largely of the rapid disposal of the public domain for private use. Significant changes occurred by the end of the nineteenth century on two fronts: a federal public land management role developed with the establishment of national parks and forests, and a federal program evolved to enhance the agricultural productivity of private lands in the arid West. The Reclamation Act of 1902 authorized the secretary of the interior to develop "irrigation works" in the West.[9] The agricultural production stimulus generated by this act remained the major federal development impact on western private land in the prewar years.

Elsewhere, however, the federal government had little direct effect on private land use until the early 1920s, when the Commerce Department under Secretary Herbert Hoover prepared and disseminated the Standard State Zoning Enabling Act to stimulate state and local land use regulation and better housing conditions throughout the country. The constitutionality of zoning laws under the state's police power was established shortly afterwards in *Village of Euclid v. Ambler Realty Co.*[10]

However, the seeds of greater federal involvement in private land regulation had been established in several early statutes concerning waters and roads. Since 1890 the Corps of Engineers had regulated against the impairment of navigation through its permit authority, which was clarified and broadened in the 1899 Rivers and Harbors Act.[11] The Corps of Engineers continued to improve navigation through the prewar years as the definition of navigable waters was expanded by the courts to include waters that were, had been, and might reasonably become navigable.

Direct federal licensing of hydroelectric developments on navigable waters was established under the Federal Power Act of 1920.[12] Hopes for federal water resource development were frustrated until the Tennessee Valley Authority established a regional model for natural resource development in 1933. Constitutional bases for water resource management expanded so that by 1940 the Supreme Court declared that "navigable waters are subject to national planning and control in the broad regulation of commerce."[13]

Another precedent-setting statute was the Federal Aid Road Act of 1916, which gave the federal government the lead in modern highway development.[14] By 1933, the federal government had spent $1 billion to help states and localities build and maintain one hundred thousand miles of rural and urban fringe roads and bridges. Annual federal aid to the states by then totaled $182 million, of which 68 percent was for highways. Only in 1944 did urban roads become eligible for this assistance.[15]

The Great Depression focused national attention on the economic distress caused by private land abuse, which Franklin Roosevelt described in his 1932 acceptance speech to the Democratic Convention:

> There are tens of millions of acres east of the Mississippi River alone in abandoned farms, in cutover land, now growing up in worthless brush. Why, every European nation has a definite land policy and has had one for generations. We have none. Having none, we face a future of soil erosion and timber famine. It is clear that economic foresight and immediate employment march hand in hand in the call for the reforestation of these vast areas.[16]

A new land use initiative was offered only weeks after Roosevelt's inauguration, when the secretary of agriculture submitted to Congress the National Plan for American Forestry, in response to Senate Resolution 175. The report strongly endorsed the national land classification survey and land use planning notions that had been recommended by President Hoover's secretary of agriculture in November 1931.[17] "Land Classification," the report noted, "taking into account physical, economic and social factors and having as its objective comprehensive land-use plans, is the most constructive proposal yet made for bringing about the highest use of land in the United States."[18]

Although Congress did not accept the land classification recommendation, the potential of national land and water planning remained attractive to President Roosevelt.[19] Land and water planning became part of the mandate for his National Resources Planning Board (NRPB), which existed from 1934 to 1944 in what Stewart Udall later described as "a brief hour of triumph for scientific planning."[20] In addition, between 1938 and 1941 the government, in cooperation with several states, sponsored the Land Use Planning Program and selected over a thousand counties in forty-seven states to inventory and classify their land resources. After intensive organizational and planning work, a unified local and county planning program was undertaken in many of these areas, but that effort encountered agency and political opposition. In 1941, federal participation ended in what had become a small but geographically broad land use planning movement.

It was the soil conservation program of the New Deal, however, that left the most enduring mark on federal management of private rural land. As far back as 1894, a Department of Agriculture (USDA) bulletin had urged farmers to contain soil losses.[21] In 1928, Hugh Hammond Bennett, who had begun his career as a soil surveyor with the Bureau of Soils twenty-five years before, wrote the influential USDA bulletin, *Soil Erosion, A National Menace*, which explained why soil problems should concern the nation and not simply farmers individually. When Congress provided money in 1930 to the USDA Bureau of Chemistry and Soils to study soil erosion causes and controls, Bennett was put in charge.

The work of Bennett and others provided the scientific information needed to justify soil conservation action. During Bennett's April 2, 1935, congressional testimony, Washington's sky grew dark with dust from the Great Plains. Only weeks later, Congress passed the Soil Conservation Act, establishing soil erosion as a national concern and creating a permanent soil erosion program under the Soil Conservation Service (SCS). The SCS was empowered to conduct surveys and research; carry out preventive controls; provide financial aid; and acquire lands, rights, or interests to achieve soil conservation goals. With that agenda established, the need shifted from research and demonstration programs (already under way on 5.5 million acres in forty-one states) to programs of direct assistance to farmers.[22]

To gain farmer cooperation, the USDA formed soil conservation districts on a state and state-subdivision basis at the recommendation of Congress. Districts aided by SCS experts developed farm conservation plans that included provisions for terracing, strip cropping, drainage, crop rotation, contouring, fertilization, pasture improvement, controlled grazing, and tree planting. The department prepared a model state law to authorize such districts, which were to be established on favorable votes of a majority of "land occupiers" and governed by five supervisors, three of whom were to be elected. President Roosevelt sent copies of the model law, recommending adoption, to governors of all the states in February 1937. The SCS's role expanded in 1939, when Congress passed the Omnibus Flood Control Act, authorizing SCS to plan small flood control projects upstream from the navigable waters under the jurisdiction of the Corps of Engineers.[23] By 1947 all the states had passed enabling legislation for soil conservation districts.

A second set of agricultural programs developed in the 1930s provided price and income support to farmers. Until just before World War II, the Agricultural Conservation Program (ACP) paid farmers to shift from soil-depleting crops to soil-conserving crops.[24] The war stimulated programs to encourage increased farm productivity through the use of fertilizer and other practices. The ACP has continued to be administered through county and state committees that determine what county conservation programs and practices are eligible for federal payments.

Thus, by the end of World War II, federal involvement in private rural land management outside of the Tennessee Valley primarily supported agricultural development and productivity. These agricultural programs continued after the war, but their support lay not with the emerging environmental movement but with the classic "iron triangle" of influential rural congressional leaders, the Department of Agriculture, and farmer organizations. Not until the mid–1970s would federal legislative and policy goals of the environmental and agricultural communities begin to converge.

FEDERAL RURAL LAND MANAGEMENT SINCE WORLD WAR II

Significant events in the evolution of the federal private rural land conservation programs can be grouped into three distinct postwar periods: (1) from 1945 to 1961, when population and urban areas grew, economic development programs flourished, and government paid limited attention to conservation; (2) from 1961 to 1975, when creative ideas for aesthetic, ecological, and land use programs were translated into innovative federal programs; and (3) from 1975 to the present, when the population in most rural areas grew faster than in most cities, when energy and economic concerns evoked more traditional resource conservation approaches, and when recession and political reaction set in.

The major federal programs emerging over these postwar years concerned the management of wetlands, coastal regions, and coal surface mines. The major failure was Congress' inability to pass the National Land Use Policy Act—a failure that reveals much about the character, motivation, power, and tactics of the postwar environmental movement.

In the mid–1960s, two books were recommended to me as required reading for young environmental enthusiasts. One was Fairfield Osborn's *Our Plundered Planet,* which had been influential in the 1950s.[25] Osborn described the environmental condition of the globe and decried the scant attention being given to soil erosion, forest protection, wildlife, and reclamation in the United States. He emphasized the need for an ethical perception that harm to resources harms the life or property of another and noted that there was "nothing revolutionary in the concept that renewable resources are the property of all the people and, therefore, that land use must be coordinated into an over-all plan." Here the conservation legacy of the New Deal asserted itself; the TVA, he suggested, was the first logical step in that direction.

The other book was Aldo Leopold's *Sand County Almanac*, which remains the classic articulation of the land ethic.[26] In the chapter on ethics, which Leopold revised many times between 1933 and 1949, he noted that we were confronted with forest and soil problems "coextensive with the map of the United States." He suggested "a universal symbiosis with land, economic and aesthetic, public and private." With social pressure, ecological education, and research findings applied to land management, "a sufficiently enlightened society, by changing its wants and tolerances, can change the economic factors bearing on land."[27]

Despite the land use concerns expressed by Osborn, Leopold, and others, federal soil conservation programs during this period ceased to be linked to political, social, or natural resource policy reform as they had been in the 1930s. Farm interests and environmental interests were motivated by different concerns. The Soil Conservation Service expanded further into small watershed protection and construction programs that pleased its constituency but that often harmed wildlife habitat and earned the enmity of environmental groups. The environmental movement neglected basic agricultural and commodity support programs, until after the mid–1970s, when soil erosion and prime farmland conservation regained attention.

Wetlands received growing conservation interest in the early postwar period. In 1956 the Fish and Wildlife Service published an inventory of the nation's wetlands that focused primarily on waterfowl production.[28] It estimated that some forty-five million acres of marshes, swamps, and bottomlands in the United States had been converted to other uses since the nation was originally settled. In 1958, with the strong support of national conservation organizations and all forty-eight state governors, Congress passed the Fish and Wildlife Coordination Act Amendments.[29] The new act required that water resource development projects and federal water resource permits integrate into their plans fish and wildlife concerns. The act was intended to benefit the estimated twenty-five million people who hunted and fished, and who contributed to the best organized conservation force of the time, as well as the sixty-six million who found "recreation and release from tension" in fish and wildlife activities.[30]

Two other congressional acts in 1958 set the stage for the land use changes and priorities destined to emerge in the 1960s: one was the Interstate Highway Act, by which the federal government was to finance 90 percent of the costs of a new forty-three-thousand-mile network of national defense superhighways. The other was the establishment of the Outdoor Recreation Resources Review Commission, which was to report on the need for new national programs for recreation.

New Programs in the 1960s and Early 1970s

A new conservation agenda for the country that reflected the views of many environmental leaders was clearly stated in 1963 by Stewart Udall, secretary of the interior, in his book *The Quiet Crisis*.[31] "Our irreverent attitudes toward the land and our contempt for the Indian's stewardship concepts are nowhere more clearly revealed than in our penchant to pollute and litter and contaminate and blight our once-attractive landscapes," Udall declared. He urged enactment of the recommendations of the Outdoor Recreation Resources Review Commission for a Land and Water Conservation Fund, controls of strip mining, ways to balance social costs and benefits, ways to control chemical and radioactive contamination, and ways to respond to the collision of incompatible land use proposals. In time this effort was to achieve significant success.

Aesthetic and recreational issues attracted national attention following the passage of the Land and Water Conservation Act in 1964. The White House Conference on Natural Beauty publicized Lady Bird Johnson's energetic efforts to establish natural aesthetics within the traditional conservation agenda. In 1965, the Court of Appeals for the Second Circuit held that the Federal Power Commission must consider natural beauty and historic interest in its proposed licensing of a pumped storage facility at Storm King Mountain on the Hudson River.[32] Publication of *With Heritage So Rich* by the U.S. Conference of Mayors proved sufficiently influential to lead to passage of the Historic Preservation Act of 1966.[33] This statute gave historic and archeological sites added protection against adverse federal actions.

A particular concern of Lady Bird Johnson and a coalition of state roadside councils, the Garden Club of America, and other national and local citizen groups was the elimination of billboards from federally supported (especially interstate) highways. In May 1965, President Johnson urged Congress to establish a road system to allow Americans to "touch nature and see beauty" and "to improve and broaden the quality of American life." The Highway Beautification Act that Congress passed established federal aid for landscaping highways and a loss of 10 percent in highway funding to any state that the secretary of commerce determined had not effectively controlled billboards and screened auto junkyards along primary and interstate highways.[34] Congress, after bitter debate, also required that "just compensation" should be paid for such removal. The act, at passage, was hailed as an overall success, but since it required compensation (regardless of what state or federal constitutional law required) the act was afterwards criticized as an environmental failure, delaying more than expediting billboard removal.[35]

Highway beautification soon became less critical to national and local conservation groups than highway construction. The next year (1966), conservationists lobbied and obtained passage of the Department of Transportation Act, which required new plans and procedures to protect natural beauty and public parks. Section 4(f) of that act became an important influence on highway programs and a precursor to the environmental impact statement by requiring that the secretary "shall not approve any program or project which requires the use of any land from a public park, recreation area, wildlife and water fowl refuge or historic site unless there is no feasible and prudent alternative."[36]

Wetlands Protection

Meanwhile the conservation of wetlands and estuaries gained increasingly strong political support at the national level. The Fish and Wildlife Coordination Act Amendments in 1958 had given the Fish and Wildlife Service legal authority to review and seek mitigation measures for Corps of Engineers navigational permit actions.

A strong hunting and fishing constituency, organized through the National Wildlife Federation, the Izaak Walton League, and the International Game and Fish Commission (which represented state agencies), supported the service. The newer, more ecologically oriented organizations added support as well, and this coalition became more potent in Washington as it employed legal and scientific arguments in behalf of wetland and estuarine protection.

Passage of the first state coastal wetland protection act in Massachusetts in 1963 and another to protect inland wetlands two years later evidenced vital local and state support for their cause.[37] The Massachusetts wetland regulations were cited regularly in the growing number of environmental, scientific, and legal symposia; educational materials; and other public information efforts of the late 1960s and 1970s. Wetlands and estuaries became transformed in the public mind from unwanted swamps and sumps to valuable resources.

The Fish and Wildlife Service and its growing constituency had powerful support from Congressman John D. Dingell of Michigan, chairman of the Subcommittee on Fisheries and Wildlife of the House Merchant Marine and Fisheries Committee. Dingell was a persuasive and dedicated supporter of wetland and estuarine protection and an increasingly energetic overseer of federal agency implementation of the Coordination Act Amendments of 1958.

Throughout the 1960s, the service recommended that the Corps of Engineers deny or modify the permits under Section 10 of the 1899 Rivers and Harbors Act that private parties requested for dredging or filling estuaries and wetlands. Because the Corps would do so only for navigational (not conservation) reasons, Congressman Dingell introduced a bill in 1967 to authorize the secretary of the interior to deny a permit for dredging, filling, or excavating within an estuary if it impaired natural resources or water quality. The bill that passed in 1968 as the Estuary Protection Act simply directed the Interior Department to prepare a national estuarine study, but the committee hearing focused public attention on the need to expand the permit criteria used by the Corps of Engineers.[38]

Changes in the Corps of Engineers policy occurred rapidly thereafter. In December 1968, the Corps unilaterally expanded its permit evaluation under section 10 to include "the effect of the proposed work on navigation, fish and wildlife, conservation, pollution, aesthetics, ecology, and the general public interest."[39] The next year, the Corps of Engineers denied a trailer park developer a permit to dredge and fill eleven acres of his tidal property in Boca Ciega Bay, Florida, on ecological grounds, even though the activity would not interfere with navigation. The state, hundreds of citizens, and the Fish and Wildlife Service supported the denial. The developer sued the Corps, won in the district court, and then in 1970 (after passage of the National Environmental Policy Act) lost before the court of appeals. Said the court: "In this time of awakening to the reality that we cannot continue to despoil our environment and yet exist, the nation knows, if Courts do not" that the destruction of fish and wildlife in estuarine waters had substantial interstate commerce effects. The Corps of Engineers now had the clear legal authority to deny a section 10 permit based on nonnavigational considerations.[40]

Corps regulation of wetlands lying outside section 10 jurisdiction (beyond navigable waters and their tributaries) emerged by a further circuitous route after passage of section 404 of the Clean Water Act in 1972.[41] The act gave the Environmental Protection Agency basic water quality permit authority but authorized the Corps of Engineers to issue permits "for the discharge of dredged or fill material into the navigable waters at specified disposal sites." EPA received authority to set guidelines for such permits and overrule them if the discharge adversely affected "water supplies, shellfish beds and fishery areas . . . wildlife or recreational areas."

Wetland regulation was not mentioned by the act or clearly contemplated by Congress, but not long after passage, environmental groups (notably lawyers of the Natural Resources Defense Council [NRDC] and the Environmental Defense Fund [EDF]) saw the possibility of applying its section 404 provisions to the hitherto unprotected inland wetlands. EPA agreed that under the new act all discharge permits covered the tributaries of navigable waters and intrastate streams used by interstate travelers for recreation or traveled by fish sold in interstate commerce—that is, wetlands. EPA Administrator Russell Train and the House Government Operations Committee separately urged the corps in 1974 to exercise permit jurisdiction beyond its navigable waters limits to cover most freshwater wetlands and streams.

In May 1973, the corps issued proposed regulations that broadly defined section 404 to cover "waters of the United States," but in its final regulations restricted coverage to navigable waters again.[42] The EDF and the NRDC followed with separate legal strategies, and the NRDC, along with the National Wildlife Federation, obtained a court order directing the corps to establish permit jurisdiction beyond the navigability limits to the extent of federal authority over water resources under the interstate commerce clause of the Constitution.[43]

Now launched on a regulatory program it had never envisioned when the Clean Water Act passed, the corps dug in its heels, despite lack of support for its position from the Interior Department, EPA, the Council on Environmental Quality, or the Justice Department. Under orders from its leadership, the corps carefully drafted and broadly circulated a press release to stimulate alarm among farmers, ranchers, foresters, and developers, stating that federal permits might be required for the most ordinary activities and that millions of people could be violating the law under risk of serious penalties.[44]

The predictable result was a storm of protest for what the *New York Times*, falling into a trap, called a corps "power grab." Environmentalists organized a major campaign against the misleading release; the administration and political leaders of the Department of the Army condemned the corps. Chastened, the corps developed interim permit guidance, clarifying its position that normal agricultural and forestry activities would be excluded from permits. The controversy that had become the most intense conflict between environmentalists and the corps was resolved with good relations, and the corps' authority was restored. As one commentator notes, "throughout the controversy . . . environmental groups rarely tried to reduce the power of the corps, or to slow it down, as they had done repeatedly in the area of civil works."[45]

The aftermath of this episode was the corps' decision in 1976 to reject a permit request for a major housing development on thousands of acres of Florida wetlands at Marco Island, and in 1977, after two years of heated debate, Congress codified the section 404 program in the 1977 Clean Water Act.

Coastal Zone Management

The evolution of federal wetland protection paralleled the growing national interest in coastal zone management. Rising recreational demands on coastal areas had been highlighted by the 1961 report of the Outdoor Recreation Resources Review Commission (ORRRC). The 1969 report of the Presidential Commission on Marine Science, Engineering, and Resources called for long-range, coordinated national management of coastal resources to cope with coastal growth and conflicts in local and state coastal programs.[46] Specifically, it recommended establishing a Federal Coastal Zone Program and a National Oceanographic and Atmospheric Administration.

By 1970, coastal zone management bills had been introduced in the Congress by members of both parties. They agreed on the essential requirements for federal grants to the thirty coastal and Great Lakes states, comprehensive coastal management, state designation of responsible state agencies to develop the plan, balanced land and water use management, federal approval of the plan, and federal actions consistent with an approved plan. As had been the case with wetlands, the ingredients were promising for political action on coastal management.

At the outset, information from authoritative government and other sources on demographic, recreational, and industrial development and scientific trends all pointed toward increasing importance of coastal resources and continued conflict between growth and resource protection in coastal zones. Coastal and ocean issues, inextricably linked, brought government and academic scientists (biologists, marine scientists, geologists, and others) into coastal zone policy development through congressional testimony and contacts with members of Congress.[47]

There was nothing academic about the growing number of local controversies that gained national attention in the late 1960s and early 1970s. One of the most important was the conflict between the development of a resort community at Hilton Head on the South Carolina coast and a proposal for a nearby refinery by BASF, a West German company. Strong local feelings for and against the refinery, Senator Strom Thurmond's request for federal assessment of its pollution effects, and Secretary of the Interior Walter Hickel's concern about pollution captured media and congressional attention.[48]

Industrial development interests were attracted to programs likely to enhance investment certainty and to avoid unproductive siting conflicts. Sensing the growing local opposition to refineries and power plants along the coasts from Maine to Florida, developers helped broaden the political support for coastal management within the administration and in Congress.

Political leadership for the national coastal program came from Senator Ernest Hollings of South Carolina, chairman of the Senate Commerce Committee's Subcommittee on Oceans and Atmosphere. Throughout the 1970s, he remained the most effective congressional advocate of coastal zone management, in part because of his ability to appeal to the concerns of diverse political and economic interests.

At first this support made the program politically attractive to the administration, but in 1972, when the program was placed in the National Oceanic and Atmospheric Administration (NOAA), the Interior Department urged a veto of the Coastal Zone Management Act.[49] The Council on Environmental Quality and the Office of Management and Budget (OMB) favored the bill, and it was signed, according to John Whitaker of the White House domestic policy staff, because of the obvious development problems along the coasts, the strong state role, the water orientation of the program, the opportunity for the Interior Department to veto any inconsistent coastal plans if the land use bill were to pass, and the fact that "Nixon concluded that the overriding need to control land-use planning in ecologically fragile coastal zones should not be delayed even though he preferred a national land-use approach."[50]

The administration nevertheless impounded the program's funds during the promised budget cutting after the 1972 election, and the fate of the program then depended on its continued strong congressional support and on internal bureaucratic ingenuity. The latter was forthcoming under NOAA's new administrator, Robert White, who appointed Robert Knecht, a deputy research director from a federal laboratory in Colorado with political experience as mayor of Boulder, to head the program. White allocated $250,000 from his administrative budget to Knecht to begin a coastal information program, which functioned for over a year to build support for the program and to establish links with the states. By the time program funds were appropriated in fiscal 1974, Knecht had broad support for the program among environmental and state government leaders and a well-developed implementation plan.

Throughout the 1970s, while an increasing number of states with initial planning grants received federal program approval and management grants, the federal role was delicately but surely carried out. The Office of Coastal Zone Management (OCZM) had to develop regulations and approve state programs that accommodated two diverse interests. The petroleum and electric power interests considered the act to be a neutral planning, management, and energy facility siting mechanism, whereas the environmentalists believed the act was intended to favor ecological protection and management. OCZM also had to work closely with powerful federal agencies likely to be affected by the consistency provision of an approved state plan. Conflicts with the Interior Department were particularly difficult and inevitable, especially regarding offshore oil and gas leasing. OCZM also had to keep close ties with congressional supporters, while playing straight with lukewarm administrations under Nixon and Ford.

National Land Use Policy Act

Like Sherlock Holmes's "curious incident" of the dog that did nothing in the night,[51] the conclusive failure of Congress to pass the National Land Use Policy Act in 1974 is curious and significant. Unlike the wetland and coastal acts that passed and developed further in the executive branch, federal land use policy proposals had insufficient support among environmentalists.

Senator Henry M. Jackson introduced his land use bill in January 1970, as a logical next step after the 1969 passage of the National Environmental Policy Act (NEPA) that he had sponsored. Enactment of the NEPA had been a response to the need recognized in the early 1960s to keep the federal government from causing undesirable environmental effects by establishing an environmental policy. The NEPA also provided a procedure for agencies to "look before they leap" and to inform the public, and it authorized the Council on Environmental Quality (CEQ) to oversee environmental programs and to report on trends and problems. The NEPA combined statements of environmental awareness with government accountability standards that transcended the environmental movement.

Jackson's desire for a land use act also had these two distinct elements: he believed that if real progress were to be made in achieving national environmental quality regulation, "control of land in the larger public interest is essential . . . because the land is the key to insuring that all future development is in harmony with sound ecological principles and environmental guidelines."[52] He sought balanced and sound decision making, not necessarily ecological goals. The National Land Use Policy Act he proposed included a federal grant-in-aid program to help state and local governments improve their land use inventories, planning, and management capabilities; encouragement for states to prepare a land use plan for future recreational, industrial, and conservation needs; and provisions to coordinate federal land use planning and federal-state relations and to improve federal data on land problems and trends. The stick, to ensure that states prepared plans, was presidential authorization to reduce federal funding (for highway, water resource, and other public works programs) by up to 20 percent if states failed to comply with the bill.[53]

Between 1970 and mid-1974, the prospects for some national land use legislation appeared auspicious. During the late 1960s, land use disputes involving energy development (such as the Storm King pumped storage proposal on the Hudson, the proposed Bridge Canyon dam on the Colorado, and the development of large steam generation plants in the Four Corners region) had become commonplace. Scores of highway and airport construction controversies erupted around the country. Of special concern to Jackson's committee was the proposed Miami Jetport adjacent to Everglades National Park.[54] These kinds of development conflicts appeared likely to continue.

Senator Jackson relied for demographic support on projections that demands for recreation areas would double by 1975, that new housing units would increase by twenty-six million by 1978, that urban areas would grow by five million acres, and that electrical energy demand would double by 1980.[55] Many large industries, particularly in energy, had a strong interest in fostering planning and decision making above the local level to facilitate their projected needs for construction of industrial plants, refineries, electricity-generating facilities, and transmission lines.

Substantial legal thought was being given to the gaps, conflicts, and adverse environmental and social consequences of local zoning and planning. Land use legislation in several states such as Maine, Vermont, Hawaii, and others, offered support and some models for a national land use law. Information on the effects of these laws added significant support for a national bill.[56] In addition, the American Law Institute's Model Land Development Code, in the works since the late 1960s, proposed greater state roles in planning decisions.

Impetus for national land use legislation had dedicated, able, and influential political leadership—in Congress, under Senator Jackson and Congressman Morris Udall, and in the executive branch, under former land use attorney John Ehrlichman, President Nixon's domestic policy adviser, and Russell Train, chairman of the Council on Environmental Quality.

President Nixon himself, until near the end of his tenure, maintained his support. In submitting to Congress CEQ's first annual report on the environment, he declared that "throughout the nation there is a critical need for more effective land use planning, and for better controls over use of the land and the living systems that depend on it. . . . The time has come when we must accept the idea that none of us has a right to abuse the land, and that on the contrary society as a whole has a legitimate interest in proper land use. There is a national interest in effective land use planning all across the nation."[57]

Although President Nixon endorsed the concept of a national land use policy, it was not until early in 1971 that the administration presented to Congress its own land use bill as part of the president's environmental program.[58] The administration bill, drafted by CEQ and approved over OMB's objections because of Ehrlichman's support, included federal grants for state development of land use plans and regulations for areas of critical concern, such as important ecological resources and lands around key transportation and other facilities.[59] It did not include funding sanctions against states that failed to participate, and it put the program in the Interior Department, not the Department of Housing and Urban Development.

The compromise administration-Senate Interior Committee bill that emerged in the fall of 1972 included the support for state inventories and projections of land uses and trends, aid for state planning and management of critical areas, and a provision for a National Advisory Board on Land Use Policy to help the Interior Department manage the program and to coordinate federal policy. The bill thus combined the concepts of land use planning and federal coordination that Jackson had sought with the concept of critical areas that developed in the 1960s.

The ultimate failure of the bill, which passed the Senate easily in

1972 and 1973 but failed finally in the House in 1974, has been attributed to two major causes. First, the Chamber of Commerce waged a vigorous, negative, and alarmist lobbying campaign that raised fears in the House concerning federal zoning and taking without compensation. These opponents cited publication of the Rockefeller Task Force's *The Use of Land* and the CEQ's *The Taking Issue* as evidence of an organized assault on traditional notions of private property.[60] Second, the departure of John Ehrlichman, who had consistently maintained administration support for the bill, and President Nixon's likely desire to shore up his conservative support in the midst of Watergate reduced White House enthusiasm for the bill.[61]

Many other acts appeared to meet several of the bill's original purposes. Passage of the Coastal Zone Management Act in 1972 and the Surface Mine Control and Reclamation Act in 1977, the evolving wetlands regulation program, and the growing importance of NEPA as a coordinating, public information tool concerned with land use conflicts removed much of the impetus for national land use planning legislation after 1974.

The National Land Use Policy Act was simply not an "environmental quality" bill that fulfilled the strongest goals of the environmental movement. Movement leaders were not motivated to reform planning and legal theory; they wanted to devote their resources to the concrete subjects that fired the imagination of their members and constituents, such as specific wetlands and coastal areas. Whether in fact the movement had more grassroots support for land use planning than its Washington leaders believed remains unclear.

Political Actions from 1975 to the Present

With the failure of Congress to pass the National Land Use Policy Act, the private land agenda of the environmental movement turned to more identifiable reforms, the most significant being federal regulation of coal surface mining—the major land conservation issue that remained unresolved since Stewart Udall had laid out his agenda more than ten years before.

The Surface Mine Control and Reclamation Act of 1977 was a long time coming. Public education about the problem really began with publication in 1963 of Harry Caudill's *Night Comes to the Cumberlands: A Biography of a Depressed Area.*[62] It was, as Secretary of the Interior Stewart Udall noted in his introduction, "a story of land failure and the failure of men." Caudill's book, in describing the destructive environmental and human history of the Cumberland Plateau of Kentucky, made a unique contribution to the evolving environmental ethic of the 1960s. Caudill's call for a Southern Mountain Authority modeled after TVA was never realized or fully appreciated, but the plight of Appalachia became a subject of new policy debate in Washington. In 1965 Congress established the Appalachian Regional Development Act to help plan economic development,[63] and it directed the Department of the Interior to study the strip mining problem and to recommend responses to the president two years later.

The Interior Department report portrayed, in large color photographs, the devastation of coal strip mining in Appalachia.[64] It recommended federal solutions only if state regulations and enforcement failed (as Caudill said they had and always would). Although no specific surface mine legislation was sent to Congress until the Nixon administration, the issue, with its dramatic environmental and human dimensions in Appalachia, captured the imagination of environmentalists in Washington.

The evolution of the surface mine act was marked by considerable lack of consensus regarding its proper coverage, the appropriate state-federal management relationship, and the economic and energy effects of controls. Data on the extent and importance of the environmental problem were not, however, in serious dispute.

The original Nixon proposal, send to Congress in 1971 after a CEQ study showed that state regulations were ineffective, included underground as well as surface mines, and it favored a strong state regulatory role under federal guidelines. The first bill to pass either house established a federal permit program, but subsequent bills enhanced the state role in line with the Nixon-Ford position.[65] The legislative structure for the statute was becoming more and more detailed and demanded increasing dedication and technical expertise from Washington's environmental lobby, led by the Environmental Policy Center. Strong congressional moves in 1971 to ban strip mining altogether lacked conclusive support but brought about even more precise slope, geographical, mine design, reclamation planning, and enforcement controls in the legislation that Congress finally passed.

President Ford pocket-vetoed the bill in late 1974, and a bill passed in 1975 met a similar fate when Ford vetoed it and Congress could not override the veto. The environmental impetus for the bill remained strong, but the Ford administration and the coal industry were concerned about the energy implications of surface mine controls. The Interior Department's Bureau of Mines and the Federal Energy Administration were convinced that the act would cause significant job losses and coal production curtailments. Although CEQ's analyses did not support that conclusion, it lacked persuasive influence within the Ford White House. It was not until 1977, during the Carter administration, that Congress passed and the president signed the Surface Mine Control and Reclamation Act, which included compromises regarding controls over steep slope and mountaintop stripping.

Passage of the bill was an important element of President Carter's environmental agenda. It was also the prime goal of some of his earliest and most vigorous environmental supporters, notably Louise Dunlop, an environmentalist with the Environmental Policy Center, who had doggedly and effectively fought for the act since the early 1970s.

The surface mine act created problems for subsequent regulators by virtue of its tortuous history. The act had become extraordinarily detailed. Its performance standards were to apply not only to the wet, steep-sloped Appalachian region but to the flatlands of the Midwest and the dryland of the West, with its low sulfur coal. Specificity was required because many states (particularly in Appalachia) were not trusted to regulate strictly. Moreover, when Congress failed to approve an outright strip mine ban, it turned to specific control language to compensate.[66] The legislative compromises seemed to add more complexity, so that "additional variances granted to the industry are balanced by more precise specificity in both language and legislative history."[67]

These problems haunted the new Office of Surface Mining when in March 1979 it developed the permanent implementing regulations that Secretary of the Interior James Watt and many states found excessively intrusive. When the act was found constitutional by the Supreme Court in 1981, however, the Reagan administration determined not to seek legislative revisions.[68]

THE CURRENT STATE OF LAND MANAGEMENT

The need to manage private rural land has become far more widely understood throughout the country than it was twenty years ago, in large part through the efforts of the environmental movement. At the federal level, environmentalists have achieved most of the specific conservation measures affecting private rural land that they developed in the 1960s. As a result there are federal laws and executive orders to reduce the adverse land use effects of federal public works and grant-funded projects, to protect certain critical wildlife habitats, to regulate coal surface mining and wetlands, and to manage coastal regions.[69]

Since the late 1970s, and especially under the Reagan administration, federal programs affecting private rural land have shifted direction or emphasis. Environmental groups and congressional supporters have played a significant role by ignoring, resisting, or supporting the changes.

One of the most sweeping and rapid changes has affected federal support for local, state, and federal land and water planning programs, including the federal river basin and regional economic development commissions. Federal support for these programs, never encouraged by the Office of Management and Budget even under the Carter administration, virtually ceased by the end of 1981. Environmentalists took little interest in these programs and hardly resisted their demise in the battles over the early Reagan budgets. An exception was the Coastal Zone Management (now Resources) Program. It exists today because it had developed strong environmental, state agency, and congressional support that has kept the program well funded.

Another significant change affecting federal involvement in private rural land issues stems from budget and staff reductions of the Council on Environmental Quality. Once a strong internal voice for federal concern about private land management, it now exercises no significant policy role. Environmental groups that worked vigorously to save the council in the Carter administration proved unable to prevent its reduction in 1981.

Federal efforts to relax regulatory controls over private land use—for example, through the Corps of Engineers permit programs and the surface mine controls of the Interior Department—were vigorous in the early 1980s. Environmental groups and congressional allies used litigation and lobbying to prevent significant changes in federal regulations affecting strip mining and wetlands. The Environmental Protection Agency greatly strengthened its oversight of the Corps of Engineers wetland permit and established a new Office of Wetlands Protection in 1986.[70]

Federal concerns about national losses of prime farmland have diminished since they were raised by CEQ and the Department of Agriculture during the Carter administration. Their joint National Agricultural Land Study (NALS) was given to the Reagan administration in January 1981, and Secretary of Agriculture James Block indicated early support for its recommendations. The recession and reduced land development, however, coupled with the reluctance of the administration to interfere with private land use, pushed the study aside. However, in the Farmland Protection Policy Act of 1981, Congress enacted one key NALS recommendation requiring federal agencies to assess and, if practical, to reduce the impact of their proposed actions on prime farmland.[71]

The concept of private land conservation through limitations on federal expenditures became politically attractive by the late 1970s, and it received more attention after passage of the Coastal Barrier Resources Act of 1982.[72] That act, which restricts federal expenditures for growth-inducing development on undeveloped barrier islands, became one of the few land use measures in the early 1980s to gain enthusiastic environmental, taxpayer group, and Reagan administration support. It passed after more than five years of economic and scientific studies supported by CEQ and the Interior Department, strong backing from a coalition of environmentalists and scientists, and special attention to the barrier island problem in President Carter's 1977 environmental message.[73]

The rural land use agenda of the environmental movement in the 1980s concentrated on agricultural and wetland issues and eventually coalesced with Reagan administration budget concerns by focusing on conservation through expenditure control.

After strong environmental lobbying, Congress passed the "sodbuster" provision of the 1985 farm bill,[74] which made farmers ineligible for most U.S. Department of Agriculture crop support and insurance programs if they brought into crop production land highly susceptible to soil erosion. A similar "swampbuster" provision applied these penalties to farmers who sought to drain, fill, or otherwise convert natural wetlands to cropland. A year later, Congress included provisions in the Tax Reform Act to eliminate farmers' ability to deduct expenses for cultivating highly erodible land and draining wetlands.

Other provisions of the 1985 farm act tied federal controls over highly erodible land to future savings in Department of Agriculture commodity programs. The Conservation Reserve Program authorized the secretary of agriculture to enter into contracts with farmers to convert up to forty-five million acres of highly erodible cropland to grassland or forests. Such retired lands would then not be eligible for agricultural commodity support. A conservation compliance provision required farms to begin in 1990 to implement an approved five-year soil conservation plan for highly erodible land.[75] These provisions had been unsuccessfully urged upon Congress by environmental groups during the Carter administration.

The long-standing goal of the environmental movement to curtail federal water resource projects achieved fruition with passage of the Water Resources Development Act of 1986.[76] The act established new cost-sharing formulas to increase contributions of nonfederal beneficiaries of Corps of Engineers water projects, and other provisions to curb environmental damage and federal spending.

With these measures, environmental interests achieved important controls over adverse impacts of federal programs on rural lands that had not been possible during the 1970s. The approach was piecemeal and limited by links to federal budget-cutting priorities. Further opportunities existed to exploit this political connection for public land management activities, such as the Forest Service's extensive sale of national forest timber for less than the sale cost taxpayers. Opportunities for new land acquisition or broader regulation of private rural lands for conservation purposes were limited during the 1980s.

Given continued high federal deficits, the land use agenda of the environmental movement and the federal government during the late 1980s and early 1990s will be modest, but changes are possible over the longer term. For example, scientists are increasingly concerned about the need to modify and respond to the effects of global climate change on forests and farmlands, and rising sea levels along the coasts. EPA and environmental groups have new programs to address the issue, which has immense implications for future infrastructure development, energy use, agricultural policy, and coastal zone management. Environmental organizations have rising concerns about the loss of biological diversity in species, ecosystems, and genetic resources at home and abroad. New federal habitat conservation programs and biological diversity regulations are distant, but distinct, possibilities. In rapidly developing suburban and rural areas throughout the country there are new experiments in cooperative local, state, and federal planning for development that protects wildlife, wetlands, and other resources. Each of these concerns focuses on needs to ensure sustainable development patterns. In time this issue may result in more comprehensive national attention to conservation practices on private rural land.

NOTES

1. Of the 2.263 billion acres in the American land base, federal public land includes about 751 million acres, and urban land or Standard Metropolitan Statistical Areas (SMSAs) include 69 million acres. Private land addressed in this paper includes undeveloped land within SMSAs. See Council on Environmental Quality and U.S. Department of Agriculture, *National Agricultural Lands Study*, Final Report (Washington, D.C.: Government Printing Office, 1981), 9.
2. Significant effects of federal tax laws include the allowed deductions for home-

owner mortgage interest payments and property taxes and the capital gain provisions for land sales. Richard P. Fishman, ed., *Housing for All Under the Law*, A Report of the American Bar Association Advisory Commission on Housing and Urban Growth (Cambridge, Mass.: Ballinger, 1978), 18. On the conservation side of the ledger, the tax code has authorized charitable deductions to encourage acquisition and protection of open space, which has fostered more than four hundred private land trusts in forty-three states. Kingsbury Browne, "Treasury and the Land Trusts—a Weakening Alliance for Conservation," *American Land Forum Magazine*, Winter 1983, 66.

3. Council on Environmental Quality, *The Growth Shapers—Land Use Impacts of Infrastructure Investments* (Washington, D.C., 1976); Calvin Beale and David Brown, "Sociodemographic Influence on Land Use in Non-Metropolitan America," in Senate Committee on Agriculture, Nutrition, and Forestry, *Agricultural Land Availability*, 97th Cong., 1st sess., July 1981, vol. 2.
4. 7 U.S.C. secs. 4201–4209.
5. 16 U.S.C. sec. 1451 *et seq.*
6. 32 U.S.C. secs. 4321–4347.
7. 33 U.S.C. sec. 1344; 33 U.S.C. sec. 403.
8. 30 U.S.C. secs. 1201–1328 (Supp. III 1979).
9. 43 U.S.C. sec. 391.
10. 272 U.S. 365 (1926).
11. 26 Stat. 454; 33 U.S.C. sec. 401 *et seq.*
12. 16 U.S.C. secs. 791(a)–793, 795–818, 820–825(r).
13. U.S. v. Appalachian Electric Power Co., 311 U.S. 377 (1940).
14. 39 Stat. 355.
15. Fishman, *Housing for All*, 29.
16. See Stewart Udall, *The Quiet Crisis* (New York: Holt, Rinehart and Winston, 1963), 140.
17. The National Land Use Planning Committee, made up of agricultural college representatives and federal employees and led by the Bureau of Agricultural Economics, was actively planning ways to encourage local and state land inventories as a first step toward systematic land use planning. Marion Clawson, *New Deal Planning—The National Resources Planning Board* (Baltimore: Johns Hopkins University Press, 1981).
18. Letter from the Secretary of Agriculture, Mar. 13, 1933, transmitted in response to Senate Resolution 175, 72d Cong.
19. Arthur M. Schlesinger Jr., *The Age of Roosevelt—The Coming of the New Deal* (New York: Houghton, Mifflin Co., 1959), 350.
20. Udall, *Quiet Crisis*, 145. Clawson has observed that the NRPB's Land Use Committee was less effective and influential than its Water Committee. Clawson, *New Deal Planning*, 120.
21. Wayne D. Rasmussen, "History of Soil Conservation, Institutions and Incentives," in *Soil Conservation Policies, Institutions and Incentives*, ed. Harold G. Halcrow, Earl O. Heady, and Melvin L. Cotner (Ankeny, Iowa: Soil Conservation Society of America, 1982), 5.
22. Dallavalle and Mayer, *Soil Conservation in the United States: The Federal Role*, Congressional Research Service Report no. 80–144 (Washington, D.C., Sept. 1980), 10.
23. Aug. 11, 1939, 53 Stat. 1414.
24. Feb. 29, 1936, 49 Stat. 1151.
25. Fairfield Osborn, *Our Plundered Planet* (Boston: Little, Brown and Co., 1949).
26. Aldo Leopold, *Sand County Almanac* (Madison, Wis.: Tamarack Press, distributed by Oxford University Press, 1977; first published in 1949 under title: *A Sand County Almanac*).
27. Aldo Leopold, "The Conservation Ethic," American Land Forum Magazine, Winter 1983, 58, 62. First published in *Journal of Forestry*, Oct. 1933.
28. U.S. Department of the Interior, Fish and Wildlife Service, *Wetlands in the United*

States, prepared by S. P. Shaw and C. G. Fredine, Circular 39 (Washington, D.C., reprinted 1971).
29. 16 U.S.C. secs. 661–667(e).
30. U.S. Congress, Senate Committee on Interstate and Foreign Commerce, *Fish and Wildlife Conservation and Water Resource Developments*, 85th Cong., 2d sess., July 28, 1958, S. Rept. 1981.
31. Stewart Udall, *The Quiet Crisis* (New York: Holt, Rinehart and Winston, 1963).
32. Scenic Hudson Preservation Conference v. FPC, 354 F.2d 608 (2d Cir. 1965).
33. U.S. Conference of Mayors, *With Heritage So Rich* (New York: Random House, 1966); National Historic Preservation Act of 1966, 16 U.S.C. sec. 470 *et seq.*
34. 23 U.S.C. secs. 131, 136, 319.
35. See John J. Constonis, "Law and Aesthetics: A Critique and Reformulation of the Dilemmas," *Michigan Law Review* 80 (1982): 355.
36. 49 U.S.C. sec. 1653(f).
37. 1963 Acts, ch. 426; 1965 Acts, ch. 116.
38. Pub. L. No. 90–45.
39. 33 C.F.R. sec. 209.120(d) (1968).
40. Zable v. Tabb, 430 F.2d (1970).
41. 33 U.S.C. sec. 1344.
42. 36 Fed. Reg. 12217–30 (May 10, 1973); 46 Fed. Reg. 12115–37 (Apr. 3, 1974); Charles D. Ablard and Brian Boru O'Neill, "Wetland Protection and Section 404 of the Federal Water Pollution Control Act Amendments of 1972: A Corps of Engineers Renaissance," *Vermont Law Review* 1 (1976): 51, 63. The authors make a strong case that Congress intended to keep the traditional jurisdiction of the corps.
43. Natural Resources Defense Council v. Calloway, 392 F. Supp. 685 (1975).
44. Jeffrey K. Stine, "Regulating Wetlands in the 1970s," *Journal of Forest History* 27 (Apr. 1983): 67.
45. Ibid., 71.
46. *Our Nation and the Sea*, the Stratton Commission Report, United States Commission on Marine Science, Engineering and Resources (Washington, D.C.: Government Printing Office, Jan. 1969).
47. John C. Whitaker, *Striking a Balance: Environment and Natural Resources Policy in the Nixon-Ford Years* (Washington, D.C.: American Enterprise Institute for Public Policy Research, 1976), 151.
48. Christopher J. Duerksen, *Environmental Regulation of Industrial Plant Siting* (Washington, D.C.: The Conservation Foundation, 1983), 67. BASF subsequently withdrew its proposal, ibid., 7.
49. Pub. L. No. 92–583, Act of Oct. 27, 1972.
50. Whitaker, *Striking a Balance*, 154.
51. Arthur Conan Doyle, "Silver Blaze," in *The Memoirs of Sherlock Holmes* (New York: The Heritage Press, 1950), 534.
52. Henry M. Jackson, "A View from Capitol Hill," Policy Memorandum No. 37 (Princeton, N.J.: The Center For International Studies, Princeton University, 1970).
53. S. 3354, 91st Cong., 2d sess., 1970.
54. Hearings on this proposal spurred interest in a national land use bill. See Noreen Lyday, *Law of the Land: Debating National Land Use Legislation, 1970–1975* (Washington, D.C.: The Urban Institute, 1976), 10.
55. Jackson, "View from Capitol Hill."
56. Fred Bosselman and David Callies, *The Quiet Revolution in Land Use Control* (Washington, D.C.: Council on Environmental Quality, 1971).
57. Council on Environmental Quality, *Environmental Quality* (Washington, D.C., Aug. 1970), xii.
58. *Environmental Quality* (Washington, D.C., Mar. 1971). This was the first of three annual environmental messages to Congress by President Nixon.
59. Whitaker, *Striking a Balance*, 158.

60. William K. Reilly, ed., *The Use of Land* (New York: Thomas Crowell Co., 1973); Fred Bosselman, David Callies, and John Banta, *The Taking Issue* (Washington, D.C.: Goverment Printing Office, 1973); Lyday, *Law of the Land*, 46.
61. Whitaker cites but does not credit the Watergate theory, *Striking a Balance*, 164, 165.
62. Harry Caudill, *Night Comes to the Cumberlands: A Biography of a Depressed Area* (Boston: Little, Brown and Company, 1963).
63. Pub. L. No. 89–4, 1965.
64. U.S. Department of the Interior, *Surface Mining and Our Environment*, July 1967.
65. William M. Eichbaum and Hope M. Babcock, "A Question of Delegation," *Dickinson Law Review* 86 (1982): 625.
66. Ibid., 629.
67. Louise Dunlop, "An Analysis of the Legislative History of the Surface Mining Control and Reclamation Act of 1975," in *Rocky Mountain Mineral Law Institute 21*, proceedings of the 21st Annual Institute, July 17–19, 1975 (New York: Matthew Bender and Co., 1976), 11–58.
68. Hodel v. Virginia Surface Mining and Reclamation Ass'n, Inc., 101 S. Ct. 2376 (1981).
69. Executive orders promulgated by President Carter in May 1977 directed federal agencies to avoid actions that modified or destroyed wetlands (Exec. Order No. 11,990, May 24, 1977) or that were located in floodplains (Exec. Order No. 11,988, May 24, 1977). They have remained in effect under the Reagan administration.
70. 49 Fed. Reg. 1387–1399 (Jan. 11, 1984).
71. Pub. L. No. 97–98, 7 U.S.C. secs. 4201–4209. Act of Dec. 22, 1981.
72. Pub. L. No. 97–348, 16 U.S.C. secs. 3501–10 (1982).
73. Robert R. Kuehn, "The Coastal Barrier Resources Act and the Expenditure Limitation Approach to Natural Resources Conservation: Wave of the Future or Island Unto Itself?" *Ecology Law Quarterly* 11 (1984): 583.
74. The Food Security Act of 1985 (P.L. 99–198), Dec. 23, 1985.
75. For a discussion of these provisions of the 1985 farm act, see *State of the Environment: a View Toward the Nineties* (Washington, D.C.: The Conservation Foundation,1987), 378–384.
76. P.L. 99–662, Oct. 17, 1986.

6

THE FEDERAL GOVERNMENT, WILDLIFE, AND ENDANGERED SPECIES

Thomas R. Dunlap

Since the mid-1960s there has been a dramatic shift in the nature and extent of government protection for wildlife. The states traditionally managed wildlife, and active measures were limited to encouraging a few game species. The federal government had only a small role, and federal law was a short collection of statutes relating mainly to refuges, parks, and the hunting of migratory birds. Now the federal government has a very active role and a complex set of rules that, in some cases, completely preempt state management. The laws, too, go well beyond simple prohibitions on the killing of species. Definitions of "taking" include harassment and in some cases even close observation; habitat and actions that would affect habitat also fall under federal scrutiny. Wildlife policy is now a matter for continuous administrative decision making, involving many interests in society. In addition, the reach of the law is much greater. Instead of a few game species, policy embraces, at least in theory, every animal above the microscopic level (except insect pests) and covers plants as well. There is a major effort to save species in danger of extinction. Between 1966 and 1973, popular pressure led to a far-reaching program to identify, protect, and restore all species in danger of extinction. Since then, there has been an almost continuous debate over the implementation and administration of that policy.

The policy of saving all threatened forms as parts of a connected whole is the result of a revolution in American attitudes toward nature. Although ultimately a reaction to the transition from a rural, agricultural society to an urban, industrial one, it is more immediately the product of modern science and of social and political changes since World War II. Americans now accept, as they did not forty years ago, that nature is a system, with each species playing a part in forming and balancing the whole. They also believe, far more than they did then, that animals have feelings, emotions, and a social life, and therefore rights that man is bound to respect. All this is justified by science. Although modern defenders of wildlife have not discarded appeals to sentiment, aesthetics, or utility, they rest their case on ecology, which points to the interdependence of all forms of life; on evolution, which presents each species as a precious biological heritage and part of earth's history; and on ethology, physiology, and behavioral studies showing how much we are like the other animals (or they, like us).

The social and political circumstances of the postwar period shaped the way this picture of animals, nature, and man was translated into policy. Prosperity created, or greatly enlarged, the class of people with the education, resources, and leisure to be concerned with environmental amenities.[1] Organizations devoted to wildlife protection mobilized that class through education and propaganda. They took their case to Congress, taking advantage of the circumstances of the late 1960s—the environmental enthusiasm that gripped the nation, the opportunities for direct participation in the political process, and the respect for science that made the Congress willing, if not eager, to use science as a guide to policy.

In tracing the movement toward comprehensive wildlife protection, we must begin before World War II, for administrative and legal developments between 1900 and 1940 laid the groundwork for the policies of the postwar period. Then, we can consider the intellectual and social base of the movement and the laws that make up the modern program: the Endangered Species Acts of 1966, 1969, and 1973; the Marine Mammal Protection Act of 1972; and the Convention on International Trade in Endangered and Threatened Species (CITES).[2] Finally, we can look at wildlife policy as part of the adjustment of people in an urban, industrial society to their new relationship with nature.

WILDLIFE IN AMERICA—FORMING A FEDERAL POLICY

Into the nineteenth century nature was, for most Americans, something to be conquered, and the abundance of wildlife a passing phase of frontier existence. Animals were destined, like the Indians and the forests, to vanish before civilization. With the end of the frontier and the rise of an urban, industrial civilization, Americans came to value nature, including wildlife, as a link to the pioneer past, and for the spiritual, moral, and even physical benefits it conferred. Contact with nature refreshed jaded sensibilities, lifted people into a higher realm, and reminded them of their connections to the world. Nature was a source of aesthetic delight; beautiful birds and mammals inspired the artist, the poet, and the writer, as well as lifting the heart. Hunting and fishing formed manly character and provided healthy outdoor recreation. Americans now sought to preserve what they had tried to destroy.

The movement to save the glories of outdoor America brought protection to some animals. In 1894, Congress banned hunting in Yellowstone National Park (to preserve the buffalo), and in 1903, President Theodore Roosevelt began the practice of setting aside parts of the public domain as wildlife refuges. There was no coherent ideology of nature protection or ecological understanding behind these impulses. They were rooted in immediate human needs and wishes—sport, aesthetics, and sentimentality. Most birds and large, "noble" species like elk, deer, and moose were generally popular, but few people were concerned with the smaller animals or with reptiles, fish, or invertebrates. Even within the favored groups, there were exceptions. Songbirds were table fare into the early twentieth century, and predators were beyond the pale. Even people who ardently defended wildlife encouraged the killing of hawks, owls, wolves, and coyotes—the "vermin." Man was the conqueror; he could, and would, pick and choose what he would save and what he would destroy.

As wildlife dwindled, state protection seemed less and less satisfactory. Federal authority, though, was limited by legal precedent, and wildlife's protectors found it difficult to get action from Congress. Jurisdiction over wildlife was one of the powers that the independent colonies had assumed upon their break with Great Britain and never ceded to the federal government.[3] Congress therefore had to appeal to constitutional powers it had in other contexts and apply them to wildlife. Its first venture, the Lacey Act of 1900, protected plume birds by outlawing interstate shipment of game killed in violation of state laws, an appeal to congressional power to regulate interstate commerce. The second major statute, the Migratory Bird Treaty Act of 1918, used the supremacy and treaty-making clauses to control duck hunting. The law enforced a treaty with Britain and Canada that provided for international cooperation to preserve populations of migratory waterfowl.[4] Wildlife preservation in national parks and wildlife refuges, the government's major responsibility at that time, rested on federal ownership of the land.[5]

Limits on federal power were not self-imposed by a Congress careful of precedent; the states and hunters' groups fought vigorously against changes in traditional arrangements. When Congress first tried to protect ducks, with the 1913 Migratory Bird Act, there were immediate legal challenges, as there were to the 1918 act enforcing the treaty of 1916 (which had been concluded partly to avoid the possibility of an adverse decision on the 1913 act). When, in the 1920s, the Department of Agriculture wanted to reduce the deer herd in the Kaibab National Forest, the state of Arizona promptly sued. The dispute was of more than theoretical interest. Wildlife was, in many areas, a significant part of the economy, and people were unwilling to surrender control to Washington, seeing little merit in having regulations and policies set thousands of miles away, possibly with little consideration of local affairs. Advocates of stronger preservation efforts, new management practices, or control of interstate species (such as migratory wildfowl) sought federal relief against the unsympathetic states. Under these circumstances, wildlife policy disputes easily became matters of constitutional law, and the Supreme Court has been involved in wildlife matters to an unusual degree.[6]

Legislative and executive action in the 1930s rationalized existing wildlife programs, policies, and agencies; expanded their scope; provided better funding; and laid the foundations of modern wildlife work. Administrative actions were an important component of this drive. The Forest Service had started wildlife research and censuses of game on its lands in the late 1920s, and it began a refuge program in 1934. The National Park Service (NPS) established a wildlife research unit in 1929 at the urging of George Wright, and in 1933 it started a new publication series, *Fauna of the National Parks of the United States*. By the end of the decade Park Service research was making major contributions to American wildlife work; Adolph Murie's studies of the ecology of wolves and coyotes are generally regarded as the beginning of ecological research on large mammalian predators.[7] Reorganization in 1939 merged the Department of Agriculture's Bureau of Biological Survey and the Department of Commerce's Bureau of Fisheries into a new agency, the Fish and Wildlife Service, in the now conservation-minded Department of the Interior. The New Deal also included game production in the work of soil and water conservation districts, further increasing the federal reach in wildlife management.[8]

The Congress also took an active role. The Senate appointed a

Special Committee on the Conservation of Wildlife Resources in 1930, the House appointed a similar committee in 1934, and legislative initiatives in the next decade reinforced executive action. The Wildlife Coordination Act of 1934 required that certain federal actions, where consistent with the primary mission of the concerned agency, aid or at least not harm wildlife. Other laws helped free long-term wildlife programs from complete dependence on the appropriations process. Funds from the sale of hunting stamps under the Duck Stamp Act of 1934 created a fund for waterfowl refuges, which also aided other species. The Pittman-Robertson Act of 1937 levied an excise tax on arms and ammunition for state wildlife restoration projects, which as a by-product imposed uniform federal standards on the work.

An informal structure of interest groups, organizations, and concerned individuals supplied information, political support, and criticism to the wildlife agencies. Scientists were a prominent and longstanding part of this network. The American Ornithological Union had pressed for the legislation establishing the Office of Economic Ornithology and Mammalogy (which became the Bureau of Biological Survey) in 1885, had helped form the National Audubon Society in 1905, and had written a model bird protection bill. Mammalogists, who had close personal and professional ties with the Biological Survey, supported its work and tried to influence policy, particularly the predator control program. The Ecological Society of America began working to preserve wild areas for study in 1919, and it pressed the Forest Service and the National Park Service to set aside wild areas and to preserve wildlife.[9]

In the 1930s, wildlife managers came to constitute a distinct professional group. In 1933 Aldo Leopold, holder of the first chair of game management in the country, published the first textbook, *Game Management.*[10] By 1936 there was a professional organization, the Wildlife Society, and, two years later, the *Journal of Game Management*. Graduate programs, if they did not exactly spring up from the arid financial soil of the Depression, at least came into existence, and trained professionals began to displace political hacks on state game and fish commissions and to fill the new jobs created by the Roosevelt administration's integration of wildlife management and production with federal conservation efforts.[11]

Private organizations, which had been an important element in wildlife work since the late nineteenth century, continued to press for their causes.[12] Bird protection groups, led by the National Association of Audubon Societies, were the best organized, and they had a major say in migratory waterfowl regulation and in establishing and setting policies for wildlife refuges. Hunters were active but were influential largely on the state level. Their main method of national communication was the sporting magazines and papers; not until the formation of the National Wildlife Federation in 1936 did they get a lobby that would serve their interests on the national scene.[13]

The arms and ammunition manufacturers also worked to advance their cause. They had tried to join forces with the Audubon Society in 1911, but the directors rejected the move as inappropriate—too many members were against hunting or suspicious of the companies.[14] They concentrated on state action to encourage game production, lobbying for game commissions, wildlife management, and research. The Sporting Arms and Ammunition Manufacturers' Institute, for example, hired Aldo Leopold to study game conditions in the Midwest in 1928 and supported the ecological research of Leopold's students when he became professor of wildlife management at the University of Wisconsin.[15]

The intellectual underpinning of modern wildlife policy emerged as the institutional framework was built. Animal ecology became a distinct field of study in the late 1920s and was integrated into game research and policy in the next decade. The now-familiar ideas of food chains, the food cycle, trophic levels, and niches (often expressed in terms of food and food preferences) provided concepts to organize the study of animal communities. Combined with quantitative techniques, which introduced new elements of rigor, these concepts allowed ecologists to attack fundamental problems and, ultimately, to present a detailed picture of the relationships among animals and between plants and animals. Charles Elton's influential textbook, *Animal Ecology*, written in 1926, presented much of the new research and research tools in accessible form, and they became common intellectual currency of ecologists and game managers in the next decade.[16]

The fundamental insight that this research impressed on scientists and observers was the complexity and interdependence of natural systems. Even the least significant species, it seemed, played a part in balancing and directing the system. People had seen nature as a collection of parts. Ecology pointed forcefully to a different vision—nature as a whole, all species linked in an intricate "web of life." Ecology also grounded vague ideas about the "balance of nature" in detailed, quantitative studies. People had always assumed that there was a "balance," but no one knew how it worked. Predation was usually thought to be the key—carnivores ate the "surplus" herbivores—but this idea was an extrapolation from the obvious *fact* of predation to the completely unknown *effect* of predation.[17] Now scientists could see to what extent predation, or other factors, really affected wildlife populations.[18] Elton said of this period that he and his colleagues were entering into a "new mental world of populations, inter-relations, movements, and communities . . ." as they began to study "living nature in the field" and to reject "easy generalizations about adaptation and the balance of life."[19]

Ecology caused preservationists to shift their intellectual ground and change their tactics. It provided foundations for romantic ideas about the connections between all things, but it also fundamentally altered the way in which preservationists saw nature. If all species were interdependent, then there was no distinction possible between "good" and "bad" ones, and one could not be saved without saving all. The strategy of refuges, protected breeding grounds, and hunting regulations, all designed to save a few species, now seemed mistaken, as did the economic arguments for wildlife—based on hunters' dollars and the toll exacted by insects and pest rodents. Into the 1920s, the Audubon Society had defended birds on the basis of studies of stomach contents, most of them by the Biological Survey, which showed that birds contributed more by destroying harmful insects and rodents than they cost in chickens. The group opposed shooting the exceptions—the bird-eating hawks—only because this would lead people, unable to distinguish between the "good" and the "bad" ones, to shoot all raptors.[20] As ecological studies accumulated, the society moved to a defense of all species on the grounds that all were part of a connected system. In 1935, in an article on "Feathered and Human Predators," *Bird-Lore* cited the studies one of Aldo Leopold's students at Wisconsin, Paul Errington, had done on bobwhite. This research, it said, showed that predator control was not good management. It also reported Leopold's remarks to the American Game Conference on the futility of most predator control work.[21] Later that year, it backed away from a long-standing policy of advocating "vermin" control on bird preserves. Predators, it said, should only be shot when it had been shown by scientific evidence that control was needed.[22]

On the eve of World War II, then, the framework existed for a policy of wildlife preservation that would include all species. Policy-makers were concerned about the steady loss of wildland and the scarcity of wildlife. An ideology that justified protecting all of nature was coming into existence. Scientific studies showing the complexity and interdependence of nature would have supported protection. The government had agencies to do the work and laws that could have supported an extensive program, and organized groups were beginning to support such a stand. What was lacking was public concern about preserving wildlife as a whole. There was interest in some useful or beautiful species, but not in all or in the system that sustained them. The change in wildlife policy would come only in the next generation, as the picture of nature supplied by ecology became the common wisdom, as people became alarmed at the damage industrial civilization was causing to the "web of life," and as they came to value more highly environmental "amenities," including wildlife. Social and political change would open the processes of legislation and administration to citizen groups. Environmentalists and wildlife defenders would then transform the agencies established in the prewar period and write new laws to carry out a new program.

THE POSTWAR WORLD

The economic boom of the postwar period and the diffusion of environmental ideas to a large part of the public created new interest in wildlife preservation. A rising standard of living and inexpensive transportation made outdoor recreation and visits to national parks and refuges less a luxury for the few than a common vacation trip. Increasing knowledge made Americans more aware of their links to the natural world, and they came to see it as essential to their own survival. Economic development was rapidly encroaching on wilderness, and visitors flooded formerly isolated areas (the Sierra Club dropped from its bylaws the phrase about encouraging more access to the mountains). Worldwide pollution, most dramatically fallout in our rain, pointed to the interdependence and closeness of the world. All these trends added urgency to the movement for preservation. What had been a distant problem was now an emergency. If nature was to be saved, it would be done now.

As they had in the period before World War II, scientists played an important role in informing and mobilizing the public. Now, though, the problem was seen in more than national terms, and the International Union for the Conservation of Nature and Natural Resources (IUCN), formed in 1948 under the auspices of the United Nations Educational, Scientific and Cultural Organization (UNESCO), played a key role. It coordinated efforts to protect wildlife and served as a scientific clearinghouse, and its publications, particularly its lists of endangered and threatened species, became a guide for world wildlife protection.[23]

Private organizations translated scientific information for the public and mobilized support for preservation. Older groups, like the Audubon Society and the National Wildlife Federation, were prominent in this effort, but an active radical humane movement joined them, with goals and larger aims different from those of traditional humane societies. It was concerned less with cruelty than with the exploitation of animals, less with the evil effects on man of his inhumanity than with the suffering of animals. It relied less on sentiment and sensitivity; it had a coherent philosophy of animal rights. Animals were, the new defenders of wildlife said, sentient beings like us, capable of feeling pain and other emotions, of forming attachments to others of their species, and of complex social behavior. Ethics required that we treat them as having rights—the first of which was the right to live.

The radical movement opposed not just cruelty but hunting and trapping, and, because its goals were fundamentally different from those of the National Wildlife Federation or the Audubon Society, it came into conflict with other wildlife groups almost as much as with agencies or opponents of preservation measures. Even other humane groups came under fire.[24] This produced not only a day-to-day friction, as news releases and charges of bad faith and slander went back and forth, but a division of effort and hostility within the wildlife movement. The divisions were not new—the Audubon Society and the National Wildlife Federation had suffered internal disagreements for years—but their scale was.[25]

The new groups had much more power and influence than their predecessors. Consider Defenders of Wildlife, which became a major force in forming wildlife policy by the mid–1960s.[26] It was a descendant of the Anti-Steel Trap League of the 1920s, but where the league had been a small, badly funded organization—basically the forum and vehicle for a few zealots—the Defenders of Wildlife had fifty thousand members, a magazine to keep them informed, a Washington office and staff to present its views to policymakers, and an aggressive stand. The league had relied on education and mobilization of women's groups and had worked for state legislation. Defenders worked at the national level as well, and it sued. In 1971, for example, it took the Department of the Interior to court, claiming that the National Environmental Policy Act (NEPA) required an environmental impact statement for the predator poisoning campaign.

Humane groups had an effect on policy because they were larger than their predecessors (the result of changing attitudes), were better organized, and understood the political system, but most importantly because they grounded their appeals and their program in science. Humanity to animals convinced few, and it was not an acceptable basis for policy. Science was, and animal rights groups used its cachet. The chief lobbyist and executive officer for Defenders of Wildlife, for example, was a wildlife ecologist, John Grandy (now with the Humane Society of the United States), who was careful to cite his academic qualifications when testifying before congressional committees. Even when classically humane goals were reached for—the abolition of the leg-hold trap—scientific studies were brought to the fore. Physiology demonstrated that animals felt pain. Ethology, animal rights groups argued, showed that animals had a social life and formed communities—which were disrupted when the members were shot. Scientists came forward to support these contentions.[27]

PROTECTING ENDANGERED SPECIES

Endangered species were a natural concern for environmentalists. Extinction, as the bumper stickers said, was forever, and the rates of extinction and habitat destruction that would cause extinction were rising. By the early 1960s, legal protection for endangered species was coming to be a political issue for the conservation community, and it would develop into a campaign for the enactment of a program of comprehensive protection. It was not a central philosophical concern of humane groups, but it was important, and it was an issue they could use for their own ends. Proponents of protection had all the advantages at this point—organization, funds, public support, a rationale for saving all wildlife, and at least the outline of a comprehensive program. These proponents could point to a variety of benefits—scientific, aesthetic, and even practical—from such a program.[28] Opponents could hardly mobilize against the goal of preserving wildlife. Even if they had been willing to take such a stand, with such a negative goal they would have attracted little support. Nor did there seem any reason, in the abstract, to oppose wildlife protection.

Legal action on endangered species began in a small way. In the 1964 Land and Water Conservation Act, the Congress established a fund to support federal and state outdoor recreation and wildlife projects, including the purchase of land for the preservation of native wildlife and game—although it did not define the term. Congressional appropriations committees balked; they thought there should be a bill specifically giving the secretary of the interior this power. The first endangered species act, then, was simply to clarify the existing authority of the secretary to carry on a comprehensive wildlife protection program. It directed him to use his statutory authority to buy land to purchase areas needed to protect native wildlife threatened with extinction and appropriated $15 million from the Land and Water Conservation Fund for that purpose (but allowed no more than $5 million to be spent in any given year and no more than $750,000 for any one project). It also enjoined the secretaries of the interior, agriculture, and defense to protect threatened wildlife on lands under their jurisdiction—insofar as protection was consistent with the primary purpose of their agency.[29]

The Endangered Species Preservation Act of 1966 was not designed to expand the scope of federal power, and it did not. The secretary of the interior could buy land to protect endangered species, but the power to purchase was not new. He could not regulate taking or commercial use of endangered species, and protection, even on federal lands, was limited by the primary mission of the agency in charge of the area. There were no definitions of endangered species, and the process of listing them was vague. Beyond the requirements that the secretary protect native wildlife threatened with extinction, that he make his decisions in consultation with the affected states and with the advice of "interested persons," and that he publish his findings in the *Federal Register*, there were no formal procedures. On the crucial question of working with the states there was only the vague directive that the secretary cooperate "to the maximum extent practicable."

Wildlife advocates wanted more—stronger management tools, standards for listing, and, above all, worldwide protection, preferably by controls on commercial traffic in endangered species. This last would seriously affect industries dealing in animals or animal products, from pets and pet food to whale oil and scrimshaw (carved whale teeth), and pressure on Congress mounted. In 1965 only the Wildlife Management Institute had sent a witness to House hearings, and only four other groups—the Audubon Society, Defenders of Wildlife, the National Wildlife Federation, and the Sport Fishing Institute—had submitted statements for the record. In 1969, by contrast, there were thirty-nine witnesses at the final House hearings, ranging from representatives of the Animal Welfare Institute and the Sierra Club to Gators of Miami and Brooklyn Better Bleach, Inc.[30]

The fur and pet industries objected strongly to controls on imports and exports, particularly if the United States were to determine what species were endangered. Local authorities, they argued, should have that job. Unilateral action by the United States, they argued, would not protect animals; it would only result in lost jobs and profits in this country. They called instead for an international agreement. Wildlife and environmental groups insisted on the ban to deny the American market to these products. They also wanted the listing of endangered species by the United States, not the country of origin. They did not oppose an international agreement, but they wanted the United States to move first and quickly; others, they said, would follow.

Environmental groups carried the day, and the Endangered Species Conservation Act of 1969 was a major step toward a comprehensive wildlife program. It gave the secretary of the interior authority to list species threatened with extinction and to control their passage into and out of the United States. It defined "wildlife" broadly, including crustacea and mollusks, amended earlier laws (the Lacey and Black Bass acts) to expand their scope, and placed the listing process under the provisions of the Administrative Procedures Act. It increased funding and allowed the expenditure of up to $2.5 million from the Land and Water Conservation Fund on any single project.[31] There were several things, though, that it did not do. Domestic species received no new protection, and of foreign species only those entering into commerce fell under the law. There was no habitat protection. The 1969 act, for environmentalists, was good, but it was not enough.

Affected businesses were now aroused and active, but subsequent legislation—the 1972 Marine Mammal Protection Act and the 1973 Endangered Species Act—passed easily. By the early 1970s, even politicians with little or no demonstrated interest in the environment were "getting right" with nature, and what might have been a contested program passed on the flood tide of environmental enthusiasm. President Nixon, for example, gave an environmental State of the Union address in February 1972, calling for a variety of new measures, including greater protection for endangered species. There were hearings on several bills that year and one, H.R. 4758, passed the House. Although it did not reach the Senate floor it seemed certain that some legislation would pass.[32]

What passed in 1972 was a related bill, the Marine Mammal Protection Act. It was a dramatic departure from earlier statutes dealing with these species. Such statutes had been concerned with managing single species for commercial harvesting; this act covered all marine mammals and had as its goal not species as such but the stability of the marine ecosystem. Previous statutes had been concerned with production, the "maximum sustainable harvest," this act with "optimum sustainable populations" and "the adverse effect of man's actions." Species, the law said, "should not be permitted to diminish beyond the point at which they cease to be a significant functioning element in the ecosystem of which they are a part."[33] To achieve these sweeping ends, the Congress completely preempted state authority over all species of marine mammals, put an indefinite, immediate moratorium on their taking (very broadly defined), and wrote an elaborate set of rules regulating commercial traffic. The Congress also proposed an ambitious program of international protection and made its policies the official American negotiating position in future international conferences.[34]

Users of marine resources as well as preservationists had long realized the need to regulate the use of this "commons," and they generally supported a program to regulate the taking of marine mammals. The bill that emerged, though, was far more comprehensive and restrictive than they wanted. The moratorium made some products illegal (which led to a flood of applications for "hardship" exemptions to allow firms to sell them), and left the fishing and whaling industries facing difficult or unpredictable legal situations. Tuna fishermen, for example, set their nets on schools of porpoises to catch tuna, and inevitably drowned some of the mammals. The act put their livelihood in a legal limbo. Killing dolphins in the course of tuna fishing (incidental taking) was allowed for a two-year period; it was then to be reduced "to the lowest practicable level," a phrase that provided neither comfort nor guidance, but led instead to legal and political battles and even a tuna fishermen's strike.[35]

The vague requirements for optimum sustainable populations produced maximum confusion, as administrators ruled with what evidence they had and were challenged in court, or waited for scientists to produce population estimates, and found themselves under fire for not taking action. Even minor users of marine mammals and their products found the act a problem. Dealers in scrimshaw had to show that their stock had been in circulation before the date of the act, and Aleuts and other Alaska natives who hunted whales for food or made handicrafts for sale fell under federal scrutiny.[36] The act proved to be a lawyer's dream and a scientist's and administrator's nightmare.

In the spring of 1973, the United States moved closer to a program of worldwide protection with the signing of the Convention on International Trade in Endangered Species. It was not a new initiative—the IUCN had put forward a draft in 1962—and the final product, signed a week before Congress began hearings on the 1973 Endangered Species Act, was the product of lengthy negotiations. It divided endangered species in two major groups. One was those in immediate danger of extinction; trade here was to be held to the absolute minimum, and the exporting country was to certify that what trade did go on was not a danger. The second category (listed in Appendix II) was species in less danger or "look-alikes," animals so closely resembling endangered ones that trade in them had to be regulated to protect the others. The convention provided complex machinery for enforcement—a system of export and import permits, and, within each country, a scientific authority to pass on population levels and trade and a management authority for enforcement. Regular international meetings of the signatories would provide a chance to discuss listings and propose amendments.[37]

A few months later, Congress passed the Endangered Species Act, after a lengthy debate over its provisions. Environmentalists had wanted enforcement by the Department of the Interior; business favored the Commerce Department. Then there were the crucial definitions of "wildlife" and "taking," which would limit, or expand, the scope of the act. Other disputes arose over the inclusion of plants and exemptions from the act for "economic hardship." (This last was a sore point with environmentalists, who thought that the exemptions from the Marine Mammal Protection Act had been granted far too liberally.)[38] The act wrote into law the wide-ranging program environmentalists wanted. It lifted restrictions on use of the Land and Water Conservation Fund to acquire habitat and, following the Marine Mammal Protection Act, defined "taking" very broadly to include almost any action that threatened a species or its critical habitat. It protected all forms of animals—invertebrates, crustacea, and insects—and included plants as well. Following CITES it set up separate lists for "threatened" and "endangered" species. The secretary of the interior (responsible for most species) had very broad authority to review the lists, and could, as under CITES, list species similar to endangered ones. The act also opened the administrative process; private citizens could petition the secretary to list a species. Perhaps most important, the act had as its explicit purpose the protection of "the ecosystems upon which endangered species and threatened species depend," and subordinated other things to that goal. It directed federal agencies "to insure that actions authorized, funded, or carried out by them do not jeopardize the continued existence of such endangered species and threatened species or result in the destruction or modification of [critical] habitat."[39]

The 1973 act broke decisively with past practice. Rather than protecting a few species of mammals and birds, it mandated protection of everything above the microscopic level (except pest insects). Rather than addressing immediate human needs and wishes or focusing on economically important species, it was concerned with the long-term health of the land and explicitly sacrificed a certain degree of economic development to this goal. The act was new, too, in asserting broad federal authority over resident wildlife. Only the Marine Mammal Protection Act had been so sweeping in its preemption of state power, and it had dealt with an area and a set of resources remote from most Americans' concerns. The new law would strike much closer to home.

With the program in place, many of the environmentalists' advantages disappeared. It was clear to professionals that the program would stand or fall on such mundane details as the criteria for listing species as endangered or threatened. It would be hard to sustain public interest in these dreary day-to-day battles, but this was the place where organized interests with a clear economic stake in the outcome would work the hardest. Opponents also found it easier to attack the program in place than to oppose its passage. It seemed to place wildlife values above all others and to threaten needed economic development. Far from ending the controversy over endangered species, passage of the 1973 act set the stage for a national debate over our relation to nature, how and how far we should protect it, and how we should balance its claims against other values.

The first concern was implementing the law. Environmental groups pressed for quick and drastic action; some wanted a blanket listing of all animals on the IUCN lists. Officials in Interior and Commerce, though, felt they had first to build up the scientific case on each species before they could act, and in response to complaints they protested that the process could not be rushed. They also pointed out (and here they had the environmentalists' support) that the act had placed major new responsibilities and complex administrative burdens on them without providing a corresponding increase in funding.[40]

More serious problems arose and business opposition increased as the act went into effect. The most dramatic problems arose over section 7, which required all federal agencies to consult with the secretary to assure that their programs harmed no endangered species, and to take "such action necessary to insure that actions authorized, funded or carried out by them do not jeopardize the continued existence of such endangered species and threatened species or result in the destruction or modification of habitat which is determined by the Secretary . . . to be critical."[41] By 1975, three cases had arisen under this directive, which placed, or seemed to place, protection of endangered species in direct conflict with economic development—a dam in Indiana would flood the cave of an endangered species of bat, a highway interchange in Mississippi crossed an area crucial to the Mississippi sandhill crane population, and the Tellico dam on the

Little Tennessee River would destroy the stream habitat of the endangered snail darter. Two of these problems were eventually settled, but the last, with its dramatic picture of a $100 million dam held up by a three-inch fish, provided the occasion for a major debate over the values and statutory requirements of the Endangered Species Act and a full-scale attack on the law.

Ironically, the Endangered Species Act only accidentally became an issue in the Tellico controversy. Valley residents dispossessed by the dam, conservationists who believed it was not needed, and the Cherokee Indians, for whom the valley was a sacred and historic place, had fought the TVA from the time it announced construction plans in the late 1960s. They had used a variety of tactics, including a demand for an environmental impact statement under section 102 of the National Environmental Policy Act. When a biologist discovered a new species of fish, which he named the snail darter, in the area to be flooded, opponents of the dam simply saw it as another chance to halt the project. They petitioned to have the darter placed on the endangered species list. Three weeks after the Fish and Wildlife Service listed the darter as an endangered species and its habitat in the Little Tennessee River "critical" for the species (in October 1975) opponents filed suit to halt construction, claiming the project was in violation of the Endangered Species Act.[42]

In court, the TVA contended that because work had begun on the dam before the Endangered Species Act was passed, the project was exempt. It also said that Congress had implicitly exempted Tellico by continuing to vote appropriations after the passage of the Endangered Species Act, that section 7 had never been intended to place an absolute value on wildlife, and that transplanted populations of the darter, now in other streams, made it possible to finish the dam without exterminating the darter. The district court agreed with the TVA, but the circuit court of appeals did not. Protection of an endangered species, it held, took precedence over even an almost completed federal project. The issue dominated congressional hearings in the spring and summer of 1978, and publicity only increased when, in the middle of the hearings, the Supreme Court ruled for the darter over the dam. Citing congressional expressions of concern for wildlife and the value of the ecosystem from hearings and debate, and noting earlier acts to protect endangered species, the Court held that the Congress intended to provide full protection for all species. Section 7 meant what it said.[43]

The problem, as Congressman Robert L. Leggett put it, was this: "Is the Tellico case an anomaly, or is it an example of the kind of conflict that we can expect to face in the future?"[44] The West, which depended heavily on federal money and federal land and derived much of its wealth from extractive industries, was particularly concerned. Any federal preservation program affects its economy, and this one seemed a nightmare: it appeared to allow any species, no matter how insignificant, to halt economic development, however needed. The National Cattlemen's Association, the Public Lands Council, and the National Woolgrowers Association made a joint statement blasting section 7 as dictatorial. Cooperation, they said, should mean genuine consultation and compromise, not the cooperation of others with the secretary of the interior. They called for some balancing mechanism that would weigh wildlife against other values. Senator Jake Garn of Utah protested against the "cynical" use of the act by environmentalists to halt development. A representative of the Upper Colorado River Commission, a Wyoming rancher, complained about the effect of the act on his business.[45]

Western economic interests, mining and lumber companies, and people generally committed to the "conquest of nature" called for major changes. Others, mainly environmental groups but also officials of the Fish and Wildlife Service and the National Marine Fisheries Service, opposed amendment. The act, they said, was working. There had been about 4,500 consultations under section 7. Only three had significantly interfered with a project. Only one, Tellico, had been resolved by court action, and that, environmentalists charged, could have been avoided if the TVA had been willing to consider alternatives to completing the dam.[46] Environmentalists were particularly insistent that section 7 not be changed. It was the heart of the act; to amend it would strike at the entire structure of endangered species protection.

Change, though, was almost certain; Tellico was too vivid an example of what might be a common problem and too good a piece of propaganda not to arouse fears. Environmentalists could only hope to limit the damage. They did. What emerged, from hearings, committee meetings, and three days of floor debate, was a compromise. The Congress established a cabinet-level committee to consider appeals from the section 7 process and directed the group to give "expedited consideration" to Tellico and Gray Rocks, two projects tied up in section 7 challenges. The final version, after several floor amendments had been rejected, passed by a vote of 384 to 12 in the House and by 94 to 3 in the Senate.[47]

Tellico did prove to be an anomaly, and pressure for major change in endangered species protection dropped. Conflict did not stop, but industries pressed more for individual remedies than for wholesale change. The National Forest Products Association, for example, pointed out that, in the West, member companies had to set aside small areas for bald eagle nests and to restrict activity in a much larger area around the nests (disturbing the eagles would, under the law, constitute "taking" of an endangered species). In the South they could not log areas of old growth deemed vital to the preservation of the endangered red-cockaded woodpecker. These limitations cost the companies millions. The association did not suggest abolishing or even amending the act, but asked for a program of compensation.[48]

Relations between the states and the federal government have been the most prominent continuing problem. The program, despite its assumption of federal supremacy, always assumed the active cooperation of the states, if for no other reason than their immense amount of manpower and expertise. Putting that machinery into place and developing uniform standards and plans proved to be a monumental job, which has continued to plague the program. The public had become interested only when something spectacular happened. In the summer of 1980, when a California condor chick died after being handled and measured by biologists, there was a major uproar, and state officials threatened to cease cooperation with the Fish and Wildlife Service and to block its program for the condor. Because the biologists had, at a stroke, wiped out 3 to 4 percent of the condor population, strong reactions were hardly surprising.

A better example of "normal" problems in this area was enforcement of CITES, which came to bear directly on state management of resident wildlife. Two species, the American alligator and the bobcat, have been particularly troublesome. When CITES placed alligators in Appendix II (their hides were likely to be confused with those of the endangered crocodile), it added a new, and confusing, layer of regulation. Trappers and state officials complained that the rules did not suit either local variations or the needs of the market. Some areas had many alligators, others almost none, and state efforts to take account of this ran into federal and international blanket regulations. Administrative delays in acting on state petitions for harvest and shifting federal policies made matters worse; even the presence of Congressman John B. Breaux of Louisiana on Subcommittee on Fisheries and Wildlife Conservation and the Environment of the House Committee on Merchant Marine and Fisheries did not bring a quick or simple resolution.

Bobcat trapping became a problem in 1977, when the international CITES authority listed all cats (except the domestic variety) as Appendix II species. This, as in the alligator situation, interjected the federal government into an area traditionally handled by the states. Defenders of Wildlife further complicated the situation by suing the Endangered Species Scientific Authority, asking for a halt to the bobcat harvest because the species was endangered. The Court of Appeals for the District of Columbia ruled that CITES required reliable population estimates of the total bobcat population of a state before permits could be granted. Population figures, though, were hard to come by, and the states finally resorted to pushing, successfully, for an amendment to the Endangered Species Act stating that the secretary was not required to make population estimates, or ask the states to do so, for Appendix II species.[49]

There have been a variety of other problems with administration. Captive breeding populations, animals used for sport (as in falconry), zoo transfers, and even specimens for scientific study required permits under the Endangered Species Act; falconers, museum scientists, and zookeepers told of frustrating delays and even threats from the Fish and Wildlife Service's enforcement branch. Low budgets and serious cuts in funding in recent years have only added to the difficulties.[50]

Many of the problems associated with the endangered species program can be traced to a single source: the lack of scientific information on which to base policy. The Congress, quite explicitly and with good reason, directed those enforcing the law to rely on science. What it did not anticipate was the paucity of facts and the difficulty of collecting them. Even finding what species there were and whether they were endangered proved to be far harder than anticipated. One ichthyologist testified during hearings on Tellico that he could go out and find an endangered species to stop any project; there were that many undiscovered organisms out there. Others disputed him, but no one was certain.[51] Even with known species, there was often too little information on population, reproduction, ecological requirements, and behavior to allow policymakers to establish clear and unexceptionable guidelines. The tuna fishermen and their problems with porpoises were the most dramatic illustration of this situation, but there were others. Lacking adequate census data and finding it difficult to decide just what the optimum sustainable population and the maximum sustainable population were, administrators have been making decisions that are open to both legal and scientific questions, and implementation of the program has suffered. The law has a reach longer than its grasp.

WILDLIFE PROTECTION IN THE FUTURE

Controversy over the provisions of the Endangered Species Act goes on for one basic reason: Americans have not reached a consensus on how wildlife should be protected or how much protection it should have. Old attitudes persist, and even among those who espouse wildlife protection, there are great differences about how animals should be protected and for what reasons.[52] Some things, however, are clear. The public is generally in favor of protection for endangered species, and finds the current laws—the Endangered Species Act of 1973, the Marine Mammal Protection Act, and CITES—a generally acceptable framework. Preservation of the natural world is now planted in the structure of the federal government and in the less formal but equally important private organization of society. We have come from the "conquest of nature," a stand that saw man as opposed to nature and that derived national identity from changing nature, to a recognition that wild nature exists at our pleasure, and that if wild areas and wild animals are to be preserved, it will be by our conscious decision and deliberate action. We have, as a society, decided in favor of preservation. Even the Reagan administration, widely viewed as unsympathetic to the program, found it difficult to do more than slow the administrative machinery, and even this provoked serious opposition—President Reagan devoted a week in the summer of 1984 to touring parks and wildlife refuges in an effort to polish his environmental record.

That we are preserving nature and doing it in the particular fashion laid down in the Endangered Species Act is in large part a consequence of scientific information disseminated to the public and the social conditions of postwar America. We see each species as a precious biological legacy, the heritage of millions of years, because we see the world in terms shaped by evolutionary biology. Our program to save all species is dictated by, and draws its inspiration from, ecology. That we have embarked on such an ambitious program is due to our affluent circumstances and relatively high level of education. The process has not ended. We have only begun to answer the old question: How are we connected with nature? We expect to find answers that will inspire our advanced industrial society to acknowledge its connection with nature in the most suitable way.

NOTES

1. Samuel P. Hays, "From Conservation to Environment: Environmental Politics in the United States since World War Two," *Environmental Review* 6 (Fall 1982): 14–41.
2. Michael J. Bean, *The Evolution of National Wildlife Law* (New York: Praeger, 1983),

part 3, traces the legal development of comprehensive wildlife protection programs. Statutes and treaties may be found in U.S. Congress, House Committee on Merchant Marine and Fisheries, *A Compilation of Federal Laws Relating to Conservation and Development of Our Nation's Fish and Wildlife Resources, Environmental Quality, and Oceanography* (Washington, D.C.: Government Printing Office, 1977).

3. The Supreme Court affirmed state authority in 1842, *Martin v. Waddell* (41 U.S. [16 Pet.] 234), developing it further in *Geer v. Connecticut* (161 U.S. 519 [1896]), where it ruled that the states held game in trust for all the people. The doctrine of state ownership enunciated by the court was, legally speaking, quite narrow; *Geer* turned upon a technical point of state power, and the decision was hedged by the qualification that state regulation not interfere with the federal power to regulate interstate commerce. Bean, *National Wildlife Law*, 12–17.
4. *Missouri v. Holland* challenged the constitutionality of this act; the Supreme Court upheld it then and later. Bean, *National Wildlife Law*, 19–21, 68–78.
5. Federal power to control wildlife on its own land, against state authority, was not entirely clear, legally speaking. Bean, *National Wildlife Law*, provides a summary of legal developments on this question, 21–25. See also, on the current situation, Gustav A. Swanson, "Wildlife on the Public Lands," in Howard P. Brokaw, ed., *Wildlife and America* (Washington, D.C.: Government Printing Office, 1978), 428–41. The predator and rodent control program, begun in 1915 and given statutory authority by the Animal Damage Control Act of 1931, was a comprehensive program, but not "wildlife" work. Killing coyotes and prairie dogs was a service to ranchers and woolgrowers. Bean, *National Wildlife Law*, 235–41, sketches the legal aspects of the situation. On the early administration of the program, see Thomas R. Dunlap, "Values for Varmints, Predator Control and Environmental Ideas, 1920–1939," *Pacific Historical Review* 53 (May 1984): 141–61.
6. Michael Bean, *National Wildlife Law*, chapter 2, discusses federal-state developments. Bean, commenting on an earlier draft of this paper, noted the unusual involvement of the Supreme Court.
7. George M. Wright, Joseph S. Dixon, Ben H. Thompson, *Fauna of the National Parks of the United States*, Department of the Interior, Fauna Series no. 1 (Washington, D.C.: Government Printing Office, 1933); Adolph Murie, *Ecology of the Coyote in the Yellowstone*, Department of the Interior, Fauna Series no. 4 (Washington, D.C.: Government Printing Office, 1941); Adolph Murie, *The Wolves of Mt. McKinley*, Department of the Interior, Fauna Series no. 5 (Washington, D.C.: Government Printing Office, 1944). That Murie's work was not completely accepted even within the Park Service is apparent in a letter from Olaus Murie to H. E. Anthony, December 5, 1945, in Murie file, Department of Mammalogy, American Museum of Natural History, New York. Olaus said that Adolph's superiors were trying to get him fired; they did not like the conclusions he had come to—that predators were not a serious problem for the parks' wildlife populations.
8. Senate Special Committee on the Conservation of Wildlife Resources, *The Status of Wildlife in the United States*, 76th Cong., 3d sess., 1940. See Bean, *National Wildlife Law*, 66, on changes in the Fish and Wildlife Service since 1939.
9. *Bird-Lore* is the best detailed guide to the Audubon Society and its connections to the scientific community. On the mammalogists see Dunlap, "Values for Varmints." The ecologists' campaign is covered in Charles S. Kendeigh, "Natural and Wilderness Areas within the National Forests," *Ecology* 22 (July 1941): 330–42, and Robert L. Burgess, "The Ecological Society of America: Historical Data and Some Preliminary Analyses," in Frank N. Egerton, ed., *History of American Ecology* (New York: Arno Press, 1977).
10. Aldo Leopold, *Game Management* (New York: Charles Scribner's Sons, 1933).
11. The charter members of the Wildlife Society were listed in 1938. The list does not give job affiliation, but from the addresses it is clear that many held government jobs. Of 295, 34 can be identified as in the Soil Conservation Service, 28 in the

Bureau of Biological Survey, 29 in state agencies, 18 in the National Park Service, and 29 in forestry. *Journal of Wildlife Management* 2 (Apr. 1938), Supplement.

12. Theodore W. Cart, "The Struggle for Wildlife Protection in the United States, 1870–1900: Attitudes and Events Leading to the Lacey Act" (Ph.D. diss., University of North Carolina, 1971); Robin Doughty, *Feather Fashions and Bird Preservation* (Berkeley: University of California Press, 1975).
13. Senate Special Committee on Conservation of Wildlife Resources, *Wildlife Restoration and Conservation*, proceedings of the North American Wildlife Conference called by President Franklin D. Roosevelt, printed for the U.S. Senate Special Committee on Conservation of Wildlife Resources, 1936. This became an annual conference; succeeding volumes appear as *Transactions of the North American Wildlife Conference*.
14. T. Gilbert Pearson, *Adventures in Bird Protection* (New York: D. Appleton-Century 1937), 232–34.
15. Aldo Leopold, *Report on a Game Survey of the North Central States* (Madison, Wis.: Sporting Arms and Ammunition Manufacturers Institute, 1931); Paul Errington, "Some Contributions of a Fifteen-year Local Study of the Northern Bob-white to a Knowledge of Population Phenomena," *Ecological Monographs* 15 (Jan. 1945): 3–34.
16. *Game Management* and the Leopold papers in the University of Wisconsin Archives testify to Elton's influence in game management. Elton's work is *Animal Ecology* (London: Sidgwick and Jackson, 1927). To see how the field had changed, compare a popular textbook of the 1930s, Royal Chapman's *Animal Ecology* (New York: McGraw-Hill, 1931) with Charles C. Adams, *A Guide to the Study of Animal Ecology* (New York: Macmillan, 1913).
17. Paul Errington, *Of Predation and Life* (Ames: Iowa State University Press, 1967), 235. Errington treats this phenomenon extensively.
18. Murie, *Ecology of The Coyote in the Yellowstone* and *The Wolves of Mount McKinley*; Paul Errington, "Predation and Vertebrate Populations," *Quarterly Review of Biology* 21 (June 1946): 221–45; Frank N. Egerton, "Changing Concepts of the Balance of Nature," *Quarterly Review of Biology* 48 (June 1973): 322–50.
19. Charles Elton, *Animal Ecology* (1927; reprint, London: Meuthen, 1966), vii.
20. A. K. Fisher, *Hawks and Owls of the United States* (Washington, D.C.: Government Printing Office, 1893), was the bible on this topic. On Audubon's defense of birds, see editorial in *Bird-Lore* 32 (Nov.-Dec. 1930): 446, "Campaign for Hawk and Owl Protection," and resolution of the directors of the National Association of Audubon Societies, May 9, 1934, *Bird-Lore* 36 (Sept.-Oct. 1934): 333–35.
21. "Feathered vs. Human Predators," *Bird-Lore* 37 (Mar.-Apr. 1935): 122–26; W. L. McAtee, "A Little Essay on 'Vermin,'" *Bird-Lore* 33 (Nov.-Dec. 1931). McAtee, one of the survey's old-line biologists, was quite interested in this question and in ecological affairs. Leopold became a director of the Audubon Society in 1935.
22. Warren F. Eaton, "Predators and Bird Preserves," *Bird-Lore* 37 (May-June 1935): 162–66. Compare this with the editorial in *Bird-Lore* 32 (Nov.-Dec. 1930): 446, or Mabel Osgood Wright, "Stories from a Bird Sanctuary, II," *Bird-Lore* 24 (Sept.–Oct. 1922): 253–55.
23. Robert Boardman, *International Organization and the Conservation of Nature* (Bloomington: Indiana University Press, 1981), 35–46.
24. Lewis Regenstein, *The Politics of Extinction* (New York: Macmillan, 1975), 30, 87.
25. There have always been tensions within wildlife organizations over animal policy. See Pearson, *Adventures in Bird Protection*, and Rosalie Edge Papers, Conservation Center, Denver Public Library. Thomas Kimball, former executive vice president of the National Wildlife Federation, discussed the problems of holding together a coalition ranging from hunters to vegetarian animal rights advocates, with the author, December 19, 1979.
26. Humane organizations particularly concerned with wild animals, like the Anti-Steel Trap League of the 1920s, had been short-lived and small. See Edge Papers,

Denver Public Library; Vernon Bailey Papers, Smithsonian Institution Archives; "Trap" file, General Files, Records of the Fish and Wildlife Service, RG 22, National Archives. On the early humane movement, see James C. Turner, *Reckoning with the Beast* (Baltimore: Johns Hopkins University Press, 1980); on anti-trapping, John Richard Gentile, "The Evolution and Geographic Aspects of the Anti-Trapping Movement: A Classic Resource Conflict" (Ph.D. diss., Oregon State University, 1983).

27. Regenstein, *Politics of Extinction*; House Committee on Merchant Marine and Fisheries, *Painful Trapping Devices: Hearings before the Subcommittee on Fisheries and Wildlife Conservation and the Environment*, 94th Cong., 1st sess., 1975.
28. Practical reasons to save wildlife were often used. See, for example, testimony of Christine Stevens, Society for Animal Protective Legislation, House Committee on Merchant Marine and Fisheries, *Endangered Species—Part 1: Hearings before the Subcommittee on Fisheries and Wildlife Conservation and the Environment*, 95th Cong., 2d sess., 1978, 301–2. The same kind of claim appeared well before this; mammalogists protesting against what they saw as the extermination of predators in the west in the 1920s said much the same thing. C. C. Adams, "The Conservation of Predatory Mammals," *Journal of Mammalogy* 6 (Feb. 1925): 83–95.
29. Bean, *National Wildlife Law*, 319–21; Senate Committee on Commerce, *Conservation, Protection and Propagation of Endangered Species of Fish and Wildlife: Hearings before the Merchant Marine and Fisheries Subcommittee*, 89th Cong., 1st sess., 1966.
30. *Conservation, Protection, and Propagation of Endangered Species of Fish and Wildlife*; House Committee on Merchant Marine and Fisheries, *Endangered Species: Hearings before the Subcommittee on Fisheries and Wildlife Conservation*, 91st Cong., 1st sess., 1969.
31. Bean, *National Wildlife Law*, 321–24; *Endangered Species*, House, 1969.
32. Senate Committee on Commerce, *Endangered Species Conservation Act of 1972: Hearings before the Subcommittee on the Environment*, 92d Cong., 2d sess., 1972.
33. 16 U.S.C. sec. 1361, in *Compilation of Federal Laws*, 494.
34. Bean, *National Wildlife Law*, 283–317; *Compilation of Federal Laws*, 494–507.
35. Bean, *National Wildlife Law*, 307–12, reviews this problem.
36. Ibid., 283–317.
37. House Committee on Merchant Marine and Fisheries, *Endangered Species: Hearings before the Subcommittee on Fisheries and Wildlife Conservation and the Environment*, 93d Cong., 1st sess., 1973, 7–74; Bean, *National Wildlife Law*, 324–29.
38. *Endangered Species*, House, 1973; Senate Committee on Commerce, *Endangered Species Act of 1973: Hearings before the Subcommittee on the Environment*, 93d Cong., 1st sess., 1973.
39. 16 U.S.C. secs. 1531, 1536; Bean, *National Wildlife Law*, 329–34.
40. House Committee on Merchant Marine and Fisheries, *Endangered Species Oversight: Hearings before the Subcommittee on Fisheries and Wildlife Conservation and the Environment*, 94th Cong., 1st sess., 1975; Senate Committee on Environment and Public Works, *Endangered Species Act Oversight: Hearings before the Subcommittee on Resource Protection*, 95th Cong., 1st sess., 1977.
41. 16 U.S.C. sec. 1536, in *Compilation of Federal Laws*, 388.
42. A chronology of actions on Tellico is in *Endangered Species—Parts 1 and 2*, House, 1978, 764–68.
43. Court ruling in *TVA v. Hill et al.* in *Endangered Species—Parts 1 and 2*, House, 1978, 565–627.
44. *Endangered Species—Parts 1 and 2*, House, 1978, 52.
45. *Endangered Species—Parts 1 and 2*, House, 1978, 47–49. Senate Committee on Environment and Public Works, *Amending the Endangered Species Act of 1973: Hearings before the Subcommittee on Resource Protection*, 95th Cong., 2d sess., 1978, 44–67.
46. S. David Freeman, director of TVA, admitted in testimony that not until May 18, 1978 (well after the controversy had started), did TVA allow the staff to consider

alternatives that did not involve the completion of the dam as planned. *Endangered Species—Parts 1 and 2*, House, 1978, 731–32.

47. Senate Committee on Environment and Public Works, *Legislative History of the Endangered Species Act of 1973, as Amended in 1976–1977, 1978, 1979, and 1980*, 97th Cong., 2d Sess., 1982, 643–46. Tellico continued to be a source of contention. The review committee rejected TVA's plan to finish the dam, ruling that there were "reasonable and prudent alternatives." Congressional supporters then attached a specific amendment for Tellico to the Energy and Water Development Appropriations Act of 1980, and the dam was saved—after several rounds of conferences and arguments. *Legislative History*, 725–1235.
48. House Committee on Merchant Marine and Fisheries, *Endangered Species Act: Hearings before the Subcommittee on Fisheries and Wildlife Conservation and the Environment*, 97th Cong., 2d sess., 1982, 228–64.
49. *Defenders of Wildlife v. Endangered Species Scientific Authority* (No. 79–2512), U.S. Court of Appeals for the District of Columbia; *Endangered Species Act*, House, 1982, 3–11; Bean, *National Wildlife Law*, 380–82.
50. House Committee on Merchant Marine and Fisheries, *Endangered Species: Hearings before the Subcommittee on Fisheries and Wildlife Conservation and the Environment*, 96th Cong., 1st sess., 1979; Senate Committee on Environment and Public Works, *Endangered Species Act Amendments of 1982: Hearings before the Subcommittee on Environmental Pollution*, 97th Cong., 2d sess., 1982; Senate Committee on Environment and Public Works, *Endangered Species Act Oversight: Hearings before the Subcommittee on Environmental Pollution*, 97th Cong., 1st sess., 1982.
51. *Endangered Species Act Oversight*, Senate, 1977, 133.
52. Stephen R. Kellert, *Public Attitudes toward Critical Wildlife and Natural Habitat Issues* (Washington: Government Printing Office, n.d.); Stephen R. Kellert and Joyce Berry, *Knowledge, Affection, and Basic Attitudes Toward Animals in American Society* (Washington: Government Printing Office, n.d.).

7

ADVERTISING THE ATOM

Michael Smith

Compared with the other subjects in this volume, nuclear power is a relative newcomer to environmental policy. Officially, no civilian nuclear program existed before the Atomic Energy Act of 1954. In its short lifetime, however, nuclear power has become one of the most controversial issues in the history of American environmental politics. For roughly the first fifteen years after the 1954 act, the peaceful atom enjoyed widespread popular support, enthusiastic press coverage, and optimistic government predictions of imminent growth. In the late 1960s, questions began to surface, and over the next decade, virtually every aspect of nuclear-generated electricity attracted environmental concern. The mining and processing of uranium; the location, construction, and operational safety of nuclear power plants; thermal pollution from those plants; disposing of radioactive waste; and emergency evacuation plans for communities near nuclear plants all became contested terrain among policymakers, regulators, and their critics. The national news media, once uncritically enthusiastic toward nuclear power, had become openly antagonistic by the time of the accident at Three Mile Island in March 1979. An unofficial moratorium on new nuclear plant orders has remained in place ever since.

What caused the peaceful atom's fall from grace? Is it permanently discredited, or only temporarily deflected? Is nuclear power an environmental blessing, a nightmare, or something in between? Although opposing sides in the nuclear debate tend to give equally brief replies to these questions, the answers are embedded in an intricate history of policy objectives, controversies, and crises that led to the current impasse. Since no brief essay could treat this history in adequate detail, I have chosen instead to address the social context of nuclear power in the United States: to trace the broader political and social pressures that shaped its early development, and to suggest how that pattern of development contributed to the environmental crisis that followed. Examined in this context, nuclear power is above all a product of federal promotional objectives in the 1950s and 1960s. To understand the clash between the atom and the environment in the 1970s, we must begin with the assumptions and goals that shaped early nuclear power policy.[1]

ORIGINS OF FEDERAL NUCLEAR POLICY

On July 18, 1945, Major General Leslie Groves, officer-in-charge of the Manhattan Project, wrote a top secret memorandum to Secretary of War Henry Stimson, who was with President Truman at the Potsdam conference. Groves's memo was a detailed account of the atomic bomb test at Alamogordo two days earlier. "For the first time in history there was a nuclear explosion. And what an explosion!" Groves reported enthusiastically. "The test was successful beyond the most optimistic expectations of anyone."[2]

On that same date ten years later, Atomic Energy Commission (AEC) Chairman Lewis L. Strauss presided over a ceremony marking the first commercial transmission of nuclear-generated electricity. On the velvet-trimmed table beside him, a huge electrical switch stood poised upright between two opposite encasements. If he threw this "two-way switch" in one direction, Strauss explained, it would launch a nuclear submarine. "But when I throw it in the other direction, as I am about to do, it will send atomic electric power surging through transmission lines to towns and villages, farms and factories—power not to burst bombs or propel submarines, but to make life easier, healthier, and more abundant." Reminding his audience that this switch represented "the great dilemma of our times," Strauss grasped the handle and threw it "to the side of the peaceful atom."[3]

The problems attending nuclear energy proved to be far more complex than this simple ceremony suggested, yet Strauss was more correct than he realized when he called his two-way switch a symbol of the social choices surrounding the atom. Despite its dramatic appearance, the switch was only a prop; both the transmission lines and the submarine had already been prepared for activation, regardless of the commissioner's ceremony. Like Strauss's switch, the commission he chaired only appeared to offer separate choices. The same agency that controlled production of the weapons had been assigned supervision of nuclear power—with many of the same aims and methods. No matter which way Chairman Strauss threw the atomic switch, the AEC was compelled to light up the towns and launch nuclear submarines—and to perform both tasks with similar urgency.

Although the environmental controversy over nuclear power did not attract sustained national attention until the 1970s, the seeds of that crisis were sown in the decade between Groves's memo and Strauss's ceremony. Like every other aspect of nuclear policy, safety standards for the development of the peaceful atom were profoundly affected by the prevailing military and political pressures surrounding the Manhattan Project and the postwar arms race. In its first commitment to nuclear technology, the federal government gambled that lavish funding, unprecedented scientific staffing, and absolute secrecy could overcome all technical uncertainties. The dramatic success of that undertaking left a powerful legacy for administrators. Postwar nuclear policy retained the Manhattan Project's penchant for secrecy, its equation of nuclear objectives with national identity, its sense of urgency, and its faith in the ability of government scientists to find last-minute solutions to any technical problems. Each of these tendencies militated against the complex and time-consuming process of establishing and enforcing reliable environmental standards for nuclear power.

After Hiroshima, Congress had to devise a new administrative framework for the development and application of nuclear technology. The Atomic Energy Act of 1946 created a civilian Atomic Energy Commission, to be composed of five presidential appointees. To provide for congressional monitoring of the AEC, the act also authorized a Joint Committee on Atomic Energy (JCAE), consisting of nine senators and nine representatives. A Military Liaison Committee would coordinate AEC policy and Defense Department objectives, while a General Advisory Committee of scientists and technical experts would contribute its own assessment and perspectives.[4]

A primary object of the 1946 act was to maintain the U.S. government's monopoly on the atomic bomb; that goal established an administrative link between nuclear weapons and the peaceful atom. Because nuclear power plants would use and generate fissionable material that could be applied to producing nuclear weapons, Congress assigned control of nuclear power to the same agency that directed the production of nuclear weapons. As a result, nuclear power became bureaucratically attached to the geopolitical status of the military bomb, and the AEC's power to classify virtually any information pertaining to the weapons applied equally to nonmilitary projects.[5]

For the first three years, the AEC concentrated on the production of atomic weapons, forbidding commercial applications of nuclear technology as a security risk. The Soviet Union's first successful detonation of an atom bomb in 1949 signaled shifts in policy emphasis. Alarmed by the rapidity with which the Soviets had broken America's atomic monopoly, and denounced at home for failing to prevent the Communists' victory in the Chinese revolution that same year, the Truman administration committed itself to a greatly accelerated defense budget, with the nation's nuclear stockpile as the centerpiece. The JCAE urged the AEC to proceed with the vastly more powerful hydrogen bomb; the General Advisory Committee strongly dissented, claiming that such a weapon would only trigger a far more deadly arms race. In 1950, the president announced the nation's intention to develop a hydrogen bomb. A renewed sense of emergency and expediency permeated nuclear policy.[6]

The Soviet bomb also set the stage for enlisting nuclear-generated electricity as a new element in atomic diplomacy. To most U.S. officials, the Soviet bomb minimized the utility of nuclear weapons as instruments for settling global disputes. Instead, both sides increasingly sought symbolic deployments, employing nuclear technology as an emblem of military and ideological superiority. The Soviets now had the capacity to develop a nuclear power program of their own. And the United States, as principal architect of the North Atlantic Treaty Organization (NATO, also established in 1949), was under new pressure to share nuclear technology with its allies. Atomic bombs as well as nuclear power plants would soon be within the grasp of Britain and France. Could the United States afford to be the only nuclear nation without an atomic power program? Between 1950 and 1953, the JCAE prodded the AEC toward a redefinition of civilian nuclear power as a necessity, rather than a threat, to national security.[7]

During the early 1950s, several disagreements persisted concerning the nature of the arms race and its relation to nuclear power; yet nearly all disputants agreed, for their own reasons, that an aggressive nuclear power program was desirable. For those who considered atomic energy to be primarily a military necessity for containing Soviet power, civilian nuclear power plants represented a possible source of fissionable material for weapons, as well as a domestic justification for an aggressive nuclear policy. One of the most prominent advocates of this point of view, Dr. Edward Teller, also sounded one of the first warnings about the environmental safety risks accompanying nuclear power. Best known as the principal developer and champion of the hydrogen bomb, Teller also chaired the AEC's Reactor Safeguards Committee. In a letter submitted to the JCAE in July 1953, Teller warned that "no legislation will be able to stop future [nuclear power plant] accidents and avoid completely occasional loss of life." Nevertheless, particularly in light of the "great and increasing need for fissionable materials in the military field," Teller concluded that "the unavoidable danger which will remain after all reasonable controls have been employed must not stand in the way of rapid development of nuclear power."[8]

A number of other administrators and scientists, however, advocated a nuclear power program as a way of transcending rather than augmenting nuclear technology's military image. David Lilienthal, the AEC's first chair, recalled the widespread conviction among early atomic administrators "that somehow or other the discovery that had produced so terrible a weapon simply *had* to have an important peaceful use." Such an attitude, he noted, "led perhaps to wishful thinking, a wishful elevation of the 'sunny side' of the Atom."[9] Glenn Seaborg, one of the codiscoverers of plutonium, was one of a number of Manhattan Project scientists who, in contrast to Teller, devoted their postwar careers to nonmilitary applications of nuclear technology. As chairman of the AEC from 1961 to 1971, Seaborg became the government's most tireless publicist for the peaceful atom.

Perhaps the greatest number of nuclear power proponents viewed nuclear-generated electricity as a propaganda weapon in the ideological struggle between the United States and the Soviet Union. In June 1953, JCAE Chairman Sterling Cole and Vice Chairman Bourke Hickenlooper released a statement calling for American supremacy in the impending "atomic power race." In the "battle for the minds of men" between the free world and "the Soviet atheistic materialists," Cole and Hickenlooper warned, atomic weapons were no longer enough. "It is urgent—and we use the term in its truest sense—for our national warfare and for our national defense" that we deploy the peaceful atom to "show ourselves and the world that the industrial vigor of America continues to lead the way to a decent standard of living."[10] Two months later, when the Soviet Union exploded its first hydrogen bomb only months after the United States had done so, AEC Commissioner Thomas Murray urged AEC Chairman Strauss to announce a civilian nuclear power program to counter the propaganda value of the Soviet bomb. In a speech before a public utilities conference in October, Murray characterized nuclear power plants as comparable to the H-bomb in strategic importance. If the Soviet Union developed nuclear power first, and offered it to developing nations in exchange for their allegiance, Americans would discover that the "nuclear power race," as surely as the arms race itself, was "no Everest-climbing, kudos-providing contest."[11]

Commissioner Strauss represented a group of administrators and members of Congress, mostly Republicans, who shared Murray's view of the importance of nuclear power, but disagreed with the contention by Murray and his allies, mostly Democrats, that the government should develop nuclear power as a public enterprise in the interest of expediency. A former Wall Street investment banker, Strauss firmly believed that nuclear power should be "opened to the genius and enterprise of American industry." The government might provide generous subsidies and access to government-developed technology, but private utilities should develop nuclear power as a showcase for the free enterprise system. The JCAE tended to share Strauss's view; its hearings on "Atomic Power Development and Private Enterprise" in the summer of 1953 were designed primarily to demonstrate the feasibility of a government-subsidized, privately developed nuclear power program.[12]

Thus by late 1953, nuclear administrators and advisers with a wide range of differing priorities agreed that nuclear power should be developed as quickly as possible. In December, President Eisenhower announced in a speech to the United Nations that the United States would embark on a new, international "Atoms for Peace" program that would reverse the common perception of atomic energy as a destructive force. In an effort "to serve the needs rather than the fears of mankind," America's nuclear technology would "provide abundant electrical energy in the power-starved areas of the world" as well as at home.[13]

The revised Atomic Energy Act of 1954 authorized the AEC to encourage, rather than forbid, the ownership, construction, and operation of nuclear power plants by private companies; such plants, however, had to be licensed by the AEC, and the government retained its technical ownership of all nuclear fuels. In effect, the 1954 act assigned two contradictory tasks to the AEC—promotion and regulation of nuclear power.

It was a lopsided mandate. For the next twenty years the JCAE, the Congress, and nearly all the AEC commissioners pushed for rapid development and promotion. Regulation might impede that development—particularly if it imposed lengthy test periods or costly safety procedures. In Commissioner Strauss's words, "The AEC's objective in the formulation of the regulations was to minimize government control of competitive enterprise."[14] The only political threat to minimal regulation was adverse public opinion toward nuclear power. The best insurance against public criticism was aggressive promotion of the benefits of the peaceful atom. Given its bureaucratic origins and the political pressures surrounding it, the AEC discovered that the public opinion was easier to regulate than the nuclear industry itself.

The 1954 act also increased the JCAE's powers. The only joint committee with legislative authority, it controlled virtually every bill in either house that related to atomic energy. The usual disagreements and alterations that occurred between committees, and between the House of Representatives and the Senate, were missing from nuclear issues. The JCAE also constituted the only informed overseers of the AEC. Members of Congress, however, were even more vulnerable to the political pressures for swift development than the commissioners. Having designed the 1954 revisions that made a nuclear power industry possible, the JCAE continued to press the AEC for an aggressive program and tangible results.[15]

For the first few years after passage of the 1954 act, hesitation over nuclear power was more likely to come from industry than from the government. Three tiers of companies were involved. Most influential were the vendor corporations, notably Westinghouse and General Electric, which had dominated earlier, defense-related contracts for nuclear reactors. These companies saw the construction and sale of nuclear power plants as a lucrative new market. Next came the private utility companies that would buy the plants and sell the electricity. With other cheap, plentiful sources of energy available, nuclear power seemed less attractive to these companies without strong incentives from the government and the vendor companies. In the early years, the utilities were drawn to nuclear power more out of fear that the government would operate it as a public utility than out of eagerness to develop it. Finally, there were the companies indirectly affected by the emergence of a nuclear power industry. Most significant among these were the insurance companies, which found the risks attending nuclear power too uncertain to warrant coverage. Without insurance, the utilities could not proceed.[16]

The AEC responded to the insurance problem by instructing its scientists at Brookhaven National Laboratory to undertake a detailed study of nuclear power plants. Over their brief period of operation, nuclear reactors had logged a good safety record, but no one had established how likely an accident might be, or how serious the consequences. In 1957, the Brookhaven staff reported its findings to the commissioners in a report titled "Theoretical Possibilities and Consequences of Major Accidents in Large Nuclear Power Plants," but commonly referred to by its document number: WASH–740. Their report hypothesized that an accidental release of radioactive material from a nuclear plant could cause 3,400 deaths, 43,000 injuries, and $7 billion in damage. The Brookhaven scientists were unable to determine the likelihood of such an accident, but deemed it only possible, not probable.[17]

No safety evaluations eventuated from WASH–740. Instead, the JCAE drafted the Price-Anderson Act of 1957, which exempted the nuclear power industry from lawsuits for injuries sustained in a nuclear accident. In place of corporate liability, Congress set aside an arbitrary sum of $560 million. By the AEC's own estimate, this amount could not begin to cover damage claims from a significant nuclear accident. The figure was large enough to remove the issue of liability as an obstacle, yet small enough to prevent public alarm at the anticipated scope of accidental damage.[18]

WASH-740 had based its estimates on a hypothetical nuclear power plant with a power capacity of 185 megawatts—about twice the size of the nation's first nuclear plant in Shippingport, Pennsylvania, which began operation in 1957. Nuclear plants in this first generation, to paraphrase Mark Twain, were to their larger successors as the lightning bug is to lightning. By the mid–1960s, Westinghouse and General Electric were proposing 1,000–megawatt plants. With the Price-Anderson Act due for renewal in 1967, the AEC asked Brookhaven in 1964 to reexamine the WASH-740 findings in light of the newer, larger reactor proposals. Expecting more favorable results than those in the original report, the commissioners found the updated estimates distressing. A major accident at one of the new reactor plants might cause 45,000 deaths, according to the Brookhaven staff; property damage could approach the $400 billion mark. And although probability was impossible to calculate accurately, the new report found "nothing inherent in reactors or in safeguard systems as they now have been developed which guarantees either that major reactor accidents will not occur or that protective safeguard systems will not fail."[19]

After discussing with industry representatives the negative public relations impact of such findings, the AEC elected to classify the WASH-740 update as secret and to approve applications for the new plants. No significant changes in reactor safety accompanied these actions. Instead, the AEC's annual reports to Congress from 1954 to 1969 reflected the political priorities and public concerns of the time. By and large, operational safety and environmental protection received much less attention than the educational and informational programs by which the AEC promoted nuclear power. These promotional efforts illustrate the political and social assumptions underlying the emblematic role of nuclear power in its formative years; ironically, they also help explain the emergence of grassroots environmental opposition to nuclear power in the 1970s.

ATOMIC ADVERTISING, 1954–1969

With the advent of a federal nuclear power program, the AEC undertook the most extensive government public relations effort since the Office of War Information had closed its doors in 1945. A positive image of nuclear power was important for several reasons. Members of Congress and their constituents had to be assured of the wisdom of continuing appropriations; utility companies needed to be persuaded to participate in the government's nuclear power program. Most important, however, was the emblematic nature of atomic energy policy in general. Like nuclear weapons, nuclear power had been assigned a crucial symbolic role by its proponents. Nuclear power plants were to serve as badges of American technological and political superiority at home, and as adjuncts of atomic diplomacy abroad. The display value of nuclear technology could be only as effective as the publicity surrounding it. From the outset, then, the question was not whether to promote nuclear power aggressively, but how best to go about it.

Officially, as well as informally, disciples of the nuclear gospel sought each other's advice on how best to proceed. Just after Eisenhower's election in November 1952, former AEC Commissioner T. Keith Glennan proposed "a national association of atomic industries" to encourage "the development and utilization of the peaceful applications of atomic energy in accordance with the best traditions of the American system of free competitive enterprise."[20] The following spring, Glennan's suggestion resulted in the formation of the Atomic Industrial Forum (AIF). In the years to come, the AIF provided one of the key settings for government-corporate cooperation in nuclear power development and promotion.

In 1956, the AIF sponsored the first national conference on public relations for nuclear power. Its proceedings offer an instructive glimpse of nuclear power advocacy in its formative stage. AIF Executive Manager Charles Robbins called the conference to order with the question of the hour: "How do we overcome the doubts and apprehensions of the wartime atom and replace these with confidence and a ready acceptance of peaceful atomic enterprise?"[21] AEC personnel addressed the conference on the progress of reactor development and discussed promotional materials provided by the federal government. Frank Pittman, the AEC's deputy director of the Division of Civilian Application, suggested that "in its civilian uses the energy of the atom should be described as 'nuclear energy' rather than 'atomic energy.' This might help eliminate the fearful feeling that is brought to the minds of many members of the public by the word 'atomic.' " Pittman also warned his audience that "it must be made clear that there is no such thing as an absolutely safe operating nuclear reactor, just as there is no such thing as an absolutely safe operating chemical plant, oil refinery, automobile, airplane, or anything else."[22]

At least one speaker commented on the diversity of motives for developing nuclear power. *Nucleonics* magazine editor Jerome Luntz noted the degree to which "atomic energy" had permeated American culture, with no clear sense of its nature or purpose. The Manhattan telephone directory for 1956, for example, contained "41 entries with the word 'atom' or 'atomic,' going from Atom Fuel Company, which is probably a coal company, to Atomic Undergarment Company." Part of the uncertainty, Luntz speculated, might be due to "the confusion among our own top level atomic energy policy makers as to why we are developing atomic power and how fast we should move in that development." Some spoke of atomic power's "propaganda" value, or of opportunities it provided to "make gains in the cold war," Luntz noted. For his own part, the editor felt that the most important issue was making nuclear power profitable.[23]

Luntz's comments provided a rare moment of candor at the public relations conference. Presumably, all the industry representatives shared his preference for profits first, geopolitics later. As is often the case when differing motives surround similar goals, the various advocates tended to speak in each other's language. Many of the participants from nuclear and public relations industries spoke as if they were in the State Department. LeBaron Foster, editor of *Public Opinion Index for Industry*, attempted to bridge economic and ideological motives, explaining that the nuclear power question was "really a problem of preserving our system of values." One sure sign of alien ideologies polluting the nuclear issue, Foster maintained, was public support for government ownership of nuclear power plants. In a public opinion study entitled "Free Market versus Socialist Thinking,"

Foster's Opinion Research Corporation had determined that blacks, unskilled workers, those with only an eighth-grade education, low-income individuals, and the foreign-born were in the "strongly socialistic" column (because they approved of public ownership), whereas business executives, upper-income individuals, Republicans, college teachers, and government employees favored the "free market" (that is, government subsidies without government control). The nuclear industry's task would be to explain its special needs to this unevenly informed public.[24]

The public relations community's message to the assembled industry and government representatives was one that would only grow in significance in the decades to follow: it is easier to develop public "tolerance for accidents" than to avoid the accidents themselves—more expedient to offer strategies for "Developing Consumer Acceptance of Radioactivity" than to establish and enforce radiation standards. Nearly all the public opinion surveys cited during the conference revealed a general public willingness to embrace nuclear power. Nonetheless, one of those same surveys warned that "'danger' lies not simply in the presence of a hazard," but also in the loss of public faith in the problem-solving ability of experts. "In all likelihood, it would take but one highly dramatic and well publicized event—a major plant catastrophe—to upset this faith."[25]

The government and the nuclear power industry thus faced potentially conflicting choices: to maintain safety standards at possible risk to public confidence, or to maintain appearances at possible risk to safety. The AEC pursued both safety and public relations, but found the latter much easier and more welcome. By the late 1950s, reactor safety guidelines and nuclear waste disposal policy remained unresolved; but the AEC and the industry were already well embarked on a massive campaign to celebrate the peaceful atom.

To create a positive public image of nuclear power, the AEC relied on two closely related techniques: secrecy and advertising. Supervision of nuclear weapons had already conditioned the AEC to classify information routinely; as the WASH-740 update demonstrates, the commission simply transferred that procedure to the peaceful atom, suppressing documents and accident reports that questioned the safety or economic feasibility of nuclear technology. What not to say seemed clear enough; the new problem was what to present to the public, and how to reach it. Through booklets, films, press releases, lectures, and exhibits, the educational arm of the AEC undertook to reach the entire nation, and much of the rest of the world.

It proved to be a highly successful campaign. The AEC's educational pamphlets enjoyed nationwide distribution. One series alone—the *Understanding the Atom* booklets—began to appear in 1952; by the end of the decade, over eight million copies were in print. Between 1960 and 1970, over forty million people attended AEC screenings of films, and nearly four times that number viewed them on television. Traveling exhibits, high school lecture-demonstrations, science fairs, and even Boy Scout merit badges in atomic energy brought the message to tens of millions of school children. In addition, the AEC participated in joint promotional efforts with the AIF and other trade associations. The major nuclear vendors were quick to mount their own publicity drives; eventually, they were joined by the private utility companies. From the mid–1950s to the late 1960s, public expectations for the peaceful atom grew and flourished in the bright light of nuclear publicity.[26]

Because the peaceful atom represented a new and complex technology, atomic advocates were free to characterize it in almost any way they chose. The images they selected warrant scrutiny, for they had a profound effect on public attitudes toward nuclear power, on AEC policies, and even on cultural attitudes toward technology and expertise in general. Like the AIF, the AEC perceived the need to dissociate the peaceful atom from the bomb. One of the earliest—and most quoted—examples of this tendency was the speech delivered by AEC Chairman Strauss in September 1954. Within fifteen years, Strauss predicted, the peaceful atom would lead us to "transmutation of the elements—unlimited power. . . . It is not too much to expect," he claimed, "that our children will enjoy electrical energy too cheap to meter . . . will travel effortlessly over the seas and under them and through the air with a minimum of danger and at great speeds—and will experience a lifespan far longer than ours."[27] This was more than Commissioner Lilienthal's "sunny side of the Atom." Strauss was invoking a vision of technological optimism, grounded in the magical properties of the atom. Words like "magic," "wonder," and "mystery" appeared routinely in the titles of countless AEC booklets, films, and magazine articles. In 1955, President Eisenhower activated the nation's first commercial nuclear power plant by waving a radioactive "magic wand." "Unlimited power" might, however, have frightening implications if the magician ever lost control. The task facing nuclear publicists was to invoke the magical possibilities of the atom while at the same time domesticating it, reducing its mysterious powers to the palpable dimensions of daily life.[28]

One of the most vivid and most widely viewed examples of this domesticated magic appeared in *Our Friend the Atom*, a Walt Disney production that first appeared on nationwide television in 1956 and soon became one of the most popular items in the AEC film library. Disney assures his listeners that "the atom is our future," and introduces us to Dr. Heinz Haber, Disneyland's new expert on "scientific development." Haber says that the development of nuclear technology has been "almost a fairy tale," illustrating his point with a cartoon update of the Arabian Nights tale of the genie and the fisherman. In the Disney version, a fisherman finds a magic bottle in his nets. When he uncorks it, a fierce, slant-eyed genie appears in a huge mushroom-shaped cloud. Furious at having been pent up for so long, the genie announces that he will destroy his liberator. The fisherman uses verbal trickery to recapture him. Chastened, the genie reemerges to grant his new master's wishes. In case the viewers missed the point, the film superimposes the genie onto reverse footage of an atomic explosion, so that the bomb appears to implode harmlessly back into the lamp. Atomic power was thus domesticated—in the form of a powerful but docile servant.[29]

Like Strauss's two-way switch, or Ike's wand, Disney's genie converts the fury of the atom into the power to perform tasks. What should those tasks include? Readers of the August 1955 issue of *Ladies' Home Journal* learned of an impending nuclear world

> in which there is no disease . . . where hunger is unknown . . . where 'dirt' is an old-fashioned word and routine household tasks are just a matter of pushing a few buttons . . . a world where no one stokes a furnace or curses the smog . . . and the breeze from a factory is everywhere as sweet as from a rose. . . .

The author of that article was President Eisenhower's special assistant on disarmament, Harold Stassen—a man seldom praised for his ability to predict the future. Stassen carefully tailored his image of nuclear magic to include "routine household tasks" as well as global transformations.[30]

Those household tasks became the focus of the atom's domesticated magic in *The Atom and Eve*, a 1965 film produced by a consortium of New England utility companies and distributed through the AEC. *The Atom and Eve* traced the parallel development of the electrical industry and "little Eve," who was fortunate enough to be born into a modern-day "electrical garden of Eden." As she approaches puberty, Eve acquires an expanding array of electrical appliances; she also eats from the tree of electrical knowledge ("Eve learned a new word: hydroelectricity"). As Eve reaches maturity, the electrical industry grows with her, finally developing nuclear technology "to meet Eve's never-ending needs." Fully grown and draped in a flowing, low-cut evening gown, Eve twirls from one nuclear-generated electrical appliance to another, lovingly embracing the refrigerator and the electric range, lying supine against the smooth expanse of the electric washer-dryer. Eve's coming-of-age culminates in her Dance of the Nuclear Light Bulb. Destined to burn for "a million years" with its inexhaustible nuclear storehouse of energy, the bulb becomes consumer culture's eternal flame, which Eve caresses as she spins and pirouettes across the screen.[31]

Electricity for household appliances was not the only way in which nuclear enthusiasts proposed to alter Eve's domestic realm. In a feature article titled "You and the Obedient Atom," the September 1958 issue of *National Geographic* echoed Disney in its observation that "these unimaginably tiny particles work like genii at man's bidding." Visiting the AEC's various national laboratories, the magazine staff found even the smallest details of daily life were being transformed by this new servant. Brookhaven National Laboratory sought to create bold new hybrids of carnations in its radioactive "Gamma Garden." Unfortunately, some less pleasing mutations, such as two-colored "schizophrenic carnations" produced "flowers of evil" alongside the "flowers of good." Meanwhile, researchers at Argonne Laboratory were testing the food of the future. Potatoes, bread, and hotdogs left exposed to air for several months—predictably—rotted. Yet irradiated samples of these same foods remained fresh and "germ-free." Argonne reported that "changes in taste [in irradiated food] are scarcely noticeable." Samples had already been fed to rats, military volunteers, and congressmen "without harmful effect." *National Geographic* concluded that "the atomic revolution" would "shape and change our lives in ways undreamed of today—and there can be no turning back."[32]

In addition to the AEC's projects and promotional efforts, the very language it used emphasized both the power and the subservience of "the obedient atom," often through the familiar analogy of parent and child. In February 1966, the AEC commemorated the twenty-fifth anniversary of the "birth" of plutonium with a highly publicized ceremony. The commission's press release described how the element's five fathers had gone about creating their "new baby." (Clearly, this "nuclear family" remained decidedly patriarchal.) Glenn Seaborg, one of the five and now chairman of the AEC, minimized the danger and difficulty involved with plutonium by characterizing it as "the ornery element"—a "difficult child" among the older, more sanguine elements. "It may be because I am so well acquainted with plutonium's early childhood that I am inclined to view it as one might consider a 'bad' child," he observed in the press release for the commemoration, "—difficult and even exasperating in its conduct at times, but replete with fascinating possibilities." For Seaborg, the childlike qualities of the atom permitted its master to recapture his own youthful sense of wonder. "In a homely sense," he wrote, "the atom is like one of those old many-sided jackknives that can do almost anything a kid would want to do—whittling, screwdriving, bottle-opening and so on."[33]

The language of nuclear promotion deleted most of its cold war references whenever it moved from JCAE hearings and AEC meetings to press releases, information booklets, and films. Both versions of the atom were intended to represent "our system of values," as LeBaron Foster put it. Nonetheless, while most government officials continued to see nuclear power as a symbol of national prestige, the public information offices of the AEC and the nuclear industry recast those values into a more accessible setting: household consumer goods powered by cheap, abundant nuclear electricity. In part, these more personalized depictions of nuclear prestige reflected atomic energy's entry into the parlance of the marketplace: utility companies, as well as their customers, had to be convinced of the need for this new source of electricity; and the principal nuclear vendors, General Electric and Westinghouse, marketed electrical appliances as well as nuclear power plants. But the AEC gained more than corporate interest by applying one set of nuclear images on Capitol Hill and another on Main Street. Trading the geopolitical lance and shield for a washer-dryer or an atomic jackknife might sacrifice some of the global stature of nuclear technology, but it had the advantage of distancing its audience as far as possible from the bomb and its attendant lexicon of national security. For most Americans, this translation of "unlimited power" from the international arena to the household removed the atom from a social environment in which they felt helpless, to the familiar trappings of a consumer culture where their choice of products conveyed the illusion of control. For those developing nations to which the AEC brought its exhibits and films, nuclear-powered appliances provided a vision of abundance to associate with American largesse.

The domesticated magic of nuclear power promotion, however, retained a certain Sorcerer's Apprentice quality; it was as if the endless proliferation of appliances and benefits would finally allay any fears that the nuclear genie might burst from its lamp again, or that the fisherman might be incompetent. In the 1950s and 1960s, such fears focused on nuclear explosives, rather than on nuclear power. As public concern over fallout and the arms race mounted, promotional images of a nuclear Garden of Eden offered a comforting respite from the perils of the military atom. The AEC went even beyond nuclear power plants in its efforts to domesticate the atom. During the same years that it launched its promotion of nuclear-generated electricity, the government sought ways of applying the bomb itself for civilian uses through Project Plowshare. An examination of that project will help to clarify the relation between the AEC's early promotional and environmental priorities, since it demonstrates, in David Lilienthal's words, "how far scientists and administrators will go to try to establish a nonmilitary use [for the atom]."[34]

PLOWSHARE AND "PLANETARY ENGINEERING"

The first environmental issue over which the AEC confronted significant public dissent was fallout from nuclear weapons testing. The controversy began in the mid-1950s, when information questioning the environmental safety of the tests first became available, and subsided after the 1963 signing of the U.S.-Soviet Limited Test Ban Treaty, which forbade atmospheric testing. In the course of the debate, it became apparent that the AEC had underestimated, sometimes dramatically, the potential environmental and health effects of radioactive fallout. Part of the government's difficulty derived from lags in scientific procedure. The AEC based its calculations of public exposure, for example, on the assumption that fallout would disperse equally throughout the global atmosphere; more advanced understanding of world climate patterns revealed uneven concentrations in the temperate zones. In addition, the controversy challenged the credibility of the AEC itself on three counts. First, the government had not arranged for adequate monitoring and follow-up testing of radiation exposure in the areas where weapons were detonated (primarily at the Nevada Testing Grounds and on island sites in the Pacific). Second, the AEC had approached radiation danger almost exclusively as a matter of short-term, external exposure, despite growing evidence of the long-term hazards of internal exposure, such as concentrations of strontium 90 and iodine 131 in the food chain. Finally, the AEC attempted to suppress evidence of fallout-induced health problems among exposed persons and animals, and classified or tried to discredit scientific studies suggesting that official exposure standards were unsafe.[35] At the same time, at least one AEC laboratory explored civilian engineering tasks for thermonuclear explosives.

When President Eisenhower announced his Atoms for Peace program in December 1953, he characterized nuclear-generated electricity as a means of converting the "greatest of destructive forces" into "a great boon for the benefit of mankind."[36] Project Plowshare (1956–1971) essentially pushed claims for the peaceful atom one step further by approaching nuclear weapons as a tool for large-scale environmental modification, such as excavation, mining, and drilling. The idea first gained support during the 1956 Suez Canal crisis, when scientists at the AEC's Lawrence Radiation Laboratory (LRL) in Livermore, California, proposed blasting a new canal across Israel with nuclear explosives. The crisis passed before any tests could be initiated; the idea, however, survived to attract various levels of AEC sponsorship for the next fifteen years.

The evolution of Plowshare was to some degree a product of the geography of AEC decision making. Like the other AEC national laboratories, the LRL staff could submit projects to Washington for approval. They saw themselves as competing with their more conventional counterparts at Los Alamos for funding and staffing. For its successful proponents, Plowshare could open a new realm of project supervision.[37]

Beyond this bureaucratic incentive, scientists and administrators brought a variety of related objectives to their support for the peaceful bomb. Edward Teller championed a broad and vigorous thermonuclear applications policy as an issue of national security. To people concerned about fallout from Plowshare experiments, he maintained that the upcoming generation of nuclear explosives would be safer than their predecessors. (Elsewhere, Teller argued that radiation from fallout "might be slightly beneficial" and that any mutations it might induce should be welcomed for accelerating the evolutionary process.)[38]

Other scientists initially embraced nonmilitary uses of nuclear explosives as a way of redeeming nuclear technology as a social force. Speaking of Project Orion, a plan for a spacecraft propelled by nuclear explosions, Freeman Dyson wrote: "We have for the first time imagined a way to use the huge stockpiles of our bombs for better purposes than for murdering people."[39]

Even from the outset, however, an overriding concern regarding nonmilitary use of nuclear explosives was their public relations impact. Harold Brown, who first submitted LRL's proposal for Plowshare, wrote to Herbert York in 1957: "It is plain that many of the difficulties involved in non-military uses [of nuclear explosives] are not technical but have to do with questions of public opinion."

Peaceful applications of these weapons might generate local "nuisance problems"; yet "such programs if successful might produce a change in attitude." After all, who would object to nuclear weapons testing (or nuclear power plants) in the vicinity if mines, harbors, canals, and even cities were being carved from the earth with thermonuclear bombs? Plowshare might set the stage for "an altered public relations program"—one that publicized projects rather than conceal them.[40]

Perhaps the most enthusiastic defender of Plowshare was AEC Chairman Glenn Seaborg, who envisaged nuclear technology as the key to a "three-dimensional" future in which people would "cut the umbilical cord that ties us precariously to [the Earth's surface]" and begin to populate the moon, the ocean floor, and the Earth's interior ("The Nether Frontier").[41] Seaborg saw Plowshare as part of a more immediate program of "planetary engineering," whereby nuclear explosives and other nuclear technologies would irrigate the deserts, remove mountains, carve "instant harbors" and canals, desalinate ocean water, and through the benefits of "Hardiman," a nuclear-powered "man-amplifier," make an ordinary man as powerful as a forklift. Like "Paul Bunyon and Babe, his Great Blue Ox," Seaborg observed, we could transform our surroundings with a well-placed (nuclear) gesture. Nuclear excavation would be much quicker and cheaper than conventional methods, and "the dangers of radioactivity from the new, *nearly fission-free* explosives are small."[42]

Project Plowshare's various projects highlighted a number of political and environmental issues. The Suez Canal crisis prompted interest in a second, sea-level canal in Panama: a "Panatomic Canal" to absorb increased shipping traffic and to minimize the national security risk of relying on a single interoceanic canal. The AEC began investigating the Isthmus of Panama for likely canal sites as early as 1956. Participants in the Second Plowshare Symposium in 1959 were assured that "the cost of excavating a canal by nuclear means is only a small fraction—[in this case] only some 16 percent—of conventional costs." Marine biologists argued, however, that a sea-level canal would create disastrous ecological imbalances by inducing contact between the separate Atlantic and Pacific marine species. But other biologists disagreed, and Seaborg observed that "tempers are short and arguments long, and great advantages of a new canal to the human species are often underestimated or forgotten in the heat of conflict."

Seaborg conceded that there was a further difficulty with projects such as the Panatomic: "Nuclear excavation might entail the temporary removal of tens of thousands of people at the very least. Such operations," he lamented,

> combined with fears of radioactivity and of induced earthquakes, could lead to considerable apprehension, regardless of the great saving to world commerce and substantial benefits a canal could bring to the area.

After $17 million in feasibility studies, the Panatomic Canal was finally shelved in 1970.[43]

Private industries found the AEC's claims for nuclear excavation sufficiently enticing to propose some projects of their own. One of the more ambitious ideas was the North American Water and Power Alliance, a proposal submitted in 1964 by an engineering firm to create a coast-to-coast waterway across the United States by nuclear blasting, particularly through the Rockies. Nuclear-powered water pumps would sustain the transcontinental flow, with the added advantage that thermal pollution from the pumps would keep the waterways ice-free.[44]

More often than not, Plowshare projects became entangled in a web of political considerations. In 1963 the Santa Fe Railroad proposed the use of Plowshare explosives to crease a "Nuclear Right-of-Way" for rail lines across the Bristol Mountains between Barstow and Needles in California. The project (dubbed "Carryall" by the AEC), like all Plowshare undertakings after 1963, was hindered by the new Limited Test Ban Treaty, which forbade above-ground nuclear explosions. Critics claimed that such projects were disguised efforts to continue weapons testing. But Carryall's fate was further complicated by support from Secretary of the Army Stephen Ailes. In a December 1964 letter to Seaborg, Ailes linked Carryall to the success of the Panatomic Canal. "All of us involved in the sea-level canal studies have recognized from the outset that a major nuclear excavation project would have to be carried out within the U.S. before we would ask another country to permit one on its territory," Ailes explained. Presumably because of its relative isolation and private industrial support, "The Carryall Project appears to be ideal for this purpose."[45]

Project Plowshare never got beyond a series of cratering experiments. The lifespan of Plowshare, however, covered crucial years in the development of AEC policies (roughly 1956 to 1971). Its history reveals several assumptions underlying these policies. Like all peaceful applications of the atom, Plowshare was important for its public relations impact on the cold war, on military nuclear policy, and on government use of nuclear technology in general. As image and credibility became increasingly significant in postwar policymaking, all nuclear technology served as a weapon of impression management.

More important for future nuclear policy was the relationship that Plowshare revealed between the AEC's imperatives for aggressive development and its recognition of environmental hazards. In the mid–1950s, of course, information concerning the environmental risks accompanying radiation remained inadequate and poorly circulated, even within the AEC. But in 1970, the commission's chairman still looked favorably on plans to detonate thermonuclear devices in densely populated areas to reduce industrial excavation costs.[46] In such a climate, how could the commission be expected to attach much significance to the far less dramatic environmental consideration of a nuclear power plant?

Ironically, Plowshare's anomalous bureaucratic status led the AEC to conduct some of its first significant environmental impact studies, which helped to establish the project's infeasibility. Project Chariot, an LRL proposal to create a harbor on Alaska's north coast by nuclear excavation, is a case in point. Ordinarily, the proposals of the weapons scientists, couched in national security mandates, were approved with little or no resistance by the AEC. In the case of Chariot, however, the scientists could not invoke military necessity, because the project called for nonmilitary use of nuclear explosives. Against the strenuous opposition of the project's LRL sponsors, the AEC ordered a preliminary assessment of the effect of Chariot on the region's ecosystem. The environmental scientists assigned to the task were so alarmed by their estimates that they were instrumental in organizing opposition to the proposed blast. As Irvin Bupp has noted,

> neither the Commission nor the laboratory knew how to proceed when it was unable to justify an undertaking on grounds of national security. All prior Commission experience with opposition to its nuclear testing activities had been conducted in the tacit understanding that opposition could, in the end, simply be ignored.[47]

In October 1969, Alaska's Senator Mike Gravel joined Senator Edmund Muskie in introducing legislation to establish a commission to examine the environmental effects of Plowshare projects. In that same month, two LRL scientists, John Gofman and Arthur Tamplin, announced that, according to their findings, permissible radiation exposure levels were twenty times more carcinogenic than their AEC employers claimed. Less than three months later, President Nixon signed into law the National Environmental Policy Act. That inchoate cluster of social forces known as the environmental movement had caught up with Plowshare and with the rest of the government's nuclear energy policies.

THE RISE OF ENVIRONMENTALISM

In the 1970s, nuclear power stood squarely at the intersection of two of the decade's prevailing concerns: energy and the environment. Born in an era of cheap energy and a booming economy, the nuclear power industry encouraged rapidly increasing consumption of electricity, while warning of future scarcity as a result of that consumption. The 1973–74 Arab oil embargo seemed custom-made for strengthening the industry's prospects. President Nixon's Project Independence called for nuclear power to provide 50 percent of the nation's electricity by the end of the century. The industry and the AEC faced, however, a simultaneous environmentalist challenge that threatened to counterbalance the gains promised by the energy crisis. The manner in which government and industry responded to this challenge was in many ways as significant as the substance of the debate.

At an AIF-sponsored conference on "The Nuclear Controversy" in 1972, AEC and industry representatives pointed to 1969 as the year the trouble began. "If we look back at the national controversy over nuclear power in the United States," the first speaker noted, "we don't have to look far. There was none on the national level before 1969."[48] In January of that year, *Sports Illustrated* published an article criticizing the effects of thermal pollution from nuclear power plants on fish and marine life. In that same year, two books attacking reactor safety and nuclear hazards captured national attention: *The Careless Atom* by Sheldon Novick and *Perils of the Peaceful Atom* by Richard Curtis and Elizabeth Hogan. *Esquire* published an article based on Dr. Ernest Sternglass's claims that the risks of low-level radiation had been dramatically underestimated. In October, the JCAE began the hearings at which Gofman, Tamplin, and other scientists challenged the AEC's radiation standards. A public forum in Burlington, Vermont, brought AEC and industry representatives face to face with their critics for a highly publicized debate. The environmental implications of nuclear power had become a national issue.[49]

At the JCAE's hearings on "Environmental Effects of Producing Electric Power" in late 1969 and early 1970, several commissioners and members of Congress made it clear that they considered more than technical issues to be at stake. AEC Chairman Seaborg warned the JCAE that a wave of "unsubstantiated fear-mongering" accompanied the public's recent interest in the environment. "In the years ahead," he predicted, "today's outcries about the environment will be nothing compared to cries of angry citizens" in future energy-starved cities, where prolonged blackouts would leave people to "shiver while imprisoned in stalled subways" in a world of "spoiled food," "darkened skyscrapers," and paralyzed police and medical services. "Fortunately," Seaborg observed, "most people are not willing to sit in the dark." AEC Commissioner James T. Ramey noted that public meetings on nuclear power issues always attracted "some professional 'stirrer-uppers'" and "phonies" who criticized AEC policy. When a scientist testified that the long-term effects of low-level radiation were unknown and suggested that the burden of proof should rest with the AEC, JCAE Chairman Chet Holifield asked, "Can we live in a society where every fact has to be proven?"[50]

Some commissioners and nuclear industry leaders viewed the environmentalist challenges to nuclear power as part of a nationwide epidemic of distrust of technology and government in general. At a January 1970 conference on Plowshare, AEC Commissioner Theos J. Thompson warned that "a danger that is perhaps greater than that of the threat to our environment" was the "growing rejection of those American ideals that have made this relatively young country a leader in the world today." To oppose nuclear technology because low-level radiation might be detrimental struck Thompson as woefully enervated: "It is as though we decided not to get out of bed anymore because we might slip on the way to the bathroom." Just as development of nuclear power had been equated with national identity, so the commissioner saw criticism of nuclear technology's environmental safety as an attack on "the American philosophy of life."[51]

General Electric nuclear executive Bertram Wolfe elaborated on Thompson's argument, charging that antinuclear critics had a "hidden agenda," manipulating antinuclear sentiment to attack corporate capitalism, centralized power, a vigorous growth economy, and the culture of abundance. Antinuclear environmentalists, Wolfe maintained, felt "that energy should be used as a means for societal change not directly connected with energy." Not just nuclear power, but "the whole structure of our society" was the target, Wolfe explained; "it's another Jonestown they're talking about."[52]

With fifteen years of experience in inspiring public confidence, why did nuclear advocates interpret specific environmental criticisms as general attacks on technology, government, and "American ideals?" No doubt some antinuclear critics did want to see broader social changes; but the AEC itself was so accustomed to approaching nuclear power as an ideological emblem that it repeatedly did what it accused its opponents of doing: judging technical issues on political grounds. The AEC was also slow to acknowledge that the most compelling environmentalist criticism came from AEC scientists, nuclear industry engineers, and even former AEC commissioners who believed that nuclear power would have to overcome the mistakes of its past in order to provide the energy of the future.

Ultimately, nuclear advocates may have been right to sense a larger social change behind the emerging antinuclear movement, but wrong about its origins. At a time when civil rights, antiwar, and feminist movements articulated a broad agenda for social change, many defenders of nuclear power saw their detractors as just another dissident group. In chapter 1 of this collection of essays, Samuel P. Hays examines twentieth-century environmentalism in light of a broad cultural shift in consumer demands from conveniences to amenities. Perhaps the electrical Garden of Eden projected by nuclear advocates presented a prewar celebration of abundance to a postwar culture that was in the process of replacing that vision with a new concern for the environment as a commodity in its own right.[53]

Congress responded more favorably than the AEC to the public's growing interest in the environment. On January 1, 1970, the National Environmental Policy Act (NEPA) went into effect. For the nuclear power industry, this meant that the new Environmental Protection Agency would determine radiation and safety standards and would require an environmental impact statement prior to the licensing of all nuclear power plants. The AEC's initial response to NEPA was not entirely cooperative. In 1971, the AEC refused to consider citizen challenges to Baltimore Gas and Electric's Calvert Cliffs nuclear plant on Chesapeake Bay on the basis of thermal pollution; the AEC further claimed that NEPA did not apply to this case. The U.S. Court of Appeals in Washington, D.C., ruled against the AEC, instructing it to comply with NEPA requirements.[54]

Meanwhile, documentation of problems with reactor safety and radioactive waste disposal mounted. Daniel Ford and Henry Kendall led the Union of Concerned Scientists investigation of failures in the AEC-approved Emergency Core Cooling System, the power plants' major protection against a meltdown. The resulting hearings, which began in 1972 and continued for eighteen months, focused national attention on the issue of reactor safety. In 1973, Ralph Nader called for a national moratorium on nuclear plant construction, and Friends of the Earth won access to internal AEC documents exposing the AEC's cover-up of the WASH–740 update in 1964.[55]

By the early 1970s, nuclear friends and foes alike spoke of the damaging effect of the largely uncritical nuclear advocacy espoused by the AEC and the JCAE. Environmental concerns had warranted institutional innovation; now Congress sought an institutional redefinition of the federal atomic establishment. The Energy Reorganization Act of 1974 split the AEC into two new agencies: a Nuclear Regulatory Commission (NRC) to oversee licensing and regulation of nuclear power plants, and an Energy Research and Development Administration (ERDA) to cover the development and promotion of energy and the production of nuclear weapons. In 1977, President Carter's Department of Energy Organization Act incorporated ERDA into a new cabinet-level agency. Meanwhile, Congress stripped the JCAE of its legislative powers and divided its tasks among five different committees.

These new institutions, however, inherited old problems. In 1975, the new NRC released the results of a two-year Reactor Safety Study, also referred to as WASH–1400 and as the Rasmussen report (after MIT nuclear engineer Norman Rasmussen, the study's director). Although the report contained mixed reviews, it largely reaffirmed the safety of nuclear reactors, comparing the likelihood of a major accident to that of being struck by a meteor. At first, the Rasmussen report provided the favorable publicity required to prop up the utility companies' sagging confidence in the economic viability of nuclear reactors. The *Bulletin of Atomic Scientists*, however, denounced the Rasmussen report as "an in-house study" of questionable value; and in 1977, the Union of Concerned Scientists published a detailed critique, challenging its methods as well as its findings. In 1979, the NRC itself finally rejected the report's optimistic findings.[56]

On March 28, 1979, a series of malfunctions occurred at Unit 2 of the Three Mile Island (TMI) nuclear power plant near Harrisburg, Pennsylvania. In the next few days, several thousand people evacuated the area, and hundreds of millions of others watched and read updates on the worst accident in nuclear power's brief history. TMI was not the first serious accident at a nuclear plant; nor was it the first accident to present a serious risk of meltdown. In October 1966, for example, a partial fuel meltdown occurred at the Enrico Fermi Fast Breeder Reactor in Detroit. In March 1975, a worker at the TVA's Brown's Ferry nuclear power plant was checking for air leaks with a lighted candle when some electrical cables caught fire; in the seven-hour fire that ensued, all of the safety backups for one of the reactor's cooling systems were impaired.

The most important aspects of TMI were the publicity it received, and the timing of that publicity. At the time, many critics considered it the Tet offensive of the nuclear power controversy, swinging the press over to the antinuclear perspective and solidifying the de facto moratorium on new reactor orders. The industry's defenders charged—with justification—that the news media sensationalized and distorted the events and the significance of Three Mile Island. The media's sense of betrayal was, however, predictable. Metropolitan Edison first delayed reporting the accident, then misrepresented its scope. In the absence of substantive information, the press was as unprepared as the general public for fine gradations of technology assessment. Unexamined nuclear fear in 1979 was a direct product of the unexamined nuclear optimism of the 1950s and 1960s.

In the long run, however, TMI had a revitalizing effect on pronuclear publicity. The Atomic Industrial Forum responded to the accident by creating the U.S. Committee on Energy Awareness (USCEA). Sponsored by nuclear vendors, engineering firms, and utility firms, the USCEA conducted a $25– to $30–million-per-year advertising campaign to restore the image of nuclear power's legitimacy among scientists and policymakers. In May and June 1983, a congressional subcommittee conducted the first hearings to investigate the Department of Energy's (DOE) nuclear public relations campaign and its connection with private efforts like that of the USCEA. Promotion of nuclear power, if not nuclear power itself, had made a vigorous comeback.[57]

CONCLUSION: PONDERING THE NUCLEAR FUTURE

On the morning of April 28, 1986, the dust on a Swedish worker's shoes set off the alarm on a radiation detector at the Forsmark nuclear power plant north of Stockholm. Safety inspectors combed the plant for leaks, and tested the seven hundred other employees. The workers' clothing registered radiation exposure at five to ten times normal background levels; yet the plant looked clean. The source of radioactivity appeared to be outside the power plant; ironically, the workers were carrying traces of radioactivity into a nuclear facility from the streets. By that night, all the world knew the source: an explosion had occurred at another nuclear power plant, in the Soviet Union. Forsmark, like much of Europe, found itself in the shadow of Chernobyl.

The environmental dimensions of nuclear power were dramatically underscored by the Chernobyl accident. Radiation from the damaged reactor observed no political boundaries, traveling farther (over a thousand miles) and dispersing less uniformly than expected. The evacuation of 135,000 people within a nineteen-mile radius of the reactor underscored the increasing concern in the United States over emergency evacuation procedures.

Chernobyl also presented a delicate public relations problem for Washington. The Soviets failed to notify the world of the accident until Swedish reports of radiation forced them to explain what had happened. Some government officials seized upon Chernobyl as proof of Soviet duplicity and secrecy; to them, the unsubstantiated (and, as it turned out, groundless) reports of two thousand fatalities were much more credible than the Soviet Union's claim (later verified) that there had been only two immediate deaths.

For the U.S. Department of Energy and the Nuclear Regulatory Commission, Chernobyl threatened to discredit nuclear power. Even the Soviets' initial reticence was uncomfortably reminiscent of Metropolitan Edison's conduct during the TMI accident (a parallel that General Secretary Gorbachev noted during a May 14, 1986, speech on Soviet television). Thus, while NRC and DOE officials criticized the Soviet Union for withholding or delaying information, they instructed their own employees, scientists, and contractors to avoid the press, and specifically to refrain from making comparisons between American and Soviet reactors. Joined by the Atomic Industrial Forum and the U.S. Committee on Energy Awareness, NRC and DOE spokespersons also released statements that contributed to a number of misconceptions: that the Chernobyl reactor had no containment structures (it had two); that all American reactors have such structures (the DOE operates five reactors with no regular containment domes); and that the Chernobyl plant's design was too dissimilar to any American reactor to warrant comparison (a claim that dissenting NRC member James Asseltine contested before a congressional committee).[58]

Predictions of nuclear power's future in the United States vary greatly, depending primarily on how one interprets the past. Some nuclear analysts argue that the twin plagues of unresolved safety issues and escalating costs, coupled with the declining demand for electricity, will consign nuclear power to an early grave. "Hard, cold economics is now doing to nuclear power what thousands of hot-blooded demonstrators never could," claims Christopher Flavin. "It is slowly, painfully shutting down the world's nuclear industries." Others, as widely divergent in outlook as the Atomic Industrial Forum and nuclear critic Mark Hertsgaard, foresee a resurgent nuclear industry. The amount of capital already invested is great enough to suggest that the major nuclear corporations are willing to wait it out;

and the far greater scope of their nonnuclear assets indicates that they can afford that luxury.[59]

The nuclear industry has maintained that the main cause of the atom's enlarging price tag is delays in plant construction and operation that result from excessive government safety and environmental regulations. Some critics and industry analysts disagree, claiming that the main reason for escalating costs was the intense competition for control of the nuclear market during the 1970s by the four largest vendors (General Electric, Westinghouse, Combustion Engineering, and Babcock and Wilcox).[60] Like auto makers in the 1950s, these corporations produced ever-larger new models, thereby forfeiting the benefits of standard production and accumulated operational experience with a given design.

If nuclear power can weather the present economic crisis, we must look to the relationship between policy and expertise, institutions and technology, for our projections. We could argue that the AEC, a politically vulnerable agency overseeing a highly politicized technology, created such a gap between expertise and policy that the environmentalists of the 1970s were able to rush into the resulting power vacuum. Such an analysis would permit some reassurance from the fact that those who challenged the institutional expertise of fifteen years ago have since institutionalized their own environmental critique into the regulatory expertise of the present. Thus the conflict of interest between federal overseers and corporate developers would appear to have been reduced. To be sure, the traditional alliance between government and industry has been strained by the environmental revolution. Even at its best, however, the regulatory process has severe limitations. It can sound the alarm, but it cannot put out the fire. The most dangerous threat to an environmentally sound nuclear policy today may be the inability of any regulatory process to extract both safety and productivity from current nuclear technology.

The AEC and its successors, like all agents of government power, became increasingly preoccupied with credibility in the postwar era. Its bureaucratic proximity to nuclear weapons rendered nuclear power especially susceptible to emblematic, rather than substantive, policies; and as with nuclear weapons, the magnitude of risks accompanying an inadequate nuclear power policy has subjected it to far greater public concern, once the facts became available, than most arenas of government authority. Today's nuclear power plants are artifacts of a generation's misplaced trust in an unexamined technology; tomorrow's plants may encapsulate within their cooling towers the next generation's misplaced trust in the regulatory process. In both cases, a certain expediency has triumphed over the nuclear mandate for socially responsible technology.

NOTES

I am indebted to George T. Mazuzan, National Science Foundation; J. Samuel Walker, NRC History Office; and Prentice Dean, formerly with the DOE History Office, for their suggestions and assistance with the research for this paper. They are not responsible for any oversights or errors I may have made.

1. For general background, see especially Richard G. Hewlett and Francis Duncan, *Atomic Shield: A History of the United States Atomic Energy Commission, 1947–1952* (University Park, Pa.: Pennsylvania State University Press, 1969); George T. Mazuzan and J. Samuel Walker, *Controlling the Atom: The Beginnings of Nuclear Regulation, 1946–1962* (Berkeley: University of California Press, 1984); Daniel Ford, *The Cult of the Atom: The Secret Papers of the Atomic Energy Commission* (New York: Simon and Schuster, 1982); Irvin C. Bupp and Jean-Claude Derain, *Light Water: How the Nuclear Dream Dissolved* (New York: Basic Books, 1978).
2. Martin Sherwin, *A World Destroyed: The Atomic Bomb and the Grand Alliance* (New York: Vintage, 1977), 308.
3. "Commercial Electric Power from Atomic Energy," *Science*, July 29, 1966, 192.
4. Morgan Thomas, *Atomic Energy and Congress* (Ann Arbor: University of Michigan Press, 1956), 13–16.
5. Stephen Hilgartner, Richard C. Bell, and Rory O'Connor, *Nukespeak: The Selling of Nuclear Technology in America* (New York: Penguin, 1982), 60–62.
6. Mark Hertsgaard, *Nuclear, Inc.: The Men and Money Behind Nuclear Energy* (New York: Pantheon, 1983), 25–26.
7. Harold P. Green and Alan Rosenthal, *Government of the Atom: The Integration of Powers* (New York: Atherton Press, 1961), 9–12.
8. Edward Teller to Sterling Cole, July 23, 1953, *Atomic Power Development and Private Enterprise: Hearings before the Joint Committee on Atomic Energy*, 83d Cong., 1st sess., 1953, 633.
9. David E. Lilienthal, *Change, Hope, and the Bomb* (Princeton: Princeton University Press, 1963), 109, 110; Glenn Seaborg and William Corliss, *Man and Atom* (New York: E. P. Dutton, 1971).
10. JCAE Hearings, *Atomic Power Development and Private Enterprise*, 2.
11. Hertsgaard, *Nuclear, Inc.*, 26; Richard G. Hewlett and Francis Duncan, *Nuclear Navy 1946–1962* (Chicago: University of Chicago Press), 237–38.
12. Lewis L. Strauss, "My Faith in the Atomic Future," *Reader's Digest*, Aug. 1955, 17; JCAE Hearings, *Atomic Power Development and Private Enterprise.*
13. *New York Times*, Dec. 9, 1953, 2, cited in Hilgartner et al., *Nukespeak*, 43.
14. George T. Mazuzan, "Conflict of Interest: Promoting and Regulating the Infant Nuclear Power Industry, 1954–1956," *The Historian*, Nov. 1981, 1–14; Ford, *Cult of the Atom*, 42.
15. Green and Rosenthal, *Government of the Atom*, 26, 12.
16. Hertsgaard, *Nuclear, Inc.*, 20–60; Steven L. Del Sesto, *Science, Politics and Controversy: Civilian Nuclear Power in the United States, 1946–1974* (Boulder, Colo.: Westview Press), 58–59.
17. U.S. Atomic Energy Commission, *Theoretical Possibilities and Consequences of Major Accidents in Large Nuclear Power Plants*, WASH–740 (Washington, D.C.: Government Printing Office, 1958).
18. Ford, *Cult of the Atom*, 45.
19. Ibid., 71.
20. *Forum Annual Report for the Year Ended June 30, 1954* (New York: Atomic Industrial Forum, 1954), 5.
21. Charles Robbins, "Uses of Atomic Industry," *Public Relations for the Atomic Industry:*

Proceedings of a Meeting for Members, March 19 and 20, 1956 (New York: Atomic Industrial Forum, 1956), 1.

22. Robert Charpie, "Power Reactors," *Public Relations*, 4–20; Shelby Thompson, "Sources of Atomic Energy Information in Government," *Public Relations*, 66–70; Frank K. Pittman, "Safety Requirements of the Atomic Energy Commission," *Public Relations*, 94–95.
23. Jerome Luntz, "Address," *Public Relations*, 40, 41.
24. LeBaron Foster, "Address," *Public Relations*, 79, 82–83.
25. William A. Stenzel, "Developing Consumer Acceptance of Radioactivity," *Public Relations*, 110–114; Harold A. Beaudoin, "Locating a Reactor in a Populated Area," *Public Relations*, 105.
26. For the AEC's public relations efforts I have relied on the AEC *Annual Reports*.
27. AEC Press Release, Sept. 16, 1954, cited in Ford, *Cult of the Atom*, 50.
28. Hilgartner et al., *Nukespeak*, 44.
29. *Our Friend the Atom* was also published as a book by the same title in 1956, with Haber listed as its author. The original film has been superseded by an update under the same title.
30. Harold Stassen, "Atoms for Peace," *Ladies Home Journal*, Aug. 1955, 48.
31. Produced by the Connecticut Yankee Atomic Power Consortium, *The Atom and Eve* featured a scale model of the Haddam Neck plant, where "even the colors chosen for the structure were selected from the natural hues of the surrounding valley" (brown) and color-coordinated to "recall the New England tradition" (white).
32. Allan C. Fisher, Jr., "You and the Obedient Atom," *National Geographic*, Sept. 1958, 303, 311, 329, 352.
33. AEC Press Release, Feb. 20, 1966; Seaborg and Corliss, *Man and Atom*, 22.
34. Lilienthal, *Change*, 110.
35. H. Peter Metzger, *The Atomic Establishment* (New York: Simon and Schuster, 1972), 112, passim; U.S. Atomic Energy Commission, *Radioactive Fallout from Nuclear Weapons Tests*, AEC:CONF–765, Nov. 1965.
36. Hilgartner et al., *Nukespeak*, 41.
37. Irvin C. Bupp, "Priorities in Nuclear Technology: Program Prosperity and Decay in the United States Atomic Energy Commission, 1956–1971" (Ph.D. diss., Harvard, 1971), 181.
38. Edward Teller and Allen Brown, *The Legacy of Hiroshima* (Garden City, N.J.: Doubleday, 1962), 180.
39. Freeman Dyson, *Disturbing the Universe* (New York: Harper and Row, 1979), 112.
40. Bupp, "Priorities," 196.
41. Seaborg and Corliss, *Man and Atom*, 234.
42. Ibid., 183. (Emphasis in original.)
43. L. J. Vortman, "Excavation of a Sea-Level Ship Canal," *Proceedings of the Second Plowshare Symposium, May 14, 1959*, Part II: Excavation (AEC:UCRL–5671), 88; Seaborg and Corliss, *Man and Atom*, 184, 195.
44. Ibid., 113.
45. Stephen Ailes to Glenn Seaborg, December 30, 1964, Carryall folder, Plowshare Papers, Department of Energy Archives.
46. Seaborg and Corliss, *Man and Atom*, 234.
47. Bupp, "Priorities," 196.
48. Bill Perkins, "The U.S. Controversy—Issues of Concern," *The Nuclear Controversy in the USA* (Atomic Industrial Forum and Swiss Association for Atomic Energy, Lucerne, Apr. 30–May 3, 1972), 1.
49. Gofman, a graduate student of Seaborg's at Berkeley, and Tamplin were both AEC scientists who became outspoken critics of nuclear policy as a result of their research for the AEC; for an account of Tamplin's assignment to discredit Ernest Sternglass, see Metzger, *Atomic Establishment*, 276–78, n. 17. See also John W. Gofman and Arthur R. Tamplin, *Poisoned Power* (Emmaus, Pa.: Rodale Press, 1971).

50. U.S. Congress, Joint Committee on Atomic Energy, *Environmental Effects of Producing Electric Power*, 2 pts., 91st Cong., 1st and 2d sess., 1969–70.
51. Theos J. Thompson, "Improving the Quality of Life—Can Plowshare Help?" *Symposium on Engineering with Nuclear Explosives, January 14–15, 1970*, vol. 1 (AEC:CONF–700101), 104.
52. Bertram Wolfe, "The Hidden Agenda," in *Nuclear Power: Both Sides*, ed., Michio Kaku and Jennifer Trainer (New York: Norton, 1982), 240–43; Hertsgaard, *Nuclear, Inc.*, 181.
53. Samuel P. Hays, *Beauty, Health, and Permanence: Environmental Politics in the United States, 1955–1985* (New York: Cambridge University Press, 1987).
54. Metzger, *Atomic Establishment*, 270.
55. Paul Turner, "Introductory Remarks," in *The Nuclear Controversy in the USA II*: Conference Papers, Atomic Industrial Forum and Swiss Association for Atomic Energy, Lucerne, May 5–8, 1974 (New York: Atomic Industrial Forum, 1975), 1–3; Ford, *Cult of the Atom*, 115–30.
56. Ford, *Cult of the Atom*, 172.
57. Daniel Ford, *Three Mile Island: Thirty Minutes to Meltdown* (New York: Penguin, 1981); Committee on Energy and Commerce, *Nuclear Public Relations Campaign: Hearings before the Subcommittee on Energy Conservation and Power*, 98th Cong., 1st sess., 1983, 53, 87, 98–117, 192–98, 351, passim.
58. *New York Times*, May 22, 1986. Nuclear power appears to be caught in the midst of political struggles within the Soviet Union as well. Despite Gorbachev's efforts to incorporate Chernobyl into the program of *glasnost* by projecting a new image of openness and reform, antinuclear demonstrators have been arrested in Kiev, and indications of deep problems within the Soviet nuclear establishment persist. On the second anniversary of the Chernobyl accident, Valery Legasov, deputy director of the Moscow Institute of Physics, committed suicide. Legasov had been among the first government scientists sent to Chernobyl to contain the damaged reactor, and his friends claimed that his decision to hang himself on the anniversary of the accident was no coincidence. *Pravda* subsequently published a posthumous article by Legasov attacking "what he called the complacent attitude of Soviet scientists and engineers toward nuclear power" (*New York Times*, May 23, 1988).
59. Christopher Flavin, "Reassessing the Economics of Nuclear Power," in Lester R. Brown et al., *State of the World, 1984* (New York: Norton, 1984), 132.
60. Hertsgaard, *Nuclear, Inc.*, 63–65.

8

THE EVOLUTION OF FEDERAL REGULATION OF TOXIC SUBSTANCES

Christopher Schroeder

To judge federal toxics policy by the legislation Congress has written since World War II, it is complex: over thirty federal statutes regulate some aspect of the toxics problem,[1] and seven different agencies administer them.[2] As much as it is complex, federal policy also seems to have grown haphazardly. These statutes were written by various congressional committees at different times, they respond to the demands of different clusters of issues and constituencies, and they reflect the varying philosophies and ambitions of key congressional leaders.

In the face of such unsystematic development, speaking generally of a single toxics policy risks great oversimplification. Still, a survey of political activity relating to toxics in the postwar era finds at least two broad trends standing out from the welter of specific statutes and regulations: (1) toxics policy has reflected an increasing social conviction that the level of risks allowed by an unregulated market economy is too high; (2) in redefining "allowable risk" to lower levels, that policy has relied heavily upon centralized regulatory agencies, rather than decentralized corrections of either the market or the private law systems that historically have policed the market. To explain these two trends, we can identify several broad social and intellectual currents at work in the United States during the postwar era.

Toxics policy has become distinguished from other areas of environmental policy less by the social currents that have shaped it, because these have influenced other areas of environmental policy as well, than by the nature of the problems that policy has confronted. The opening section of the paper articulates the nature of the "toxics problem" as a peculiar public policy issue. It also provides a brief sketch of the existing toxics regulatory framework. The following section traces the two trends just noted, with special reference to the broad social currents that have influenced them.

In recent years, we have been given reason to question whether both these trends may have run their course. The Reagan administration touted deregulation of the market as its preferred strategy for a host of social issues, and agencies within the administration responsible for toxics policy resisted tightening further a number of toxics controls. Because this paper is primarily historical, it will not comment upon all the deregulation arguments or prospects, but the historical account it advances has been influenced, as history commonly is, by the contemporary issues upon which that history might bear. Accordingly, the paper's conclusion offers some observations about how toxics policy is likely to continue to respond to the forces that have shaped it thus far.

FEDERAL REGULATION OF TOXICS

National realization that toxics present a distinct policy dilemma has come only recently. When the Environmental Law Institute published a comprehensive analysis of federal environmental law in 1974, it did not contain a separate chapter on toxic substances.[3] Its sparse treatment of toxics (other than pesticides) was embedded in chapters on the Clean Air and Clean Water Acts, and only five pages out of two hundred in those chapters focused on toxics.[4] By 1970, only nine statutes governed toxics. In the 1970s, sixteen statutes were added to the list.[5]

Toxics come into contact with the environment and human beings as ingredients in food, clothing, toys, and other consumer goods. They are produced and used by design (pesticides), by inadvertence (Tris), and as the by-products and residuals of other activities (hazardous wastes, auto emissions, stack gases, and water discharges). Toxics regulation policy has been largely driven by the political salience of these different components of the overall problem, each arising separately from the others. Toxics in consumer goods, for example, were considered a different problem from toxics in drinking water, even though the chemicals involved were often identical.

Compartmentalizing the toxics problem has been reinforced by certain dynamics of the political process. Widening the scope of any specific piece of legislation is resisted, because such an action would widen the range of issues that must be accommodated and enlarge the number of concerned groups, thus decreasing the chances of compromise and increasing the opportunities for political losses.[6] Once specific programs are in place, constituent groups adjust their behavior to accommodate the realities of regulation. They can then become highly resistant to subsequent proposals to integrate, reform, or streamline as new organizing principles emerge. Agencies, too, become invested in their jurisdiction and will often weigh in on the side of marginal tinkerings within existing regulatory boundaries rather than comprehensive reform.

The toxics regulatory structure epitomizes this growth pattern, and it has resulted, not in a well-designed cabin, but in a pile of logs. Table 1 lists the results. The most recently enacted statutes, such as the Toxic Substances Control Act (TSCA) and the Resource Conservation and Recovery Act (RCRA), contain provisions requiring the Environmental Protection Agency (EPA) to coordinate regulatory action where statutory jurisdictions overlap,[7] but these requirements fall short of providing unifying structure to regulatory strategy, priority setting, or risk assessment and management. The coordination that exists has been established largely by executive actions, such as the creation of the Interagency Regulatory Liaison Group in the Carter administration.[8]

The current regulatory bureaucracy must face embarrassing questions because of statutory and regulatory fragmentation. Why, for example, do regulatory measures designed to protect against the same health hazards from the same toxic cost far more per life saved in the air program than in the worker safety program?[9] Why does regulation of the same toxic substance under the same program produce abatement costs in different industries that vary by an order of magnitude?[10] Why do statutory definitions of acceptable risk vary markedly from statute to statute, although similar health hazards associated with toxics are of concern in all the statutes? Why do the stipulated procedures through which the administering agency takes regulatory action vary across statutes? Why, after years of trying, does the federal government lack a consistent, coordinated policy with respect to carcinogens?[11]

Some of these questions may have answers, but this paper cannot

Table 8.1
FEDERAL LEGISLATION REGULATING TOXICS

STATUTE	*YEAR PASSED*
Food and Drug Act (FDA)	1906
Federal Insecticide Act (FIA)	1910
Food, Drug and Cosmetic Act (FDCA)	1938
Federal Insecticide, Fungicide, and Rodenticide Act (FIFRA)	1947
Pesticide Chemicals Amendment (PCA)	1954
Food Additives Amendment (Delaney Clause)	1958
Federal Hazardous Substances Labeling Act (FHSA)	1960
Clean Air Act (CAA)	1963
Motor Vehicle Air Pollution Control Act (MVAPCA)	1965
Solid Waste Disposal Act (SWDA)	1965
Clean Water Restoration Act (CWRA)	1966
Animal Drug Amendment (ADA)	1968
Child Protection and Toy Safety Act (CPTSA)	1969
National Environmental Policy Act (NEPA)	1970
Poison Prevention Packaging Act (PPPA)	1970
Occupational Safety and Health Act (OSHAct)	1970
Consumer Products Safety Act (CPSA)	1972
Federal Water Pollution Control Act (FWPCA)	1972
Federal Environmental Pesticide Control Act (FEPCA)	1972
Safe Drinking Water Act (SDWA)	1974
Hazardous Materials Transportation Act (HMTA)	1975
Toxic Substances Control Act (TSCA)	1976
Resource Conservation and Recovery Act (RCRA)	1976
Clean Air Act Amendments (CAA)	1977
Clean Water Act (CWA)	1977
Federal Pesticides Act of 1978 (FPA)	1978
Comprehensive Environmental Response, Compensation and Liability Act ("Superfund")	1980
Consumer Product Safety Amendments of 1981 (CPSA)	1981
Safe Water Drinking Act of 1986	1986
Superfund Amendments and Reauthorization Act of 1986	1986
Water Quality Act of 1987	1987

Sources: See note 196 for references to this legislation.

explore them in depth.[12] Two common characteristics occur across statutes and regulations consistently enough to note here. These concern definitions of acceptable risk and the use of highly centralized regulation to control toxics problems.

Definitions of acceptable risk. Either a statute contains an explicit definition of the degree of risk that can be generated by actions under the law's jurisdiction, or the law's standard-setting provisions permit a definition to be derived. Statutory definitions come in three basic types: balancing, health-based, and technological and economic feasibility based, known as TEF-based.[13] Each type of statute rejects the status quo ante levels of risk associated with the toxics to which they apply, but they do so through different methodologies. Balancing statutes require evaluation of the pros and cons of individual chemicals, although without mandating a formal cost-benefit analysis.[14] The Toxic Substances Control Act, for example, requires the elimination of "unreasonable risk to health or the environment" after consideration of the benefits of the uses of the chemical, the economic consequences of prohibition or restriction, and the effects of the chemical on humans and the environment.[15] Health-based statutes draw the acceptable risk line by stipulating a level of health or environmental quality protection that is to be achieved, regardless of the adverse economic effects of achieving that level. The Delaney Clause is an example of a health-based statute.[16] Under it, any food additive that may cause cancer in man or animals is prohibited. TEF-based statutes require controls of toxics to the extent that technology will allow, subject to the constraint that installing the technology be economically feasible. The mandate of the Occupational Safety and Health Administration (OSHA) to remove the risks of material health impairment to workers as a result of toxics exposure "to the extent feasible" is a TEF statute.[17] Table 2 summarizes the operative definitions of acceptable risk from the major federal toxics statutes according to these three categories.

The total toxics legislative package is both ambitious and ambiguous. It aims to extend some government superintendence of exposure to toxics into every source, every pathway, and every exposure environment. None of the statutes, however, escapes the ambiguities surrounding the definition of acceptable risk. In comparing the statutes, it is difficult to explain why balancing is appropriate for consumer products, TEF for water pollution, and health-based for food.[18] Still greater ambiguity exists within the definitions of any specific statute. Congress has intentionally made balancing an impres-

Table 8.2
DEFINITIONS OF ACCEPTABLE RISK

	STATUTE (SUBJECT MATTER)	*ACCEPTABLE RISK LEVEL*
BALANCING	FIFRA (Pesticides)	"Any reasonable risk to man or to the environment, taking into account the economic, social and environmental costs and benefits of the use of the pesticide."
	TSCA (Chemical testing—Sec. 4)	"May present an unreasonable risk of injury to health or the environment."
	TSCA (Chemical regulation—Sec. 6)	"Presents or will present risk of injury to health or the environment."
	CPSA (Consumer products)	"Unreasonable risk of injury."
	FDCA (Drugs)	Balance risks and benefits.
	HMTA (Materials in transit)	"Unreasonable risk to health and safety or property."
HEALTH	FDCA (Food additives)	Induces cancer in man or animal.
	RCRA (Hazardous waste sites)	"Necessary to protect human health and the environment."
	CAA (Hazardous air pollution)	"An ample margin of safety to protect human health . . ."
	FHSA (Consumer products—labeling)	Labels as are "necessary for the protection of public health and safety."
TEF	SDWA (Toxics in drinking water)	"To the extent feasible . . . (taking costs into consideration)."
	CWA (Toxic water pollutants)	"Best available technology economically achievable."*
	CAA (Toxics from vehicle emissions)	"Greatest degree of emission reduction achievable through . . . [available] technology."
	OSHAct (Occupational toxics)	"Adequately assures to the extent feasible that no employee will suffer material impairment of health or functional impairment."

*EPA can impose a stricter standard if warranted by a consideration of "the toxicity of the pollutant, its persistence, degradability, the usual or potential presence of the affected organism in any waters, the importance of the affected organisms, and the nature and extent of the effects of the toxic pollutants on such organisms," and the adequacy of other regulations.

sionistic process,[19] relegating the hard decisions about comparisons and trade-offs to the regulatory agencies, which in turn have not hastened to publicize the grounds for their decisions.[20] Decisions concerning health-based statutes must deal with "health" as a continuum, or more likely a matrix, of physiological, neurological, and psychological conditions.[21] Toxics can stimulate different bodily responses at different exposure levels, with the lowest levels of response probably shading out of the area that physicians would label an adverse health effect into an area where only sensitive instruments can detect a physiological or neurological response.[22] TEF statutes pose the question, "What is feasible?" both technologically and economically.[23]

Despite these ambiguities, friends and foes alike acknowledge that the objective of these statutory definitions of acceptable risk is to reduce risk exposure from the status quo ante. Analysts friendly to the regulatory objectives protest that the statutory language promises protection that regulatory implementation fails to deliver. Critics of the statutes protest that they aim at levels that are too stringent and inefficient.[24]

Highly centralized regulation. The toxics regulatory structure aims at preventing toxic injuries primarily by identifying specific risk-causing attributes of private behavior and then mandating actions to alter those attributes, including prohibiting the behavior if the attributes cannot be altered. The exact form of the mandate varies from statute to statute. For example, for toxics regulated under the hazardous air pollution control program, such as benzene or vinyl chloride, until recently EPA determined the best control technology, then mandated that air emissions be reduced to the level that technology could achieve.[25] Lead, in contrast, has been declared a "criteria" pollutant, so it is regulated by the EPA under the health-based provisions of the Clean Air Act governing such pollutants.[26] Under this regime, the EPA determines how much ambient lead is acceptable to human health and then prohibits any emissions that would produce more than that amount. Under the Federal Insecticide Fungicide and Rodenticide Act (FIFRA), if the EPA determines that a pesticide poses an unreasonable risk to the environment, it either prohibits its use entirely or restricts it to certain applications.[27]

Federal agencies thus proscribe behavior believed to be unacceptably risky. Although they frequently do not act until adverse health effects have been confirmed or until substantial exposure to risk has already taken place, in theory they are not supposed to wait until then. In the pollution control fields, license or permit procedures ensure that private conduct will conform to agency standards.[28] In the product oversight fields, approaches vary. FIFRA requires advance registration of new pesticides, and a 1978 amendment mandated reregistration of all existing pesticides. The burden is on the registrant to show that the pesticide is not unreasonably risky in its planned uses. The Toxic Substances Control Act does not require preclearance of new toxic chemicals, but it does insist on premanufacturing notification to the EPA, and the EPA is empowered to suspend manufacture that it believes is unacceptably risky.[29] In addition, the EPA can order additional testing of certain chemicals to provide better evidence concerning risk.[30] The Consumer Product Safety Act requires neither preregistration nor prenotification on consumer products, but the Consumer Product Safety Commission (CPSC) is charged with monitoring the consumer markets and can suspend sale of products it determines to be unreasonably risky.[31] All three statutes share the characteristic of opening up private activities to analysis by public agencies, with those agencies empowered to prohibit such actions before any injury occurs.

Aaron Wildavsky has contrasted this form of "anticipatory" regulation with the "resilient" approach of an unregulated economy plus the private tort remedy system.[32] Anticipatory risk prevention policy attempts to avert harm that would otherwise occur by subjecting risky action to risk analysis. A resilient policy responds to dangers after they have been discovered and harm has occurred, seeking to adjust to and mitigate the effects of those known dangers.[33] The crucial difference between these two is not that one policy is "prorisk" and the other "antirisk" or that one policy is preventive and the other not. The crucial distinction is that an anticipatory policy requires submitting one's actions for review by independent third parties who can then flatly prohibit certain actions, whereas a resilient policy relies on self-analysis and transactions among affected parties to determine what actions shall take place.

Although Wildavsky's distinction is significant, his terminology is slightly off the mark. Because humans can be held responsible for the consequences of their actions, human "resilient" systems are also "anticipatory," in a way that ecological systems, from whose study the term "resilient" is borrowed, are not.[34] What distinguishes the two is that resilient systems emphasize decisions made at a more decentralized level than anticipatory systems do. The focus on centralization versus decentralization more accurately captures the crucial distinction.[35] Even with this, one must be careful, for decentralized social systems make at least some centralized decisions. For example, collective decisions determine those consequences for which individuals are held responsible.[36] There are probably no actual examples of purely centralized or purely decentralized systems. Within the universe of mixed systems, the extent of centralization will vary, as will the particular decisions that are centralized, and it is fundamentally a dispute over the advantages of different mixed systems that separates defenders of the existing regulatory structure from deregulators.[37]

Along a continuum between centralized and decentralized, the current regulatory structure for toxics is highly centralized. It centralizes the definition of acceptable risk. It centralizes the evaluation of private activity and is not satisfied with private assessments of the propriety of private behavior in light of the perceived consequences. It centralizes the proscription of unacceptable behavior, generally not providing private parties the choice of complying or paying, or the option of trading around the proscription through private transactions.

The anticipatory, centralized features of toxics regulation have been growing. The evolution of pesticide regulation is typical, progressing from the 1910 act, which required no preregistration (only enabling the government to stop sale of dangerous pesticides after danger had been shown through actual use), to the current preregistration process. Even some of the relatively modern statutes have moved from their original versions in the direction of further centralization.

Centralization has procedural ramifications. Centralized agencies administer centralized regulations. These agencies struggle with complex issues on the frontiers of medical science, atmospheric and water-system modeling, and engineering and technological projections, always in a politically charged forum. Their work is in turn scrutinized under what has come to be called the "hard look" of judicial review. As a result, all these agencies have developed complex procedures that have contributed to an extremely cumbersome regulatory implementation process.

Progress under this complex structure of centralized regulation can only be termed painfully slow. By 1979, OSHA had issued only twenty-three permissable exposure levels (PELs) for toxics, of more than a hundred recommended to it by the National Institute of Occupational Safety and Health (NIOSH).[38] In 1977, the General Accounting Office suggested that, at the rate it was going, OSHA would take more than a century to issue PELs for the known occupational hazards, let alone get to any new ones.[39] That estimate must now appear optimistic, because OSHA has issued only two new PELs since 1979.[40] After its establishment in 1970, the CPSC debated for several years whether it even had jurisdiction over chronic toxic effects such as cancer. After it began to address toxics, it became embroiled in a few highly controversial cases, including those involving urea foam insulation and formaldehyde. Partly because of controversies over how those cases were handled, Congress amended the Consumer Product Safety Act in 1981,[41] further encumbering the rule-making process for chronic toxic effects, ensuring even slower decision making.[42]

EPA would not move forward under the toxics program of the Federal Water Pollution Control Act until sued by environmental groups,[43] and it is still working through a list of toxics under a settlement decree in that case subsequently ratified by the Congress.[44] By 1981, EPA had regulated just seven hazardous air pollutants.[45] Seven years after its enactment, RCRA remained snarled in litigation over complex rule making to establish standards for hazardous waste disposal sites.[46] Superfund has been the focus of political intrigue but little clean-up activity,[47] although EPA administrator William Ruckelshaus made improving Superfund a priority of his administration.[48] The reregistration process for pesticides has proved even more time-consuming than anticipated,[49] and EPA does not enforce the pesticide restrictions it does issue.[50] All aspects of TSCA implementation also have been roundly attacked as inadequate, virtually since the program's inception. Studies by the Office of Technology Assessment (OTA) indicate that the minimum information requirements of the premanufacture notification program routinely are not met,[51] and the EPA's powers under TSCA's section 4 to issue rules concerning further testing for suspected chemicals have been the object of litigation because the rules have not been written.[52]

The Toxics Problem: The Low-Exposure Dilemma

Government in the postwar era is broadly characterized as rejecting status quo solutions and replacing decentralized mechanisms with more centralized ones. What distinguishes toxic substance regulation from other fields is the nature of the public policy problem that toxic substances pose. Toxics can cause serious harm, even death, at extremely low levels of exposure, yet areas of society and the economy can be disrupted by attempts to eliminate the final amounts of low-level toxics. This constitutes the low-exposure dilemma: we must either tolerate some residual risks of serious harm, perhaps even death, or tolerate other significant economic or social dislocations. The next section summarizes some of what we know about the dilemma.

Health effects. Cancer, "the fright-laden disease," heads the list of human health problems identified with toxics. The number of substances known to be carcinogenic is not yet great. In 1978, the International Agency on Research on Cancer (IARC) listed 18 substances as human carcinogens, and another 18 as probably carcinogenic but without sufficient confirming data to satisfy the agency.[53] In the United States, 102 substances or substance families are regulated *as if* they were human carcinogens under one or more regulatory statutes.[54] OSHA has stated that "most substances do not appear to be carcinogenic," but any conclusions about the total number of discrete substances that ultimately will be shown to be human carcinogens are vulnerable to criticism in light of the extremely small number of substances tested and the questionable validity of many of the tests that have been conducted.[55] It is unlikely that testing will ever keep up with the number of new chemicals introduced commercially each year, estimated at approximately 1,000,[56] let alone test all of the 70,000 currently in use.

Cancer only heads the list of health problems that toxics can cause. Mutagenicity, teratogenicity, neurological disorders, and chronic lung illnesses all have been associated with various toxic substances. Some substances essential to life in small doses, such as nickel, become toxic at higher levels.[57] The criterion for regulation under the Toxic Substances Control Act is simply whether a particular substance presents "an unreasonable risk of injury to health or the environment."[58] Despite this comprehensive definition of toxicity, only chemicals that have toxic potential at low, even extremely low, exposures constitute a persistent policy dilemma. Lead, for instance, is believed to cause lasting adverse effects in children at blood levels slightly over 30 micrograms per 100 milliliters.[59] OSHA initially promulgated permissable exposure level for benzene in the workplace at ten parts per million,[60] and some scientists believe that dioxin can be dangerous at atmospheric concentrations measured in parts per billion.[61] Since OSHA and EPA subscribe to the working hypothesis that cancer can be caused by contact with a human carcinogen at any exposure level, carcinogens pose the problem of low tolerance carried to its logical extreme.[62]

Control costs. These health effects would create no serious dilemma if eliminating them were not so expensive or disruptive. In the past several years, we have seen dramatic evidence of the economic burdens of toxics control. The EPA has estimated the cost of cleaning up the worst identified hazardous waste site at $22.7 million. Estimates for the 15,000 controlled sites range between $3 million and $5 million. These figures may be dramatically low, because they are based on the possibility of transferring the hazardous wastes to secure landfills, but an OTA study concludes that the landfill techniques that formed the basis of EPA's estimates may be less than safe. Adopting EPA's assumption, the price for clean-up of 15,000 sites would be $60 billion.[63] Others have estimated as much as $260 billion for the same clean-up.[64]

As the toxics focus shifts from loss determination to loss prevention, the benefits from prevention should be considered. Estimating the benefits accruing to specific programs to improve health and safety is notoriously difficult,[65] but even though much uncertainty attends all those efforts, estimates are discouraging. OSHA and EPA cost-effectiveness figures are much higher than those of other federal agencies in the health and safety field. One study has estimated the median cost per life saved ranges from $64,000 at the National Highway Traffic Safety Administration, to $102,000 at the Department of Health and Human Services, to $50 at the Consumer Product Safety Commission.[66] In contrast, median estimates for EPA and OSHA, including some regulatory studies dealing with matters other than toxics, were $2.6 million and $12.1 million, respectively.[67]

These are large costs. What fragmentary evidence we have reveals that efforts to eliminate the last few toxic particles will be even more expensive. One study of EPA and OSHA carcinogen regulations has suggested a range in costs from less than $200,000 per annual cancer death avoided for OSHA's asbestos standard to a median estimate of $73.5 million per annual cancer death avoided for EPA's proposed 97 percent reduction in ambient benzene levels.[68] In another instance, an analysis of the incremental cost per cancer death avoided as a result of acrylonitrite regulation for one industry segment progressed from $8.12 million for improving from the status quo ante to 2.0 ppm, to $98.45 million for improving from 2.0 ppm to 1.0 ppm, to $860.23 million for improving from 1.0 ppm to 0.2 ppm.[69]

The low-level exposure problem makes these last few particles extremely important to risk management and public policy, for it is also apparent that not controlling toxics is still an expensive strategy. No comprehensive studies have calculated the medical expenses and insurance costs of treating all victims of toxic hazards, to say nothing of attempts to attribute a full economic value to lost earning capacity due to death or debilitating illness and lost industrial productivity from absenteeism or loss of skilled personnel. However, special situations have produced some estimates. For example, Paul MacAvoy has estimated the total liability accruing from the 18,000 asbestos-related claims filed so far to be $40 billion, based on "cautious estimates of death rates, caseload, and court award."[70]

THE DYNAMICS OF POLICY EVOLUTION

This section identifies several broad changes that have occurred in American society, technology, and values and analyzes how these changes have helped shape our current toxics regulatory structure.

The Growth of Toxic Substance Production

We begin with a technological imperative: in the postwar era, toxics became plentiful in the American economy to an unprecedented extent. Today, the United States produces and uses over 70,000 chemicals, of which the Environmental Protection Agency speculates that 20 percent may be carcinogenic.[71] Twenty-five percent of the 1,500 active ingredients in pesticides are suspected carcinogens.[72] The National Institute of Occupational Safety and Health lists 28,000 chemicals as toxic, and of these it designates 2,200 as suspected carcinogens.[73] These chemicals do not reside safely inside laboratories. They are the lifeblood of a major segment of the American economy. The chemical and allied products industries employed 1.065 million workers in 1984, and another 12 million worked in chemical processing.[74] The industry itself produces goods totaling $189 billion in value, amounting to 6 percent of the gross national product (GNP) in 1984.[75] Many chemicals are intermediate products, used in the manufacture of refrigerators, radios, televisions, flooring materials, pipes, and toys.[76]

Toxics are ubiquitous in waste streams as well as in manufactured products. A preliminary study of the Los Angeles air basin found 128 organic compounds or compound groups, including chloroform, carbon tetrabromide, vinyl chloride, and benzene.[77] Another analysis of the air in Los Angeles identified fourteen polycyclic aromatic hydrocarbons, including two carcinogens—benzo(a)anthracene and benzo(a)pyrene.[78] A National Academy of Sciences study reports that over 700 organic contaminants have been identified in drinking water throughout the country and that these represent approximately 15 percent of the total weight of organic matter in the water, suggesting that the total number of chemical contaminants could be much higher.[79] Estimates of the number of hazardous waste dump sites, which can become the source of further contamination both of drinking water supplies and of the ambient air, range from a low of 4,802 in an industry study[80] to a high of 50,000,[81] the upper estimate of a government study that was relied upon heavily in drafting the Superfund legislation. Of these sites, the government study estimates that about 2,000 could present serious dangers to human health, whereas the industry study put 431 sites in the health-endangering category.[82] There is no way of knowing what chemicals are in many of these sites.

Toxics enter the human body from these waste streams as well as through exposure to goods manufactured with toxic chemicals, such as pajamas, foam insulation, and lead-based paints. Other routes are still more direct. Fairly common foods contain suspected toxic substances. Caffeine, for instance, has caused chromosomal damage in laboratory animals, enough to cause the Food and Drug Administration (FDA) to warn that it is "inappropriate to include caffeine among the substances generally recognized as safe,"[83] although subsequent studies have cast doubt on this conclusion.[84] The FDA has banned one artificial sweetener, cyclamate, that used to claim a substantial market share and viewed another, saccharin, with suspicion but was restrained from banning it by congressional action.[85] Nitrites, chemical preservatives commonly used in foods such as bacon and smoked hams to control botulism, become transformed in the digestive process to nitrosamines, compounds believed to be carcinogens.[86] We also create toxics during the preparation of food, as when meat or fish is broiled, producing benzo(a)pyrene.[87] As frightening as the known or suspected toxic compounds in food may be, worse fears may come from the seemingly never-ending number of such substances as knowledge of toxic effects, superior detection equipment, and ongoing testing produce a steadily growing list of suspects.

The production, dissemination, and disposal of chemicals constitute critical control points for the reduction of toxics entering the environment, but other industrial processes also cause concern. Lead was a standard ingredient in all gasoline until production of some lead-free gasoline was mandated in 1973.[88] Petroleum refining and associated processes continually emit a complex mixture of petrochemical gases. Coal-burning power plants (as well as the currently experimental processes that produce liquid and gas fuels from coal) produce gaseous emissions that include polycyclic aromatic hydrocarbons, while other toxic substances, including arsenic, cadmium, fluoride, mercury, molybdenum, antimony, and selenium, affix themselves to particulate emissions.[89] The discharge of those substances was substantial enough to call into question EPA's original policy of controlling particulate emissions on the basis of total suspended particles,[90] a strategy that ignores the tendency of toxic material to adhere to the smaller particulates that current technology is least successful in capturing.[91] Under EPA's current regulatory strategy, the smaller, more dangerous particulates are being much less successfully controlled than the larger, less dangerous ones.[92]

Much of our awareness of the problems of toxic exposure is new since World War II, as is a great deal of the exposure itself. The war created a need to synthesize natural materials, such as rubber, when supply lines were threatened or cut off.[93] (DDT came into commercial prominence after the war directly as a result of its success in controlling malaria in the Pacific theater.[94]) After the war, the industrial infrastructure and the research and development efforts of the petrochemical industry took off in the direction of consumer goods. Between 1940 and 1960, production of synthetic organic chemicals increased 13 times.[95] Between 1960 and 1980, the increase was a still-substantial 4.5 times,[96] so that production had increased almost sixty-fold in forty years. The synthetic organics account for a significant share of toxic materials, suggesting that the postwar chemical industry, "the first modern industry," has been something of a Faustian bargain.[97]

The early court cases over toxics regulation largely concerned the control of synthetic organic chemicals.[98] Other early controversies also concerned chemical legacies of wartime or immediate postwar technological changes. Asbestos, for example, was heavily used in shipbuilding during the war, and much of our epidemiological evidence of its toxic effects comes from studies of war-era shipyard workers.[99] The quantity and dispersion of lead can be traced to its addition to gasoline as an antiknock ingredient, a direct result of the production of high-combustion engines after World War II.[100]

Protection of Traditional Health and Property Values

The existing legal system proved entirely inadequate to protect long-established individual rights from invasions by the burgeoning production of toxic substances brought on by the technological changes just described.[101] Under settled principles in the law of nuisance, trespass, and personal injury tort, actors can be held liable for the injuries they cause to the person or property of another.[102] Polluters were never an exception to this rule,[103] but by the middle of this century, the procedures of the common law were no longer adequate to control the technology that caused pollution-related harms.

Litigation at common law must be initiated by the party asserting a rights violation. The plaintiff must commit time, energy, and resources to a lawsuit; the adversarial system does not supply a champion to absorb those costs. Issues of scientific and medical complexity, common in environmental litigation, add substantially to the plaintiff's burden.[104] Modern technologies characteristically disperse relatively small damages among a considerable number of individuals. Given the dynamics of the litigation process, any one individual frequently lacks an adequate incentive to pursue even a valid claim for money damages.[105]

Before these technological harms became widely known, there was little incentive to adjust the procedures for recovery so that widely dispersed, relatively minor damages could be recovered, and such modifications would arguably have been unjustifiably costly.[106] As the number of these little insults grew, however, a point was reached where the aggregate size of such claims was so large that the expense of recovery was justifiable.[107]

This point was reached in the postwar period. The technological advances already noted facilitated the production of more and more toxic material.[108] Increases in population after World War II meant that dispersions of toxics affected more and more people. In addition, affluence influenced the value placed on that harm. It is widely assumed that demand for environmental quality increases with wealth,[109] so that as wealth increases, the avoidance of a given degree of environmental harm becomes highly valued. The combination of these factors produced a great increase in the valuation of the damage that technology causes.

The protection of traditional values of human health and property would have justified some institutional reforms to respond to the changing pressures of technological change and social advance. The class action lawsuit was for a time promoted as one such possible reform. The class action responds to the problem of individually small but collectively significant damage by providing the class action lawyer an incentive to prosecute an action on behalf of all affected individuals.[110] However, although the class action device has been occasionally used in environmental litigation, neither the courts nor the Congress has devoted much energy to improving it for use in the toxics field, and in fact a number of doctrinal obstacles emerged in the 1960s and 1970s to stifle, rather than promote, class actions.[111]

Innovation by the courts themselves was stymied in large part by the very same complexity that currently slows the efforts of administrative agencies.[112] Courts were also properly sensitive to the difficult political and distributional issues raised by decrees that impose liability or injunctions on only the defendants before the court, leaving competitors with at least a temporary advantage, and threatening the economic stability of the geographical area affected by its limited decree.[113] In sum, the courts did not generally respond to increased pressure from technological and social change on traditional values by making class suits more serviceable. This failure in turn increased pressure on the legislative and administrative apparatus of government to do so.

Where individual rather than class claims were brought, the courts have been more responsive. When exposure to toxics first takes place, the situation resembles the one of widely dispersed, low-level harm described earlier. The "harm" that has been inflicted at that stage is the risk of future injury, a risk that will frequently be a small addition to the background risks already present. Later, the harm coalesces into a much smaller number of highly concentrated damages, and the case more closely resembles a classic individual tort claim.

For a time, individual plaintiffs faced extraordinary obstacles in pursuing these claims. Statute of limitation rules, for example, did not take into account the long latency periods of toxic injuries, so claims were barred before plaintiffs could have known whether they had been injured or to what degree.[114] Courts also discredited statistical evidence linking the defendant's chemical to the plaintiff's injury.[115] When injuries were the result of the joint or cumulative effects of several defendants' actions or of the actions of only one defendant whose responsibility could not be isolated, the normal obligation of a plaintiff to allocate liability to specific defendants often prevented recovery.[116]

In recent years, courts have greatly relaxed these obstacles.[117] Plaintiffs today have greater opportunities to litigate and a much greater chance of prevailing. The rapid growth in "toxic torts" litigation testifies to the efficacy of some recent procedural reforms.[118]

One important type of case has remained largely unaffected by these reforms, however. This is the case in which the plaintiff's injury *may* have been caused by the defendant's behavior, but it may also have resulted from causes for which the defendant is not responsible, and the plaintiff cannot present convincing argument and evidence that the defendant's conduct was the more likely actual cause.

Whenever a toxic chemical does not leave its "signature" on its victims, usually by causing some esoteric disease not likely to be caused by other sources, exposure to toxics can result in this type of case.[119] The American public currently believe that modern technology has created an environment in which risk and hazards are beyond our ability to control or manage.[120] The present inability of lawsuits to provide redress in these nonsignature situations exacerbates that sense of powerlessness and further increases the pressure for legislative response.

Thus, two types of technological harms have been impervious to any satisfactory adjudication by courts: widely dispersed, low-level harms sufficient in the aggregate to justify imposing liability, and discrete, severe harms that could not be traced to a specific defendant, although statistical calculations warranted the conclusion that defendant had caused some of the total number of such harms befalling an exposed population.[121] These failures to protect traditional values supplied much of the fuel for institutional reform vesting authority in administrative agencies.[122]

Responding to Changing Social Values

The values Americans attach to wilderness preservation, to reverence for natural beauty, and to nondegradation of natural resources are not new; environmentalists have long maintained that environmentalist policies affirm principles latent in our heritage, principles to which we have always assented to some degree.[123] However, until recently these values were not so dominant as they are at present, nor were they fully recognized in law.

Our shared concept of what values should be protected by legal rules and institutions has been changing from more exclusively private values that markets are better equipped to protect to social values that require some corrections and adjustments to traditional market actions.[124] Like the more traditional pollution effects just described, these social goods can be widely dispersed, but to a potentially far greater degree.[125]

Legal rules and institutions have been adjusting to reflect the new emphasis on environmental quality and amenity values. Although cases recognizing such values as independently sufficient grounds for nuisance litigation remain difficult to find, this is largely because statutory recognition and protection of them has leapfrogged common-law developments. However, the judiciary has recognized that harms to amenity values are genuine legal injuries. In 1972, the Supreme Court declared that harm to "the aesthetic and environmental well-being" attributable to the existence of pristine public lands was sufficient "injury in fact" to allow suit to challenge federal approval of commercial development on those lands.[126] Zoning for purely aesthetic purposes used to be rejected by many courts as outside the government's interest in public health and welfare. Erosion of this principle first occurred through subterfuge,[127] but in recent years aesthetic values have been fully acknowledged as legitimate bases for land use regulation, whether the countervailing interest has been private property rights[128] or First Amendment free speech rights.[129] Open space zoning,[130] wetlands regulations,[131] and historic landmark preservation[132] all rest substantially on the social values of environmental amenities, and all have been sustained by courts.[133]

While such values have been judicially recognized, individuals still lack adequate incentives to protect widely shared interests. Legal reforms such as permitting individuals so inclined to champion those social interests can improve but not totally redress this bias.

Suits protecting environmental amenities are typically suits for injunctions. Even with the widespread availability of attorneys' fees, such suits lack the extra incentives of cases for money damages, in which the prevailing plaintiff's attorney can receive a substantial percentage of the damage award as compensation for her services.[134]

Protectors of emerging social values in the environment urged increased action upon Congress. Within the evolution of the federal legislation that resulted, one can trace the increasing emphasis on those values. For example, the Federal Insecticide Act of 1910 was written primarily in response to complaints that farmers were not equipped to evaluate the efficacy claims of pesticide peddlers, and so could not know when products were adulterated or their contents misrepresented.[135] Adequate consumer information is a prerequisite to efficient market exchange of private goods; the object of the act was thus to correct a narrow market defect by ensuring the validity of information provided by pesticide salespeople to farmers.[136] As already mentioned, the 1910 act was superseded by the Federal Insecticide, Fungicide and Rodenticide Act (FIFRA) in 1947. In the intervening years, Congress had taken some initial steps to protect consumer health from hazardous pesticide residues on foodstuffs.[137] FIFRA expanded the 1910 act's interest in efficacy to include some protections from adverse health effects due to human contact with pesticides.[138] These health-oriented protections remained consistent with a predominantly private goods valuation of pesticides, because attention was fixed on the first users of the product, the farmers, and on the consumers of the foods. Because the consumer had no adequate means of detecting hazardous residues, the rationale for some government intervention remained fundamentally identical to that of the 1910 act.

In retrospect, FIFRA was a transitional statute. It marked the shift in pesticide regulation from after-the-fact testing and sanctioning to an anticipatory policy of requiring registration of any pesticide before distribution. FIFRA, however, also implicitly assumed that pesticides did not migrate from their point of use.[139] Consequently, it failed even to appreciate the emerging social goods implications of pesticides. Rachel Carson's *Silent Spring* attacked that misunderstanding: many pesticides can and do remain in the environment for a considerable time, where they are consumed by insects, animals, and humans, causing appreciable further damage. The sphere of adverse effects is thereby tremendously expanded both in quantity (many more individuals than the farmer and consumers of specific products may be harmed) and in kind (insects and animal species that are not the targets of pesticide control can be killed).[140]

In 1972, FIFRA was amended by the Federal Environmental Pesticide Control Act (FEPCA). The extensive public and congressional debates that led up to the enactment of FEPCA focused both on human health and on environmental quality as aspects of the pesticide problem.[141] The act acknowledged these two dimensions by stipulating that pesticides should not be registered for use if they would cause "unreasonable adverse effects on the environment," defined to mean "any unreasonable risk to man or to the environment, taking into account the economic, social and environmental costs and benefits of the use of the pesticide."[142]

The Evolutionary Path of Toxics Policy

The preceding sections have identified four developments in postwar American society that have had important influences on the evolution of toxics policy: (1) awareness of the low-exposure dilemma; (2) increased production of toxic substances that exacerbate the harm from exposure to toxics; (3) threats to traditional health, safety, and property values from those toxics; (4) threats to emerging public values connected to the environment. Because the judicial system did not sufficiently respond to these influences, pressures have been concentrated on the legislative and administrative arms of government.

The history traced thus far provides initial insight into the first of the two broad trends that unify toxics policy. Federal legislation has been redefining allowable risk downward, because the combined institutions of the market plus the (insufficiently responsive) private law regulatory system permitted toxic exposures to occur that were inconsistent with the combination of traditional health and property values and our emerging environmental amenity values. By looking in somewhat more detail at the manner in which these four influences affected policy development, as well as at the timing of their influence, we can gain further insight into the second trend, the use of centralized regulation administered under a multiplicity of separate statutes.

The delayed appreciation of the low-exposure dilemma. Until the mid–1970s, legislation treated toxics largely as a subcomponent of other, more general issues, especially the media-oriented issues addressed in the major pieces of air and water pollution control legislation. In the 1980s, by contrast, we understand toxics as a distinct problem that cuts across media.[143]

Technological reasons cannot explain the late emergence of the toxics problem. Many types of toxics production accelerated rapidly in the immediate postwar period, and the growth rates of the traditional pollutants in air and water—control of which received increasing attention throughout the early 1960s—were certainly no larger than those of toxics in the same period. Similarly, our knowledge of the adverse health effects of toxics has grown dramatically since the late 1960s, but it was by no means absent in the earlier periods. The toxic effects of many heavy metals have been understood for decades, for example. The asbestos litigation has revealed that industry was aware of the toxic impact of asbestos exposure at least as early as during the war; Rachel Carson's 1962 exposé of pesticides drew on research results of the 1950s; and passage of the Delaney Clause in 1958 demonstrated a legislative awareness of the hazards of carcinogenic substances.

Damage caused to traditional human health values does not explain the pattern of toxics policy growth, because any focus on such effects attempts to read our current emphasis on such effects back into a historical record that reflects different priorities. The environmental movement developed initially because of perceived threats to the public values connected with the environment, not because of the direct threats of industrial activity to health.[144] Although health has always been prominent among the concerns raised by environmental disruption, the movement placed even more emphasis on man's potential to destroy needed ecosystems as the mechanism through which human well-being would be harmed.

To express its concerns, the rising environmental movement drew on the science of ecology. For example, the Council on Environmental Quality's first annual report emphasized ecological principles in stating environmental issues,[145] and the entire movement stressed the protection of natural systems and the connections between human activity and environmental disruption. In many respects, the emerging public amenity, preservationist, and ecological values drove the movement and defined its agenda.

The focus on ecology led naturally to formulating environmental concerns in terms of the basic components of the ecological cycle—air, water, and land. Furthermore, in taking this approach, the environmental movement was able to build on an existing statutory structure. Federal water pollution legislation had existed since before the war, and air pollution legislation was sought after the Donora deaths in 1948 and the realization in the 1950s that auto exhausts and smog were related.[146] Congress had passed water pollution control acts in 1948, 1956, 1961, 1965, and 1966, and air pollution control measures in 1955, 1963, 1965, and 1967. The existing federal presence in control of air and water pollution laid a foundation for the new environmental measures enacted through the Clean Air Act Amendments of 1970 and the Federal Water Pollution Control Act Amendments of 1972.

Under a media approach to environmental legislation, various pollutants can be distinguished, but the critical role they all play is essentially the same: they disrupt natural systems. For purposes of organizing the political agenda, specific pollutants are subsidiary to natural media. Thus, toxics were not immediately recognized as posing problems that could benefit from an approach that started with the toxic substance itself and assessed all exposure routes, rather than dealing with toxics in land, air, and water separately.

Over time, direct health effects of toxics demanded much more attention, until today public fear of health effects dominates the news after each new incident of exposure or suspected exposure. Increasingly, the feared health problems are chronic and long-term effects caused by low-level exposure, rather than acute and immediate effects caused by high doses. This shift in emphasis further explains the late recognition of the toxics problem. Throughout the construction of our original environmental legislation, incidents of high visibility, such as the Donora smog, the proposal to dam the Grand Canyon, and the fish kills in the Great Lakes, played an important catalytic role. In contrast to such drama, toxics at low levels are often nearly imperceptible, and they can cause harm years after exposure has ceased. It has taken longer to turn to the less immediate, frequently latent problems posed by long-term exposure to low levels of toxics.

Even after chronic effects have been acknowledged, only time and experience can reveal the dimensions of the low-exposure dilemma. We can trace an evolution within the toxics statutes themselves, as the definitions and interpretations of allowable risk, although still ambiguous, have been transformed over time in appreciation of the low-exposure problem. To trace this evolution, it is helpful to treat product regulation separately from pollution regulation.

Pollution statutes address the presence of toxics in waste streams from industrial, agricultural, commercial, and municipal courses. Early statutory formulations of acceptable amounts of toxic chemical pollution were extremely strict and generally health-based. The Clean Air Act Amendments of 1970, for example, required toxics to be reduced to a level sufficient to protect human health with an ample margin of safety.[147] The 1972 Clean Water Act Amendments imposed a similar requirement for water-borne toxics.[148] The implication of the health-based standards was that toxics that may cause adverse effects at the lowest levels of exposure must be entirely eliminated. As federal agencies began to administer these early standards, they confronted the low-exposure dilemma that some of these toxics can be eliminated only with great social or economic dislocation.

The low-exposure dilemma placed great internal pressure on the absolutist health-based standards. Through a combination of actions by federal agencies, courts, and the Congress, in the 1970s these standards were progressively modified to permit some non-zero levels of toxics exposure.[149] This process has not been easy, nor have the revised standards avoided the ambiguities noted earlier, but it has been occurring. In both air and water regulations, technological and economic feasibility standards have replaced exclusively health-based ones. TEF standards are the regulatory embodiment of "doing everything humanly possible" to avoid toxic exposure. Because an underlying premise of these standards is that technologies will continually improve, and thus eventually will allow the total elimination of toxics exposures, these standards have the virtue of not explicitly acknowledging that some risk must be tolerated. On the other hand, by stopping regulation at the point of feasibility, such standards avoid massively dislocating economic effects. They thus present a temporary political accommodation to the dilemma posed by low-exposure risks and high avoidance costs.

Where economic or social dislocations do not appear likely, on the other hand, absolutist health-based standards for toxics represent the current politically acceptable risk level. There is some evidence that when the absolutist air and water standards were originally written, it was assumed that most toxics did *not* present a low-exposure dilemma. The legislative history of the 1970 Clean Air Act Amendments, for instance, suggests that most members of Congress believed that toxics provisions would apply to only a handful of chemicals and that controlling those would not be severely disruptive.[150] Once the low-exposure dilemma was confirmed, however, the regulatory apparatus quickly recoiled from the absolutist position. In contrast, the absence of an enduring low-exposure dilemma probably explains the longevity of the absolutist Delaney Clause. Despite persistent criticisms of the strict application of the clause, it may be that controlling food additives in this way simply does not involve substantial dislocations, because additives generally have substitutes or serve largely cosmetic functions.[151] As an exception testing the rule, Congress instructed the FDA not to ban saccharin when some research suggested its carcinogenicity. At the time, the lack of a substitute artificial sweetener made the threatened ban potentially disruptive.[152]

Product regulation statutes govern such things as food, drugs, consumer products, new chemicals, and pesticides. Their risk-level standards have also evolved under the pressure of the low-exposure dilemma, although textual changes in the statutory standards are less evident. Product regulation statutes from their inception required the administering agency to make decisions on the basis of a comparison of the product's benefits to its hazards. Their language has proven sufficiently flexible to accommodate changes in the valuation of hazards and in what counts as a hazard.

Technological alternatives enter into the balance through the consideration of plausible substitutes for the subject product. Roughly comparable substitutes for a product, if they can reduce risk at a tolerable cost, perform approximately the same role in product regulation that feasibility assessments play in pollution legislation. Such substitutes enable an agency to eliminate or restrict the use of the product without being unduly disruptive.

The Growth of Centralized Regulation. Complete historical investigation of how highly centralized regulation of toxics came to dominate state and local regulation as well as private lawsuits is much too complex to permit more than some preliminary observations here, yet it is too crucial an aspect of toxics policy to go entirely without attention.[153]

Immediately after World War II, the United States had great confidence in administrative government.[154] During the New Deal era, central planning by government had won its battle for intellectual and political respectability. The battle for judicial approval had also been won in the 1930s, in the face of counterarguments that a more radically unregulated economy and social structure were constitutionally mandated. The war itself had been successfully waged through a highly concentrated apparatus headquartered in Washington. Although it had pinched domestically, the effort had succeeded and was accorded the respect that America routinely grants to success. Dean Landis's encomium to administrative expertise had been published just before the war,[155] and in passing the Administrative Procedure Act in 1946, Congress ratified the considerable value of administrative agencies. Scholars examining administration in the period immediately following the war generally took for granted the propriety of substantial controls on the economy by independent commissions and executive agencies.

Over the same period during which the environmental movement grew, confidence in administrative agencies ebbed dramatically, a story that has been amply chronicled elsewhere.[156] The impact of that decline on the environmental movement has been decidedly schizophrenic. Consistent with sharply critical analyses of administrative performance, environmentalists argued that federal agencies were a major component of the environmental problem. The National Environmental Policy Act, which marks the legislative coming-of-age of environmentalism, aims specifically at altering the irresponsible promotion by federal agencies of large-scale development projects.[157] Early environmental litigation, which played such a critical role in organizing public opinion on environmental matters, concentrated on suits against federal agencies for undertaking or approving such large-scale projects.[158]

In great measure, however, environmental advocates have sought better administration, not less administration, to control the problems of administrative agencies themselves and the problems of environmental degradation. In the 1970s, federal toxics policy lurched toward more and more administration, and more and more administration of the anticipatory variety. For this development to be something other than a flat disregard of the failure of administration, it must be explained either by a conviction that the evils of administration are curable, or by a belief that even flawed administration is the lesser of evils. Undoubtedly, the explanation partakes of both.

One congressional response to administrative failure has been to reduce the opportunities for agencies to be captured by their regulatory clients.[159] Through amendments to the enabling legislation, as in the case of the land management agencies, Congress has redirected the substantive mandates and goals of agencies. Increased oversight hearings and investigations are thought to give Congress other control points over agency actions.[160] The "hard look" style of judicial review also responds to the phenomenon of agency failure and has a large base of support in the Senate for this reason.[161]

Statutes have imposed nondiscretionary duties on federal agencies, as well as strict statutory deadlines for performing specified regulatory tasks. Congressional dissatisfaction with the performance of environmental agencies under the Reagan administration has resulted in increased use of these strategies, as well as producing newer ones, including the legislative "hammer," pursuant to which an agency is given to act on a measure a stipulated time, after which a statutory measure goes automatically into effect.[162]

Congress has also created new agencies to address environmental problems. EPA, OSHA, and CPSC are the primary regulators for toxics. All are modern "mission-oriented" agencies created in the 1970s.[163] Because their mandates are organized by problem rather than by industry, they appear less vulnerable to some of the more obvious forms of capture by a single regulated constituency, although they still suffer from a need to rely heavily for information on regulated groups, and regulated groups still have a staying power before these agencies that gives them substantial advantages.[164]

The newer agencies also benefit from much stronger economic justifications. The critiques of agency performance in the 1960s and 1970s concentrated on the older, economic regulatory agencies operating under broad mandates to regulate a specific industry "in the public interest." The economic rationale for such agencies relies on monopoly tendencies within the industry or on other perceived ills, such as "destructive competition," that would materialize if the industry were deregulated.[165] The validity of these rationales has been seriously challenged in recent years on both empirical and theoretical grounds.[166] Because of this challenge, the deregulation movement has achieved some of its most notable successes in deregulating, or partially deregulating, industries within the jurisdiction of the older economic regulatory agencies.[167]

These attacks on the theoretical justification for the economic agencies cannot be extended so easily to the newer environmental, health, and safety agencies. Even assuming "that an unregulated marketplace is the norm and that those who advocate government intervention must justify it by showing that it is needed to achieve an important public objective that an unregulated market cannot provide,"[168] the general missions of these agencies survive that test. Environmental problems can be interpreted as a species of "market failure," so that agencies might legitimately attempt (1) to make industrial sources bear the full costs of their production of pollutants, a result that will improve economic efficiency under assumptions broadly applicable to technological risks;[169] (2) to supply appropriate levels of the social goods associated with environmental quality, because neoclassical economic theory concedes that the unregulated marketplace will not do so.[170]

The modern environmental, health, and safety agencies represent one of two responses the environmental movement has made to technological risks and their association with such large-scale organizations as the nuclear power industry, petrochemical firms, and automobile manufacturers. To control the threats of such organizations to individuals and environmental values, they embody an effort to construct organizations of countervailing power.

> In a highly technological society such as ours, the need for increased regulation is manifest. It is inconceivable to think of "lessening the regulatory burden," as some put it, at a time when private industry has the power to alter our genes, invade our privacy, and destroy our environment. Only the government has the power to create and enforce the social regulations that protect citizens from the awesome consequences of technology run amok.[171]

The opposite response to large-scale institutions is to dismantle them. Amory Lovins's eloquent plea that public utility companies be allowed to wither and die in favor of "soft energy paths,"[172] as well as the appropriate technology movement, which asserts "small is beautiful," epitomize this polar response to threats associated with large-scale technologies.[173] The deeper vision of the appropriate technology movement prefers small-scale, communal living patterns—and in this regard it resonates with certain visions in utopian literature that also have a heritage within American thought.[174]

The soft paths tradition illustrates that the link between the substantive program of environmentalism and centralized agencies is not ineluctable. In its aversion to large-scale institutions, the soft paths movement shares some common ground with the deregulation movement and that movement's emphasis on decentralized decision making. In their social orientations, however, the two fall far apart. The land ethic social structure animating much of the soft paths branch of environmentalism stresses communalism and interconnection, not the neolibertarian, isolationist philosophy that drives much of the deregulation literature.[175] What is more, when soft path proponents have turned to practical policy advocacy, they frequently have embraced the strategy of countervailing power. Public utility commissions are supposed to ensure that utility companies explore energy conservation measures, for instance.

One reason for the dominance of the approach of countervailing power held by centralized agencies is clearly strategic. Environmental advocates must be concerned with maximizing the impact of their scarce advocacy resources, and they can usually accomplish this with national-level legislation. Working legislation through a temporarily receptive Congress promises much greater returns than decentralized efforts at the state and local levels. Setting federal agencies to the task of implementing national legislation further enables national environmental advocacy groups to increase their influence. Instead of policing the activities of fifty state agencies or litigating common-law actions in fifty different court systems,[176] environmental organizations can monitor the work of one central bureaucracy. If the rules and general operating instructions that agencies formulate to direct their work can be monitored and influenced, agencies can to some degree be expected to carry through on them. Thus, for example, the formulation of a definition for the best practicable technology for waste removal from liquid waste streams is a much more significant event than the subsequent application of that definition to specific polluters. This strategic view explains why *Dupont v. Train,*[177] establishing EPA's authority to issue nationally uniform effluent standards for water pollution control, was such a critical case, and why environmental groups seldom participate in EPA's rule-making proceedings for specific industrial subcategories.[178]

In the Congress, the reliance on administrative agencies to carry out the environmental agenda has partially been attributable to institutional inertia. Although both the criticisms of agency activities and the possibility of alternative mechanisms for intervention, such as the creation of marketable permits for pollution or legislating changes in common-law standards of liability, now seem firmly established, such was not the case from the mid–1960s to the early 1970s, when the basic structure of governmental response was being laid down. Even in the academic community, the property rights theory that underlies many such regulatory alternatives was relatively new. R. H. Coase's seminal article appeared in 1960,[179] and it was decidedly oblique in articulating alternatives, being more concerned with discouraging legislative action altogether. It took writing by Harold Demsetz, Guido Calabresi, J. H. Dales, Garrett Hardin, and others through the 1960s to develop the policy relevance of property rights theory.[180] None of these ideas could be expected to migrate instantaneously from academe to Congress, especially because strong objections to the entire property rights approach had also developed.[181]

Thus, in the formative years of environmental legislation, alternatives to traditional regulation never received a full hearing. The reasons for centralized regulatory development run deeper than failure to consider alternatives, however. The key idea behind "resilience" is to permit private parties to learn from their mistakes. Initiating change occurs at the individual level; control by society over the problem is therefore highly indirect, operating through individual actions. This basic mechanism is much too indirect to satisfy the demands either of public concern, or of the political system, in which members of Congress need symbolic and dramatic results for which they can claim credit.[182]

The political dangers of exercising control come when control shades into meddling. Alfred Kahn used to urge his colleagues in the Carter administration "to keep the hand of government as invisible as possible."[183] Bureaucratic management through anticipatory mechanisms seems wholly to fail that test. So far as congressional politics is concerned, however, administration provides a way to distance many criticisms. Members of Congress can claim full credit for dramatic legislation. The events that truly chafe on the private sector, the specific acts of enforcement and prescreening, will come later, and they will primarily be the responsibility of the agency. When those events occur, members of Congress retain the option of sharply criticizing the agency.[184]

To recapitulate, centralized regulatory agencies began to administer new definitions of allowable risk in an atmosphere of public and intellectual confidence in administrative agencies, their use was consistent with the application of the principle of countervailing power as applied by the environmental movement to its own agenda, they had strategic advantages for the modestly funded and staffed advocacy arm of environmentalism, and they presented certain political advantages for the Congress. The interests of one more affected constituency deserve consideration. The growth of environmental legislation cannot be understood without appreciating the role played by opponents of stringent environmental controls.[185] Most environmental legislation has been the product of a clash of proponents and opponents, not of a well-defined national consensus on specific issues.[186] Initial toxics provisions, such as the first versions of toxics regulation in the air and water acts, may be exceptions, as they were enacted with little organized opposition or debate. The harsh realities of the low-exposure dilemma were not then palpable enough to support resistance to a very popular environmental groundswell. Subsequently, however, these realities were brought home, and industrial organizations and other spokesmen for less regulation strongly resisted any further federal intrusions. Since that time, a pattern has emerged.

Consider the Toxic Substances Control Act as an example. In 1971, the Council on Environmental Quality's (CEQ) *Toxic Substances* report had called for legislation embodying the essential parts of what eventually became the Toxic Substances Control Act of 1976. For several years, the chemical industry flatly opposed such legislation. By 1975, however, industry was taking what Congressman John Murphy could subsequently call a "very constructive approach" to the legislation, attempting more to mold a bill they could live with than to oppose passage of any bill at all.[187] Industry's changing attitude contributed to a climate in which the bill could pass the Congress by substantial margins, 73 to 6 in the Senate and 360 to 35 in the House.[188]

This is not the occasion to analyze the strengths and weaknesses of the bill as enacted. What does bear noting is the pattern of opposition behavior, which shifted from initial recalcitrance to increasing conciliation. The history of environmental legislation as far back as the 1910 pesticides law,[189] as well as more recent bills,[190] reflects this recurring pattern. At some point, opponents of legislation realize that some form of regulation is going to pass and attempt to improve their position through cooperation and work on compromise language. One more reason for such shifts is that national legislation may well be an advantage to national industries once an issue has achieved such notoriety that state and local governments are considering regulation of their own. At that point, industries see that they will be better off with national uniform standards administered by federal regulatory agencies than with a melange of different local approaches. The history of the 1910 pesticide bill indicates that the infant chemical pesticide industry supported national legislation partly from a concern that state legislation might hamper growth.[191] The toxic substances bill reflects the same considerations. When the CEQ report was first issued, concern at the federal level appeared to be ahead of public sentiment and information, and little activity was occurring at the state and local levels. By 1975, this situation had changed. Toxics were becoming an extremely visible public issue, sufficient to motivate states and localities to consider regulatory responses.

National enterprises continue to urge federal uniform standards as opposed to multiple state and local controls in the toxics field, defying the Reagan administration's professed emphasis on turning regulation back to the states and localities. National oil companies and chemical associations successfully argued that the original Superfund legislation preempted state power to tax for Superfund purposes.[192] Industry has also mounted successful challenges to worker "right to know" state legislation, contending that the Hazard Communication Standard of the Occupational Safety and Health Administration preempts such laws.[193] Arguments favoring federal standards governing products liability are stimulated primarily by national industry, and turn largely on the interest in avoiding victimization by fifty different state products liability laws (and the accompanying forum-shopping practices of the products liability plaintiffs' bar).

CONCLUSION

Given its objective of illuminating the basic themes that have run through toxics policy and that help explain its evolution, this paper perhaps inevitably has underemphasized disagreements and dissenting points of view. Along the way and from this point on, each significant change in the structure or the implementation of toxics policy has sparked controversy and will continue to do so. Indeed, this may be particularly true for the future, for we have moved past the easiest targets of opportunity to improve the level of toxics exposure throughout the country. Progress from here will entail either increasingly severe controls or major alterations in lifestyles and in the processes that produce toxics.

The pressure for toxics regulation will continue to be generated by the forces examined here: threats to health and property values and threats to environmental amenity values that Americans increasingly hold. Whether the policy response predominantly employs centralized regulations depends on whether such regulation maintains the comparative institutional advantages outlined above. There are some reasons to question whether it will.

First, many toxics issues are locational and site-specific. Adversely affected groups will have some organizational advantages at the state, regional, and local levels in, for example, resisting a toxic waste disposal site, that they will lack at the national level. Second, an inventive plaintiff's bar has been probing the limits of tort law to provide compensatory damages for various types of toxics exposure; their successes will generate litigation-based pressure on the activities of toxics generators that is not directly dependent upon federal legislative or regulatory action.

Third, the process of centralized regulation has suffered near-paralysis in attempting to discharge its current burdens.[194] Toxic chemicals and chemicals that may be toxic are far too numerous to be addressed under existing administrative procedures and norms. Substantial movement via centralized agencies seems likely only if Congress and the president can agree on a series of reforms to simplify the regulatory process. Undoubtedly this will require sacrificing a degree of analysis and of process in exchange for regulatory output, because current administrative formats continue to err on the side of inaction in the face of uncertainty or protest.[195] Thus far, however, neither manufacturers nor environmental organizations have viewed such changes as being sufficiently in their own interests to promote them consistently.

Despite these considerations, the deregulation tide of the 1980s has left toxics regulation nearly untouched.[196] Federal administrative proceedings remain nearly the sole forums within which the debate over the appropriate levels of control for specific toxic chemicals can bring together the scientific evidence, analytical expertise, and interest group representations necessary to air the issues. If federal toxics policy is going to continue to evolve, it will have to reach some new working consensus on ways to end the regulatory bottleneck.

NOTES

1. The precise number of statutes depends in part on how the subsequent amendments to existing legislation are counted.
2. The Environmental Protection Agency (EPA), the Occupational Health and Safety Administration (OSHA), the Consumer Product Safety Commission (CPSC), the Food and Drug Administration (FDA), the Mine Safety and Health Administration, the Department of Transportation, the Department of Housing and Urban Development.
3. Environmental Law Institute, *Federal Environmental Law*, ed. Erica L. Dolgin and Thomas G. P. Guilbert (St. Paul: West Publishing Co., 1974).
4. Ibid., 709–12, 1103–4.
5. In anticipating the importance of the toxics problem, the Council on Environmental Quality's (CEQ) 1971 monograph, *Toxic Substances*, was truly pathbreaking. Although others, notably Rachel Carson in *Silent Spring* (Boston: Houghton-Mifflin, 1962), had articulated the problems associated with specific toxics, the CEQ study appears to be the first to treat toxics as a generic problem. The study presents no express definition of toxic substances, but it expresses concern over "many chemicals . . . because of their potentially toxic effects at extremely low levels of exposure and their presence in many media." *Toxic Substances*, 1. The

multimedia character of the problem largely drives the study, because the insufficiency of then-existing laws and regulations to deal with toxics is largely attributable to this characteristic. This essay defines the toxics problem somewhat differently.

6. Insofar as the political environment responds to catastrophes that seem to require dramatic and targeted responses, the chances that the regulatory regime for a problem like toxics will be enacted only after deliberate consideration of all the interconnected aspects of the problem are further reduced. See, e.g., J. Krier, "The Pollution Problem and Legal Institutions: A Conceptual Overview," *U.C.L.A. L. Rev.* 18 (1971): 429, 461 (suggesting this as a factor in the development of national air pollution legislation).
7. Under TSCA, if the EPA administrator determines that a risk to health or the environment associated with a chemical substance or mixture could be eliminated or reduced by actions taken under another federal law administered by the EPA, the other law must be used. 15 U.S.C. secs. 2605(c)(1), 2608(b) (1982). However, the administrator still has the discretion to take action under TSCA if he believes that it is in the public interest to do so. 15 U.S.C. sec. 2608(b) (1982). If the administrator determines that the risk could be eliminated or reduced by action taken under a federal law administered by another agency, then the administrator should notify that agency of the risk and defer to the actions taken by the notified agency. 15 U.S.C. sec. 2608(a) (1982). The administrator is further required to employ the least burdensome requirements to protect against risks regulated under TSCA, as well as to consult with other agencies in order to prevent duplicative requirements on those subject to regulation. 15 U.S.C. secs. 2605(a), 2608(d) (1982).

   RCRA establishes a regulatory program to manage and control hazardous waste. EPA is required to integrate all the provisions of RCRA with any other acts of Congress that grant regulatory authority to the administrator in order to avoid duplication in regulation. 42 U.S.C. sec. 6905(b) (1976). Furthermore, the secretary of the interior, with the concurrence of EPA, must integrate the regulations dealing with coal mining wastes, promulgated under RCRA, with the regulations developed under the Surface Mining Control and Reclamation Act of 1977. 42 U.S.C. sec. 6905(c) (1980). These added-on requirements to integrate and coordinate fall substantially short of integrating toxic programs, priorities, procedures, and staff.
8. The Interagency Regulatory Liaison Group, composed of representatives of EPA, FDA, OSHA, and CPSC, was established on August 2, 1977, as a program of cooperation to help eliminate waste and duplication in regulation by the federal government. *Env't. Rptr.-Curr. Develop. (BNA)* 8 (Aug. 5, 1977): 511–12. In 1979, the Food Safety and Quality Service of the Department of Agriculture joined the group. 44 Fed. Reg. 39,858 (1979).
9. For instance, the implicit cost per life saved of OSHA's benzene standard of 1 ppm has been estimated at $18.9 million (the standard was upset by the Supreme Court before implementation, Industrial Union Dept. v. American Petroleum Inst., 448 U.S. 607 [1980]), while the EPA's proposal to mandate 97 percent removal of benzene under the Clean Air Act is estimated to cost between $66.8 and $80.2 million per death avoided. Ivy E. Broder and John F. Morrall, III, "The Economic Basis for OSHA's and EPA's Generic Carcinogen Regulations," in *What Role for Government?* ed. Richard F. Zeckhauser and Derek Leebert (Durham, N.C.: Duke University Press, 1983), 250. While all estimates such as these are fraught with uncertainties, no one should be surprised by wide discrepancies, because the statutes have no provision mandating equal cost-effectiveness, the agencies generally resist making such calculations, the health-based and technological and economic feasibility-based (TEF-based) statutes employ criteria other than cost, and the balancing statutes do not require cost-benefit analysis.
10. OSHA's 1 ppm benzene standard was estimated to cost benzene refiners $103.8

million per death avoided, while tire manufacturers would invest $1.1 million per death avoided. See Broder and Morrall, "Economic Basis," 252. Studies of marginal abatement costs per quantity removed show no more uniformity. For instance, copper mines are calculated to pay an incremental cost of $184,000/kg of mercury removed under the Clean Water Act, while lead and zinc mills expend $24,000/kg. D. M. Jenkins et al., "Draft Final Report to the EPA: Incremental Cost Effectiveness of Water Pollution Abatement C–37" (Dec. 28, 1979). Meanwhile, the chlor-alkali industry pays $19/kg for removing mercury from its air emissions. D. M. Jenkins et al., "Draft Final Report on Incremental Cost Effectiveness for Airborne Pollution Abatement" (June 5, 1979), table 7.

11. The Interagency Regulatory Liaison Group (IRLG) suggested a common carcinogen policy in 1978, but it was never adopted by all the member agencies. Under the Carter administration, EPA and OSHA developed widely divergent carcinogen policies. In 1986, the Environmental Protection Agency issued a generic cancer risk assessment policy to govern EPA activity under all the statutes it administers. EPA Carcinogen Guidelines, 51 Fed. Reg. 33, 992 (1986). If adopted by other federal agencies, the guidelines could become the basis for coordinated cancer risk assessment. They have received considerable criticism, suggesting that they might not enjoy general acceptance after a change in administrations. See Howard Latin, "Good Science, Bad Regulation, and Toxic Risk Assessment," *Yale J. of Reg.* 5 (1988): 89, 95–106.
12. Some of them are discussed further in Christopher H. Schroeder, "Legislative Treatment of Multi-Media Pollutant Exposures," report to the National Commission on Air Quality (1981).
13. Under TSCA, "a formal cost-benefit analysis under which a monetary value is assigned to the risks" is not required. House Committee on Interstate and Foreign Commerce, *Toxic Substances Control Act: Report to Accompany H.R. 14032*, 94th Cong., 2d sess., 1976, H. Rept. 94–1341, 14. Neither does the CPSC have to conduct formal cost-benefit analyses. E.g., *Consumer Product Safety Act: Report to Accompany H.R. 15003*, 92d Cong., 2d sess., H. Rept. No. 92–1153, 1972, 33. ("[T]here should be no implication, however, that in arriving at its determination the Commission would be required to conduct and complete a cost-benefit analysis prior to promulgating standards under this act.") The Fifth Circuit ruled that the Occupational Safety and Health Act (OSHAct) as a balancing statute requiring a reasonable relationship between costs and benefits, but even then did not require formal cost-benefit analysis. American Petroleum Institute v. Occupational Safety & Health Admin., 581 F.2d 493 (5th Cir. 1978), *aff'd on other grounds*, American Petroleum Institute v. Occupational Safety & Health Admin., 448 U.S. 607 (1980). The Supreme Court subsequently overruled the Fifth Circuit's interpretation of the OSHAct's definition of acceptable risk, holding it to be a TEF-based statute instead. American Textile Mfrs. Inst. v. Donovan, 452 U.S. 490 (1981).
14. Thomas O. McGarity, "Media-Quality, Technology, and Cost-Benefit Balancing Strategies for Health and Environmental Regulation," *Law & Contemp. Prob.* 46 (1983): 159–60.
15. 15 U.S.C. sec. 2604(b)(4)(A) (1982).
16. P.L. 85–929, 68 Stat. 511 (1958) [at 21 U.S.C. 348(c)(A) (1970 ed.)].
17. 29 U.S.C. sec. 655(b)(5) (1982).
18. Some efforts have been made to justify a specific definition, often in the face of criticism that it ought to be one of the other varieties. E.g., David D. Doniger, Liroff, and Richard A. Dean, *An Analysis of Past Federal Efforts to Control Toxic Substances* (Baltimore: Johns Hopkins University Press, 1978), 37–42 (health-based Delaney Clause defensible as a *per se* rule under conditions where benefits of substance unlikely to outweigh risks, and costs of determining this balance case by case outweigh the gains to be achieved from such fine-tuning); Christopher H. Schroeder, "Foreword: A Decade of Change in Regulating the Chemical Indus-

try," *Law & Contemp. Prob.* 46 (1983): 1, 22–28 (TEF-based water regulations justifiable because media-based regulation of water had proven impossible, given difficulties of modeling and of attributing responsibility for pollution to specific sources). See also D. Bruce LaPierre, "Technology Forcing and Environmental Protection Statutes," *Iowa L. Rev.* 62 (1977): 771 (analyzing the efficacy of statutes from different categories in accomplishing the goal of technology forcing). Economics-based criticisms of the regulatory fragmentation generally argue that *all* statutes should use the balancing approach, with full cost-benefit calculations. I know of no study contending that the entire array of health-based, TEF-based, and balancing statutes are approximately right under some normative theory of toxics regulation. Professor McGarity, however, has recently analyzed media-based and technology-based regulation of pollutants according to the functional criteria of (1) efficiency; (2) administrative feasibility; (3) survivability; (4) enforceability; (5) fairness and equity; and (6) ability to encourage technological advance. See McGarity, "Media-Quality," 200–233 (n. 1).

19. See n. 14.
20. Court opinions frequently identify the indeterminacy of agency decision criteria by stating that the agency's ultimate decision turns on "policy judgments," a code phrase indicating that further search for clarity of criteria will be unavailing. See, e.g., Industrial Union Dept. v. Hodgson, 499 F.2d 467, 499 (D.C. Cir. 1974) (OSHA's determination of exposure level "rests in the final analysis on an essentially legislative policy judgment, rather than a factual determination, concerning the relative risks of underprotection as compared to overprotection"); Tanners' Council of America v. Train, 540 F.2d 1188, 1192 (4th Cir. 1976) (EPA's determination of adequate technology almost "entirely judgmental"). Under "hard-look" review, agencies must more fully articulate their factual predicates, and sometimes the intermediate models and assumptions upon which they rely, but these restrictions still confine agency choices within a considerable range, rather than compel the agency to identify with precision the weights being supplied to the various elements of their analysis.
21. See, e.g., Lead Indus. Ass'n v. EPA, 647 F.2d 1130 (D.C. Cir. 1980) (reviewing EPA's standard for airborne lead, summarizing debate in the proceeding over whether the elevation of erythrocyte protoporphyrin in the blood of people exposed to low concentrations of ambient was a "sub-clinical" effect not signifying any adverse health effect, but simply showing a reaction to lead that current instruments are sensitive enough to detect).
22. EPA's latest carbon monoxide regulation is based on a study showing that angina patients experience the onset of angina pain in light treadmill exercise experiments 16 percent sooner in an atmosphere with low CO concentrations than in a CO-free environment. "Carbon Monoxide: Proposed Revisions to the National Ambient Air Quality Standards," 45 Fed. Reg. 55,066, 55,073–76 (1980). The most plausible interpretation of this experiment is that it represents one stop along a continuum of responses that angina patients will have to varying levels of exposure, rather than some sharp dividing line between health and illness. See also Lead Indus. Ass'n v. EPA (n. 21).
23. On the difficulties of interpretation of technological feasibility, see LaPierre, "Technology Forcing." Justice Rehnquist has thought the definition of "feasible" so vague as to make the OSHAct unconstitutional, Industrial Union Dept. v. American Petroleum Inst., 448 U.S. 607, 683–86 (1980) (Rehnquist, J., concurring); American Textile Mfrs. Inst. v. Donovan, 452 U.S. 490, 544–48 (1981) (Rehnquist, J., dissenting), but the concept has become sufficiently well defined to surmount that extremely undemanding standard. For an interpretation of the concept, see United Steel Workers v. Marshall, 657 F.2d 1189, 1263–1308 (D.C. Cir. 1980).
24. See William R. Havender and Elizabeth M. Whelan, "Cancer's Uncertainties and Public Policy," *Cato J.* 2 (1982): 543, 555–62. See also Lester B. Lave, ed.,

introduction to *Quantitative Risk Assessment in Regulation* (Washington, D.C.: Brookings Institution, 1982), 3 ("Rather than attempting to determine the most probable estimate of risk and the range of uncertainty, the risk assessment [of agencies] generally has been structured to be arbitrarily conservative"); Peter J. Aranson, "Pollution Control: The Case for Competition," in *Instead of Regulation: Alternatives to Federal Regulatory Agencies*, ed. Robert W. Poole, Jr. (Lexington, Mass.: Lexington Books, 1982), 339, 380. (The bureaucrat's incentive to minimize detectable errors results "in the EPA's overriding concern to regulate pollution emissions to a level of zero risk. . . . Of course, a policy goal of creating zero levels of environmental risk merely shifts risk to other margins, such as business and economic failure.")

25. Schroeder, "Foreword: A Decade of Change in Regulating the Chemical Industry," *Law & Contemp. Prob.* 46 (1983): 31–32.
26. 42 U.S.C. sec. 7409(a) (Supp. V 1981).
27. 7 U.S.C. sec. 136(bb), sec. 136a(c)(5), sec. 136a(d)(1)(c) (1982).
28. E.g., 42 U.S.C. sec. 7410(a)(1) (Supp. V 1981); 33 U.S.C. sec. 1342(a) (1982).
29. 15 U.S.C. sec. 2605(a)(1) (1982).
30. 15 U.S.C. sec. 2605(a)(4) (1982). The EPA can order more tests if it "finds that there is a reasonable basis to conclude that the manufacture, processing, distribution in commerce, use or disposal of a chemical substance or mixture, . . . presents or will present an unreasonable risk of injury to health or the environment." 15 U.S.C. sec. 2605(a) (1982). Under 15 U.S.C. sec. 2604(f)(2)(b) (1982), the EPA can order additional tests during the notification period if the EPA believes that the chemical presents "an unreasonable risk of injury to health or environment before a rule promulgated under 2605 of this title can protect against such risk." 15 U.S.C. sec. 2604(f) (1982). 15 U.S.C. sec. 2604(f)(1) (1982).
31. 15 U.S.C. sec. 2957 (1982).
32. Aaron Wildavsky, "Foreword: Rights and Regulation: Ethical, Political, and Economic Issues," in *Rights and Regulation*, ed. Tibor R. Machan and M. Bruce Johnson (Cambridge, Mass.: Ballinger Publishing Co., 1988), xv.
33. Ibid., xv-xvi.
34. Ecological systems have feedback mechanisms, to be sure, but they do not rely upon the cognitive powers of the members of ecological communities.
35. Cf. Frank I. Michelman, "Pollution as a Tort: A Non-Accidental Perspective on Calabresi's Cost," *Yale L. J.* 80 (1971): 647, 663.

> There are two distinct functions performed by a legislative body when, with a view to optimizing the cost output of some activity, it establishes rules for allocating liabilities and evaluating costs. We can call these "centralizing" and "decentralizing" functions. The centralizing function arises out of the possibility we have discussed that the legislature . . . will perceive in certain kinds of activities costs which will not otherwise be recognized or monetized either by those who control the means of economically avoiding such costs or by others with whom they transact. The centralizing function, then, consists of coercing potential cost avoiders into acting in accordance with the legislature's . . . perceptions of costliness.
>
> The decentralizing function arises out of the opposite possibility that certain costs can be most accurately valued by the most decentralized set of choice-makers able and motivated to recognize and monetize them; and that sacrifices economically necessary to avoid such costs can be most accurately valued by those who would have to make the sacrifices. The decentralizing function, then, is to provide a framework for voluntary transactions designed to maximize the probabilities that individuals or relatively small groups will be motivated to recognize and monetize costs, with the result that those who could avoid costs with the least sacrifice will be motivated to consider whether the sacrifice is worthwhile.

36. Then, too, tort liability is a more centralized action than enforcing a requirement on contracts entered into before polluting actions being taken, because the tort system uses the more centralized mechanism of judge or jury to determine damages, rather than relying upon individual bargains to determine value. Thus, within the putatively resilient market-plus-tort system there are gradations in the amount of centralization employed. See Guido Calabresi and Douglas A. Melamed, "Property Rules, Liability Rules and Inalienability: One View of the Cathedral," *Harv. L. Rev.* 85 (1972): 1089.
37. See Michelman, "Pollution as a Tort," 664 (we are not interested in studying either extreme possibility).
38. R. Jeffrey Smith, "Toxic Substances: EPA and OSHA Are Reluctant Regulators," *Science* 203 (Jan. 5, 1979), 28.
39. U.S. General Accounting Office, *Delays in Setting Workplace Standards for Cancer-Causing and Other Dangerous Substances* (Washington, D.C., 1977), 9.
40. In 1984, OSHA issued a PEL for ethylene oxide; in 1986, it revised downward an existing PEL for asbestos. 59 Fed. Reg. 25796 (1984); 51 Fed. Reg. 22733 (1986).
41. Telephone conversation with Susan Haywood, May 25, 1984.
42. Richard A. Merrill, "CPSC Regulation of Cancer Risks in Consumer Products: 1972–81," *U. Va. L. Rev.* 67 (1981): 1261, 1375.
43. NRDC v. Train, 8 Envt'l Rep. Cas. (BNA) 2120 (D.D.C. 1976) (Flannery Decree). The EPA's reluctance to regulate is reviewed in Khristine L. Hall, "The Control of Toxic Pollutants Under the Federal Water Pollution Control Act Amendments of 1972," *Iowa L. Rev.* 63 (1978): 609.
44. See 33 U.S.C. sec. 1317(a)(1) (1982).
45. Office of Technology Assessment, U.S. Congress, *Assessment of the Technologies for Determining Cancer Risks from the Environment* (Washington, D.C., 1981), 178, table 35.
46. See, e.g., Chemical Mfrs. Ass'n v. EPA, 673 F.2d 507 (D.C. Cir. 1982).
47. See, e.g., *Oversight of the Comprehensive Environmental Response Compensation and Liability Act of 1980: Hearings Before the Subcommittee on Environmental Pollution of the Senate Committee on Environment and Public Works*, 97th Cong., 2d sess., 1982, Rept. No. 97–459, 13 (remarks of Senator Moynihan, EPA slow to implement Superfund); ibid., 145 (remarks of Khristine Hall).
48. Hinsin and Rasher, "EPA Hazardous Waste Enforcement: A Policy in Evolution," in *Hazardous Waste Litigation*, ed. R. Mott (New York: Practising Law Institute, 1984), 41, 43.
49. *EPA Pesticide Regulatory Program Study, 1982: Hearing Before the Subcommittee on Department Operations, Research and Foreign Agriculture of the House Committee on Agriculture*, 97th Cong., 2d sess., 1982, "Staff Report, Regulatory Procedures and Public Health Issues in the EPA's Office of Pesticide Programs," 146. ("Interviews with program personnel, conducted by Subcommittee staff, indicate that reviews are taking much longer than initially expected and that assembling and assessing the available safety data are particularly difficult. Moreover, many registrants have not promptly submitted required studies. Thus, although the program has now issued a number of standards, it still has a very long way ahead of it.")
50. See, e.g., U.S. General Accounting Office, *Stronger Enforcement Needed Against Misuse of Pesticides* (Washington, D.C., 1981).
51. *Toxic Substances Control Act Oversight, 1983: Report on 98–514 Before the Subcommittee on Toxic Substances and Environmental Oversight of the Senate Committee on Environment and Public Works*, 98th Cong., 1st sess. (1983), 192 (statement of Michael Gough, Office of Technology Assessment, U.S. Congress, "not all PMN's report even the TSCA specified items").
52. NRDC v. Costle, 14 Envt'l Rep. Cas. (BNA) 1858 (S.D.N.Y. 1980). For a more extensive summary of federal regulatory actions concerning toxic carcinogens, see Office of Technology Assessment, *Federal Testing and Regulation of Carcinogens* (Washington, D.C., 1988), chap. 3.

53. OTA, *Cancer Risks*, 140, 141, table 30. IARC has since expanded its list of carcinogens to 23, and has listed 61 chemicals, groups of chemicals, or industrial processes as possibly carcinogenic in humans. IARC, *Chemicals and Industrial Processes Associated with Cancer in Humans*, Suppl. 4 of *IARC Monographs on the Evaluation of the Carcinogenic Risks of Chemicals to Humans* (Lyon: the agency, 1982).
54. OTA, *Cancer Risks*, 15. The discrepancy between these regulatory actions and IARC's list is attributable to decisions to regulate before IARC's more stringent testing requirements have confirmed the substances' carcinogen status.
55. Ibid., 130.
56. J. Clarence Davies, "The Effects of Federal Regulation on Chemical Industry Innovation," *Law & Contemp. Prob.* 46 (1983): 41, 54.
57. William M. Blakemore, "Toxic Metal and Metalloids," in *Handbook of Carcinogens and Hazardous Substances Chemical and Trace Analysis*, ed. M. Bowman (New York: Marcel Dekker, 1982), 641, 649.
58. 15 U.S.C. sec. 2601(b)(2) (1976).
59. R. G. Landsdown et al., "Blood-Lead Levels, Behavior, and Intelligence: A Population Study," *Lancet* 1 (Mar. 30, 1974): 538–41, cited in Herbert Needleman et al., "Deficits in Psychologic and Classroom Performance of Children with Elevated Dentine Lead Levels," *N. Eng. J. Med.* 300 (1979): 689–95. EPA regulation of lead in air reflected these studies. Recent findings, not yet incorporated into any EPA action, indicate that prenatal and postnatal exposures to lead at concentrations much less than 25 micrograms per deciliter have an adverse effect on early cognitive development. See David Bellinger et al., "Longitudinal Analyses of Prenatal and Postnatal Lead Exposure and Early Cognitive Development," *N. Eng. J. Med.* 316 (1987): 1037–43.
60. 29 C.F.R. sec. 1910–1000 (1982).
61. "EPA Plans to Beef Up Budget for Dioxin Research," *Chem. & Eng'g News*, July 11, 1983, 16.
62. EPA Policy and Procedures for Identifying, Assessing and Regulating Airborne Substances Posing a Risk of Cancer: Proposed Rules, 44 Fed. Reg. 58,6442 (1979); OSHA Identification, Classification and Regulation of Toxic Substances Posing a Potential Occupational Carcinogenic Risk, 42 Fed. Reg. 54,148–54,247 (1977).
63. Office of Technology Assessment, *Technologies and Management Strategies for Hazardous Waste Control* (Washington, D.C. 1983).
64. Conservation Foundation, "Clean-Up and Fix-Up Costs Rise Relentlessly," in *Conservation Foundation Newsletter*, Mar. 1983, 3. OTA's latest costs estimate for the Superfund is $100 billion over the next fifty years. Office of Technology Assessment, *Superfund Strategy* (Washington, D.C., 1985), 3.
65. Estimating benefits requires determining levels of toxic emissions or discharges, plotting dispersion of the discharges so that exposure levels can be calculated, then correlating those exposure levels, which can vary over time and location, to health effects. Each of these steps is fraught with problems in acquiring accurate data and translating that data, generally through the use of mathematical models, into results suitable for analysis. See generally McGarity, "Media-Quality," 179–87.

    When toxics are at issue, perhaps the most confounding problems are associated with the correlation of exposure levels to health effects. For instance, for carcinogens there are four primary risk-assessment methodologies: (1) epidemiological studies, (2) animal bioassays, (3) "short-term" tests, and (4) structure-activity correlations. Epidemiological studies are the sole means of obtaining health effects data directly on humans, but they suffer from fundamental defects. First, because they are retrospective studies of human populations exposed to toxics, they are unable to control for many other environmental and behavioral variables within the study population, and this impairs ability to extrapolate to other populations. Second, epidemiology can only supply point correlations which can be fit to any number of possible extrapolation curves. See James P.

Leape, "Quantitative Risk Assessment in Regulation of Environmental Carcinogens," *Harv. Env. L. Rev.* 4 (1980): 86, 92–93; Interagency Regulatory Liaison Group, "Scientific Bases for Identification of Potential Carcinogens and Estimation of Risks," 44 Fed. Reg. 39,858, 39,862 (1979) (IRLG Report). For an exploration of epidemiology's legitimate uses, see Khristine Hall and Helen Silbergeld, "Reappraising Epidemiology: A Response to Mr. Dore," *Harv. Env. L. Rev.* 7 (1983): 441. Animal bioassays face arguments about whether it is scientifically justifiable to translate conclusions about carcinogenicity in laboratory animals to human beings. Furthermore, animal experiments generally employ large doses of the substance in order to induce a statistically significant response, but then the translation of these high-dose results to the low doses germane for many toxics policy disputes must be made via mathematical models, and there is no consensus on the appropriate one. IRLG Report at 39,863–65; L. J. Carter, "How to Assess Cancer Risks," *Science* 204 (1979): 811–13; Leape, "Quantitative Risk Assessment," 93–95. "Short-term" tests subject cell populations to chemical agents and then examine for mutations or abnormalities. The correlation between the results and human health effects is controversial, and the tests probably have their greatest value as screening tools. Bruce G. Ames, *Environmental Chemicals Causing Cancer and Genetic Birth Defects: Developing a Strategy to Minimize Human Exposure* (Berkeley: Published for the California Policy Seminar by Institute of Governmental Studies, University of California, 1978), 11–22; James P. Leape, "Quantitative Risk Assessment," 95–96. Structure-activity correlations compare the chemical structures of known carcinogens to that of the test substance. These are probably the least reliable indicators, and not at all helpful for supplying the quantitative information required by policy analysis. See Louis Slesin and Ross Sandler, "Categorization of Chemicals Under the Toxic Substances Control Act," *Ecology L. Q.* 7 (1978): 359, 372. On these problems, see generally Thomas O. McGarity, "Substantive and Procedural Discretion in Administrative Resolution of Science Policy Questions: Regulating Carcinogens in EPA and OSHA," *Geo. L. J.* 67 (1979): 729; Occupational Safety and Health Administration, Identification, Classification and Regulation of Toxic Substances Posing a Potential Occupational Carcinogenic Risk, 41 Fed. Reg. 54,147 (1977).

66. Ivy E. Broder and John F. Morrall, III, "Economic Basis," 242, 250 (table 15.2).
67. Ibid., 251 (table 15.3).
68. John D. Graham and James W. Vaupel, "The Value of Life: What Difference Does It Make?" in *What Role for Government?* 176, 184–85.
69. Ibid., 18.
70. *N.Y. Times*, Feb. 14, 1982, Section C, p. 1, col. 1.
71. "Toxic Substances: EPA and OSHA Are Reluctant Regulators," *Science* 203 (1979): 28 (remarks of Douglas Costle).
72. Ibid.
73. Ibid., 29.
74. Bureau of Economic Statistics, Inc., *Handbook of Basic Economic Statistics*, vol. 37, no. 1 (Washington, D.C.: Bureau of Economic Statistics, Inc., Jan. 1982).
75. Ibid., 167, 228.
76. If the sales of these industries are added to the chemical industry proper, the impact of chemicals on the economy is much larger than 6 percent of GNP.
77. H. J. Taback et al., *Control of Hydrocarbon Emissions from Stationary Sources in the California Air Basin: Final Report* (Sacramento: California Air Resources Board, 1978).
78. Robert J. Gordon, "Distribution of Airborne Polycyclic Aromatic Hydrocarbons," *Envt'l Sci. & Tech.* 10 (1976): 370.
79. National Academy of Sciences, Safe Drinking Water Committee of the National Research Council, *Drinking Water and Health* (Washington, D.C.: NAS, 1977), 489–92.
80. See *Env. Rptr.-Curr. Develop.* (BNA) 10 (1982): 2152.

81. Fred C. Hart Associates, "Preliminary Assessment of Cleanup Costs for National Hazardous Waste Problems" (1979) (prepared for the U.S. Environmental Protection Agency), 25.
82. Ibid., 22.
83. Federation of American Societies for Experimental Biology, "Evaluation of the Health Aspects of Caffeine as a Food Ingredient" (SCOGS–89 1978, prepared for the U.S. Food and Drug Administration), 36.
84. Ibid., 38; see also M. M. Nelson and J. O. Forfar, "Associations Between Drugs Administered During Pregnancy and Longenital Abnormalities of the Fetus," *British Med. J.* 1 (Mar. 6, 1971): 523–27.
85. See Office of Technology Assessment, *Cancer Testing Technology and Saccharin* (Washington, D.C., 1977).
86. Magee, Montesano, and Preassmann, "N-Nitroso Compounds and Related Carcinogens," in *Chemical Carcinogenesis*, ACS Monograph No. 173, ed. C. E. Searle (Washington, D.C.: National Academy Press, 1976).
87. Takashi Sugimura, "Mutagen-Carcinogens in Food, with Special Reference to Highly Mutagenic Pyrolytic Products in Broiled Foods," in *Origins of Human Cancer*, ed. Howard H. Hiatt, J. D. Watson, and Jay A. Winsten (Cold Spring Harbor, N.Y.: Cold Spring Harbor Laboratory, 1977).
88. 38 Fed. Reg. 1254 (Jan. 10, 1973), codified at 40 C.F.R. sec. 80.22 (1982). EPA is currently phasing down the lead content in all gasolines. See 40 C.F.R. sec. 80.20 (1982).
89. Federal Interagency Committee on Health and Environmental Effects of Energy Technology, *Health-Effects Research and Standard Setting at EPA*, cited in SRI International III–43–44 (report to Nat'l Comm'n on Air Quality 1980). See generally D. Masselli and N. L. Dean, Jr., *The Impacts of Synthetic Fuels Development* (Washington, D.C.: U.S. Department of Energy, 1981).
90. National Commission on Air Quality, *To Breathe Clean Air* (Washington, D.C. 1981), 74 (suggesting changing from total suspended particles to a fine respirable particle standard); Environmental Protection Agency, Office of Air Quality Planning and Standards, *Review of the National Ambient Air Quality Standard for Particulate Matter: Draft Staff Paper; Summary of Staff Conclusions and Recommendations for the Primary Standards* (Washington, D.C.: Office of Air Quality Planning and Standards, 1981), 1.
91. Smaller particles, less than 15 micrometers long, are also more harmful because they are more readily inhaled than larger particles. The finest particles, below 2.5 micrometers, are suspected by some scientists of presenting the greatest dangers because only they can penetrate deeply into the lungs. Environmental Protection Agency, *Air Quality Criteria for Particulate Matter and Sulfur Oxides* (external review draft, Apr. 1980), 1.
92. The two major control technologies for particulate emissions from stationary sources, electrostatic precipitators, and bag houses, capture larger particles more easily than smaller ones. EPA has now proposed separate standards for smaller particulates: an annual average between 55 and 110 mg/m^3 for particulates 10 microns or less, and between 8 and 25 mg/m^3 for particulates 2.5 microns or less. 49 Fed. Reg. 10408 (March 20, 1984).
93. K. Knorr, *Rubber After the War* (Stanford, Calif.: Stanford University, 1944), 4–5.
94. T. Dunlap, *D.D.T.: Scientists, Citizens and Public Policy* (Princeton, N.J.: Princeton University Press, 1981), 61.
95. Fred Hoerger, William H. Beamer, and James S. Hanson, "The Cumulative Impact of Health, Environmental and Safety Concerns on the Chemical Industry During the Seventies," *Law & Contemp. Prob.* 46 (Summer 1983): 59, 82, fig. 4.
96. Ibid.
97. Daniel Bell, *The Coming of the Post-Industrial Society* (New York: Basic Books, 1976), 116 (chemistry the "first modern industry . . . since one has to have a theoretical

a priori knowledge of the properties of the macromolecules one is manipulating in order to create new products").

98. DDT and related pesticides led the parade.
99. Ronald M. Motley and Richard H. Middleton, Jr., "Asbestos Disease Among Railroad Workers: Legacy of the Laggin' Wagon," *Trial* 17 (Dec. 1981): 38 at n. 3.
100. See Council on Environmental Quality, *Environmental Quality—1980* (Washington, D.C., 1979), 206–10.
101. Richard Epstein has suggested that the fundamental common law principles of property, contracts, and tort ought properly to be "static," and that the frequent defense of reforms in common law doctrine as necessary responses to changed social conditions is not persuasive. Epstein, "The Static Conception of the Common Law," *J. Legal Stud.* 9 (1980): 251. The discussion immediately following in the text is consistent with the thesis as to substantive law principles, since it argues only that the procedures through which courts processed those substantive claims had gotten out of step with social conditions. Later the argument diverges from Epstein's thesis.
102. For a summary of these doctrines, see R. Stewart and James E. Krier, *Environmental Law and Policy* (Indianapolis: Bobbs-Merrill, 1978), 204–13; Julian C. Juergensmeyer, "Control of Air Pollution Through the Assertion of Private Rights," *Duke L. J.*, 1967, 1126. Juergensmeyer even suggested that these existing doctrines were sufficient to regulate the problem of air pollution, but financial limitations on private actions made public regulation necessary. Thus anticipating the deregulation argument, Professor Juergensmeyer overlooked the problem of proving causation, the presence of significant social values affecting the outcome, and the strong preference for direct preventive measures.

    Not all the details of private law recovery doctrine are long-settled. Of particular significance, debate continues concerning whether the appropriate standard of recovery in personal injury cases involving toxics should be negligence or strict liability, and also whether damages should be awarded in nuisance (strict liability) even when an injunction is denied because the utility of the defendant's conduct outweighs the gravity of the plaintiff's injury. Any configuration of doctrines protects against some toxics injuries, however; thus that much has been long settled.
103. E.g., William Aldred's Case, 9 Co. Rep. 57, 77 Eng. Rep. 611 (1611) (suit for damages against the stench and unhealthy odors of defendant's pigsty).
104. For a summary of the recurring issues, see Trauberman, "Statutory Reform of 'Toxic Torts': Relieving Legal, Scientific, and Economic Burdens on the Chemical Victim," *Harv. Env. L. Rev.* 7 (1983): 197–202; McGarity, "Regulating Carcinogens in EPA and OSHA," 729, 749–53.
105. Even if some of them did litigate, perhaps out of strong ideological preferences for less pollution and for making polluters pay, litigation would not predictably occur in sufficient quantity to encourage polluters to recognize the full costs of their pollution in their production decisions. This is undesirable even under a viewpoint that emphasizes allocative efficiency, because it will generally be the case in the typical setting of a relatively few polluters and a relatively large group of affected individuals that imposing liability on the polluter will lead to greater efficiency in the use of scarce resources, i.e., clean air. See, e.g., James E. Krier, "The Pollution Problem and Legal Institutions," 429, 444–49.
106. Modifications in doctrines or procedures for recovery of damages inevitably involve costs that include not only the costs of administering the system for meritorious claims, but for non-meritorious yet nevertheless litigated claims as well. If these costs are greater than the benefits to be gained by imposing them on the polluter, making the modification will not improve the efficiency of the system. See Harold Demsetz, "The Exchange and Enforcement of Property Rights," *J. Law & Econ.* 7 (1964): 11, 14. ("In asking the implications of the nonexistence of some markets, we seem to have forgotten the cost of providing

market services or their government equivalent . . . [T]he absence of a price does not imply that either market transactions or substitute services are desirable.")

107. The failure to impose those costs on the polluter will have become Pareto-relevant, in economic terms. E.g., James M. Buchanan and Roger L. Faith, "Entrepreneurship and the Internalization of Externalities," *J. Law & Econ.* 24 (1981): 95; Buchanan and Stubblebine, "Externality," *Economica*, n.s., 29 (1962): 371.

108. Barry Commoner ascribes major responsibility for environmental deterioration to technology. B. Commoner, *The Closing Circle* (New York: Bantam Books, 1971). Paul Ehrlich and John Holdren responded, claiming that population growth is the leading culprit. Paul Ehrlich and John Holdren, "Review: The Closing Circle," *Environment* 14 (Apr. 1972): 24. This debate has important policy implications: if environmental deterioration is perceived as a result of population growth, "a reduction in the population sufficient to render tolerable the environmental deterioration" might be suggested, whereas if it is due to faulty technology, appropriate action might be "to correct counter-ecological technologies and to change the economic mechanisms which generate them." Commoner, response to the review, ibid., 25. Nothing in the present article turns on the outcome of this debate, which has no objectively correct answer. American regulation of toxic substances responds as if Commoner's is the correct view, however, emphasizing technology assessment and technological pollution abatement programs, while nowhere suggesting either global or localized population management to be an explicit policy instrument.

109. See W. Tucker, *Progress and Privilege: America in the Age of Environmentalism* (Garden City, N.Y.: Doubleday, Anchor Press, 1982) (arguing that environmentalism is the product of upper- and middle-class attention to values that the poor cannot afford); see also "The Cost-Internalization Case for Class Actions," *Stan. L. Rev.* 21 (1969): 383, 392–98.

110. Although a class plaintiff can be compensated for the time and expense invested in being a class representative, the incentives to be a class representative are substantially less than are those potentially motivating the attorney in the case, who frequently stands to gain a substantial percentage of the total recovery, and whose time would in any event be compensated much more highly than the named plaintiff's. Thus it is the dynamics of plaintiffs' lawyers' case selection that must be understood in determining the effects of class action reforms. Actually, this observation is not limited to class suits. In any contingent fee litigation, of which almost all toxic cases are examples, case selection by the plaintiffs' bar is a crucial determinant of the efficacy of private recovery mechanisms. See, e.g., David Rosenberg, "The Causal Connection in Mass Exposure Cases: A 'Public Law' Vision of the Tort System," *Harv. L. Rev.* 97 (1984): 851, 889–92 (plaintiff lawyers are the tort system's gatekeepers).

111. Under the Federal Rules of Civil Procedure, the obstacles include (1) supplying notice to class members at plaintiff's expense, Eisen v. Carlisle & Jacquelin, 417 U.S. 156 (1974); (2) demonstrating that each class member has an individual claim greater than the jurisdictional minimum, which in diversity cases is $10,000, Zahn v. International Paper Co., 414 U.S. 291 (1973); (3) convincing a court that common questions of law and fact predominate over individual questions, and that the class action is a clearly superior and economical means of resolving the underlying dispute. See Advisory Committee Note to Fed. Rule of Civ. Proc. 23, 39 F.R.D. 69, 103 (1966) (discouraging the use of class action in mass accident cases because of the "commonality" requirement). In state court actions, plaintiffs face a variety of obstacles, depending upon the class action rules of the specific jurisdiction. Even generally liberal class action states, such as California, have held that considerations such as the uniqueness of property and the individual variations in damages preclude nuisance-type litigation in a class action format. E.g., City of San Jose v. Superior Ct. of Santa Clara County, 12 Cal. 3d 447, 458–63 (1974) (despite recognizing that class actions provide an important means for

redressing "group wrongs" in modern "complex society," class action for nuisance caused by noise from low-flying aircraft landing and taking off from an airport was denied because class members had insufficient community of interest and class representative could not provide adequate representation). See generally Robin Alta Charo, "Class Actions and Mass Toxic Torts," *Colum. J. Envir. L.* 8 (1982): 269.

112. See Barry B. Boyer, "Alternatives to Administrative Trial-Type Hearings for Resolving Complex Scientific, Economic, and Social Issues," *Mich. L. Rev.* 71 (1972): 111.
113. E.g., Boomer v. Atlantic Cement Co., 26 N.Y.2d 219, 257 N.E.2d 870, 309 N.Y.S.2d 312 (1970). ("[T]he judicial establishment is neither equipped in the limited nature of any judgment it can pronounce nor prepared to lay down and implement an effective policy for the elimination of air pollution. This is an area beyond the circumference of one private lawsuit. It is a direct responsibility for government and should not thus be undertaken as an incident to solving a dispute between property owners and a single cement plant—one of many—in the Hudson River Valley. . . . For obvious reasons the rate of the research [into abatement technology] is beyond control of defendant. If at the end of 18 months the whole industry has not found a technical solution a court would be hard put to close down this one cement plant if due regard be given to equitable principles.")
114. Traditionally, a statute of limitations period began to run at the time of plaintiff's initial contact with the causative agent, not at the time an injury was discovered. See generally Superfund Section 301(e) Study Group, *Injuries and Damages from Hazardous Wastes—Analysis and Impact of Legal Remedies*, 97th Cong., 2d sess., 1982, 15 ("the traditional rule caused great hardship to plaintiffs who found that their claims were time-barred before they had first discovered their injuries").
115. The reluctance of courts to entertain purely probabilistic evidence of causation spans a number of doctrinal fields. See Rosenberg, "Causal Connection," 857 (traditionally, "some 'particularistic' proof of the causal connection is required"); Donald W. Large and Preston Michie, "Proving That the Strength of the British Navy Depends on the Number of Old Maids in England: A Comparison of Scientific Proof with Legal Proof," *Envtl. L.* 11 (1981): 557.
116. Where several defendants are likely to have contributed to sustained injuries, plaintiffs have been barred recovery when required to prove the extent of injuries attributable to each defendant. See, e.g., Tidal Oil Co. v. Pease, 153 Okla. 137. 5 P.2d 389 (1983) (livestock poisoned by polluted waters). See generally William Lloyd Prosser, *Handbook of the Law of Torts* sec. 52 (St. Paul: West Publishing Company, 1971). If the damage is considered indivisible, that is to say not capable of being apportioned, defendants are held jointly and severally liable. Some, but not all, toxic tort cases would come within this rule, eliminating the apportionment obstacle.
117. Statute of limitations barriers have been eased by a shift from the traditional rule to some version of the "discovery" rule (see n. 114). Thirty-seven states, through judicial decision or statute, have adopted the rule that limitations periods only begin to run when the harm is apparent or ought reasonably to have become apparent. See, e.g., Witherell v. Weimer, 77 Ill.App. 3d 582, 587, 396 N.E.2d 268, 272 (1979); Warrington v. Charles Pfizer & Co., 274 Cal.App. 2d 564, 569–70, 80 Cal. Rptr. 130, 133 (1969); N.C. Gen. Stat. sec. 1–52 (Cum. Supp. 1981). See generally Superfund Section 301(e) Study Group, *Injuries and Damages from Hazardous Wastes*, 13–78; Francis E. McGovern, "Toxic Substances Litigation in the Fourth Circuit," *U. Rich. L. Rev.* 16 (1982), 247, 253–58. Courts are also becoming more receptive to statistical evidence. See, e.g., Union Carbide Corp. v. Industrial Comm'n, 581 P.2d 734, 737–38 n.6 (Colorado, 1978) (court should avoid "futile searches for unattainable factual certainties").

    Apportionment problems have produced four different judicial responses.

Defendants have been held jointly and severally liable for injurious acts committed in concert with others or pursuant to a common design. E.g., Bichler v. Eli Lilly and Co., 436 N.Y.S.2d 625 (App. Div. 1981); Abel v. Eli Lilly & Co., 94 Mich.App. 59, 289 N.W.2d 20 (Ct. App. 1979). Second, the theory of alternative liability allows recovery against several nonconcerting wrongdoers when plaintiff cannot prove which caused the injury or the extent of individual responsibility. E.g., Borel v. Fibreboard Paper Products Corp., 403 F.2d 1076 (5th Cir. 1973). Third, enterprise liability holds the industry as a whole responsible for injuries resulting from its activities. E.g., Hall v. E. I. DuPont De Nemours & Co., 345 F.Supp. 353 (E.D.N.Y. 1972). Finally, a market-share liability theory holds a defendant liable for a percentage of the plaintiff's injury that corresponds to the percentage of the market of the injurious substance that the defendant's sales represents. E.g., Sindell v. Abbott Laboratories, 26 Cal. 3d 558, 607 P.2d 924, 163 Cal. Rptr. 132 (1980).

118. The recent developments in toxic tort doctrine are summarized in Daniel Farber, "Toxic Causation," *Minn. L. Rev.* 71 (1987): 1291.

119. A number of commentators consider this the most substantial obstacle to the tort system's responding to toxic exposure litigation. E.g., William R. Ginsberg and Lois Weiss, "Common Law Liability for Toxic Torts: A Phantom Remedy," *Hofstra L. Rev.* 9 (1981): 859, 922; Mark D. Seltzer, "Personal Injury Hazardous Waste Litigation: A Proposal for Tort Reform," *B.C. Envtl. Aff. L. Rev.* 10 (1983): 797, 821–24; Rosenberg, "Causal Connection," 855. For a discussion of some of the toxics that leave their signature, see Kenneth Abraham, "Individual Action and Collective Responsibility: The Dilemma of Mass Tort Reform," *Va. L. Rev.* 73 (1987): 845, 860.

120. This attitude toward toxic risks is an aspect of the feeling of loss of control that afflicts our presently complex society. See generally L. Harris and Associates, *Risk in a Complex Society: A Marsh McLennan Public Opinion Survey* (Washington, D.C.: Louis Harris and Associates, 1980).

121. See, e.g., Rosenberg, "Causal Connection."

122. See James M. Landis, *The Administrative Process* (New Haven: Yale University Press, 1938) (stating that the creation of administrative bodies was due, in large measure, to "a distrust of the ability of the judicial process to make the necessary adjustments in the development of both law and regulatory methods, as they related to particular industrial problems").

123. E.g., Mark Sagoff, "On Preserving the Natural Environment," *Yale L. J.* 84 (1974): 205 (developing the argument that environmental preservation is part of our social history, our literary traditions, and our religious ideals). The various strands of intellectual history that modern environmentalism comprises are reviewed in Joseph M. Petulla, *American Environmentalism: Values, Tactics, Priorities* (College Station, Tex.: Texas A&M University Press, 1980).

124. The argument that follows is similar to that of Professor Sax, who contends that recent refusals to protect the right to develop property should be traced to loss of faith in the ability of the market to allocate property rights to appropriate uses in light of changing values placed on nonconsumptive, or social, uses of property. Joseph L. Sax, "Some Thoughts on the Decline of Private Property," *Wash. L. Rev.* 58 (1983): 481.

Social values of the kind involved here can be related to the economist's categories of common goods and public goods. Both common and public goods present difficulties for market allocation because the market's efficiency depends on the ability to exclude nonparties to transactions from benefiting or being harmed by it, for only in that way will the parties to the transaction represent the full value of the good or commodity. Common goods are ones for which it is technologically infeasible to exclude nonpayers, as in the case of fresh air. If one purchases clean air from a factory for one's own breathing consumption, he cannot prevent other people from benefiting from that purchase. Public goods are those whose consumption is nonrival—one consumer's "consumption" does not deplete the resource, at least within a range. Enjoyment of a beautiful vista is an example. On these categories, see, generally, Richard A. Musgrave and Peggy B. Musgrave, *Public Finance in Theory and Practice*, 2d ed. (New York: McGraw-Hill, 1976), 51–52; Francis M. Bator, "The Anatomy of Market Failure," *Q. J. Econ.* 72 (1958): 351, 374–75. It may be feasible to exclude nonpayers from benefiting from public goods, e.g., by fencing off the vantage points for the vista, but it is inefficient to do so because the costs of providing that good to additional people is zero, and hence its price should be zero. On the importance of this distinction, see Howard A. Latin, "Environmental Deregulation and Consumer Decisionmaking Under Uncertainty," *Harv. Env. L. Rev.* 6 (1982): 187, 215.

Social values problems of the sort discussed in the text arise when goods whose value was formerly almost exclusively private, e.g., amenable to market allocations, acquire additional values that are public or common in character. The sale of alligator skins acquires social value when the public becomes concerned over preservation of the species, for example. Alternatively, social values problems can arise when the social values at stake in the transaction shift from the sort that are generally compatible with private transactions to different values that are not. When development and broad industrial progress were paramount American values, private transactions that promoted putting property to such uses were consistent with those values. Social programs to promote those values frequently took the form of public subsidies to industrial development. Preservation, continuity, and stability, which have all been growing in prominence as social values, are in conflict with permitting private transactions to govern property disposition exclusively, in part because individuals cannot readily assemble bids for property promotive of preservation. See generally Sax, "Decline of Private Property."

125. People can place an "option value" on natural resources, reflecting the advantage they individually perceive in knowing that a resource is there, should they some day wish to use it. Additionally, preservationists exhibit an "existence value," a value on preservation "for its own sake," regardless of whether they believe they will ever have the option of directly enjoying the resource. The methodologies for calculating the magnitude of these armchair values are disputed and complex; both the nonexistence of markets in them and the difficulty their consumers will have in fixing an accurate price for them play important roles in arguments in favor of employing nonmarket methods for recognizing them.
126. Sierra Club v. Morton, 405 U.S. 727, 734 (1972). The Court required the plaintiff to be an actual user of the threatened location in order to have standing to sue, thus rejecting full recognition of "armchair environmentalists," ibid., but the decision effectively sanctioned suits by champions of nontraditional injuries.
127. See, e.g., St. Louis Gunning Advertisement Co. v. City of St. Louis, 235 Mo. 99, 137 S.W. 929 (1911) (ordinance regulating billboards upheld because they provide cover for rats, criminals, and prostitutes).
128. Metromedia, Inc. v. City of San Diego, 453 U.S. 490, 570 (1981) (Rehnquist, J., dissenting). ("[T]he aesthetic justification alone is sufficient to sustain a total prohibition on billboards within a community.")
129. City Council of Los Angeles v. Taxpayers for Vincent, 552 U.S.L.W. 4594 (May 15, 1984). (Municipal ordinance prohibiting the posting of signs on public property is constitutional since "[t]he problem addressed by this ordinance—the visual assault on the citizens of Los Angeles presented by an accumulation of signs posted on public property—constitutes a significant substantive evil within the City's power to prohibit.")
130. Agins v. City of Tiburon, 447 U.S. 225 (1980).
131. Just v. Marinette County, 567 Wis. 2d 7, 201 N.W.2d 761 (1972).
132. Penn Central Transp. Co. v. City of New York, 438 U.S. 104 (1978).
133. See cases in n. 126–32.
134. Rosenberg, "Causal Connection," 851, 900–902.

135. This paragraph and the several following draw heavily on Kenneth Kornblau, "The Evolution of Pesticide Legislation in the United States: From Consumer Protection to Risk-Benefit Analysis" (unpublished paper on file with the author, 1982).
136. Other factors influenced support for the legislation. Agriculture was a vital economic sector, and supporters of scientific agriculture, including wider use of pesticides and insecticides, feared that farmer distrust would retard the diffusion of scientific ideas into agriculture. Pesticide manufacturers encouraged federal legislation once they realized the alternative was a proliferation of potentially conflicting state laws. Kornblau, "Evolution of Pesticide Legislation," 14–16.
137. See, e.g., Food, Drug and Cosmetic Act (FDCA), P.L. 75–717, 52 Stat. 1040 (1938). See generally Kornblau, "Evolution of Pesticide Legislation," 20.
138. Kornblau, "Evolution of Pesticide Legislation," 32.
139. Ibid., 20.
140. Congress actually initiated the first full-scale government studies of "the effects of insecticides, herbicides and fungicides upon fish and wildlife resources of the United States" four years before *Silent Spring* was published, in P.L. 85–582 (1958). This was the same year the Delaney Amendment, mandating a zero residue level on foods for any chemical "found to induce cancer when ingested by man or animal," was enacted. P.L. 85–929, 68 Stat. 511 (1958), codified at 5 U.S.C. sec. 348(c)(A) (1970).
141. See Kornblau, "Evolution of Pesticide Legislation," 46–64.
142. 7 U.S.C. sec. 136(bb) (1982).
143. An early study on toxics by the Council on Environmental Quality identified their multiple pathways to human and environmental exposure—the so-called multimedia problem—as a special problem of toxics. Council on Environmental Quality, *Toxic Substances* (Washington, D.C., 1971) 1.
144. Samuel P. Hays, "Three Decades of Environmental Politics: The Historical Context," chapter 1 of this volume.
145. Council on Environmental Quality, *Environmental Quality: First Annual Report* (Washington, D.C., 1970), 6–17.
146. J. Clarence Davies and Barbara S. Davies, *The Politics of Pollution*, 2d ed. (Indianapolis: Pegasus, 1975), 26.
147. 42 U.S.C. sec. 7412(a)(1) (1976 & Supp. V 1981).
148. 22 U.S.C. sec. 1317(a) (1976).
149. The history of the evolution of toxics standards in air and water regulation is told in Schroeder, "Foreword: A Decade of Change in Regulating the Chemical Industry," *Law & Contemp. Prob.* 46 (1983):1.
150. See John D. Graham, "Failure of Agency-forcing: The Regulation of Airborne Carcinogens Under Section 112 of the Clear Air Act," *Duke L. J.* 100 (1985).
151. See Doniger, Liroff, and Dean, *Past Federal Efforts to Control Toxic Substances*.
152. Richard Merrill has argued that in fact the absolutist Delaney Clause was never applied literally by the Food and Drug Administration, which has employed a number of avoidance devices to make it workable. Advances in detection technology and expansions in the universe of carcinogens since 1958 may now have brought the full force of the low-exposure dilemma into the Delaney Clause's area of coverage. See Richard A. Merrill, "FDA's Implementation of the Delaney Clause: Repudiation of Congressional Choice or Reasoned Adaptation to Scientific Progress?" *Yale J. of Reg.* 5 (1988): 1.
153. A good summary of the vast literature on the causes of increasing federal regulation generally, together with references, is the Advisory Commission on Intergovernmental Relations, *Regulatory Federalism: Policy, Process, Impact and Reform* (Washington, D.C.: ACIR, 1984), 61–101. For environmental programs particularly, see ACIR, *The Federal Role in the Federal System: The Dynamics of Growth—Protecting the Environment: Politics, Pollution and Federal Policy* (Washington, D.C.: ACIR, 1981). See also Thomas McGraw, "Regulatory Change, 1960–79, in

Historical Perspective," in U.S. Congress, Joint Economic Committee, *Government Regulation, Achieving Social and Economic Balance*, vol. 5 of *Special Study on Economic Change*, 96th Cong., 2d sess., 1980. J. C. Davies and B. Davies, *The Politics of Pollution*, remains a classic study of the political forces motivating stricter federal air and water pollution controls.

154. E.g., James O. Freedman, *Crisis and Legitimacy: The Administrative Process and American Government* (New York: Cambridge University Press, 1978), 5. (New Deal experience with central agencies influenced their expanded use to deal with World War II problems of controlling materials, prices and production. Use of agencies continued apace after war.)
155. J. Landis, *The Administrative Process*.
156. E.g., Richard B. Stewart, "The Reformation of American Administrative Law, *Harv. L. Rev.* 88 (1975): 1667, 1681–88 (collecting sources).
157. 42 U.S.C. sec. 4321 et seq. (1982).
158. E.g., Scenic Hudson Preservation Conference v. Federal Power Comm'n, 354 F.2d 608 (2d Cir. 1965), *cert. denied*, 384 U.S. 941 (1966) (action against pumped storage facility proposed by Consolidated Edison for Hudson River); Calvert Cliffs Coordinating Council v. Atomic Energy Comm'n, 449 F.2d 1109 (D.C. Cir. 1971) (action under NEPA challenging procedures AEC would use to authorize nuclear reactor); NRDC v. Morton, 458 F.2d 827 (D.C. Cir. 1972) (action challenging Department of the Interior's plan to lease Outer Continental Shelf lands for oil exploration).
159. Whether such procedural measures can be effective is a matter of debate. See Wesley A. Magat and Christopher H. Schroeder, "Administrative Process Reform in a Discretionary Age: The Role of Social Consequences," *Duke L. J.* (1984): 301, 321–22 (collecting sources).
160. Ibid., 315.
161. Senator Dale Bumpers has proposed an amendment to the Administrative Procedure Act that would instruct courts to decide all questions of law "de novo" and not to grant any "presumption of validity" to administrative rules. Administrative Procedures Act Amendment of 1976, S.2408, *Cong. Rec.* 94th Cong., 1st sess., 1975, 29,956. The latest version of the Bumpers proposal to be voted on by the Senate was contained in S.1080, a proposal comprehensively to reform the administrative procedures for rule making, that passed the Senate by a 94–0 vote. *Cong. Rec.* 97th Cong., 2d sess., (Mar. 24, 1982) S.2718.
162. For a review of recent enactments that employ various devices to control of agency action, see Sidney Shapiro and Robert Glicksman, "Congress, the Supreme Court and the Quiet Revolution in Administrative Law," *Duke L. J.*, 1988, 819–78. Congressman James Florio has quite candidly stated that Congress has been assuming the role of agency regulator with respect to the major toxic waste statutes, Superfund and RCRA, through very detailed legislative drafting. See James Florio, "Congress as Reluctant Regulator: Hazardous Waste Policy in the 1980's," *Yale J. of Reg.* 3 (1986): 351.
163. The translation of postwar criticisms of administration into a program for creating mission-oriented agencies that were expected to be less vulnerable to these defects and failures of traditional agencies is reviewed in Alfred A. Marcus, "Environmental Protection Agency," in *The Politics of Regulation*, ed., James A. Wilson (New York: Basic Books, 1980), 267. To some extent, these expectations were fulfilled through the 1970s. Although EPA, OSHA, and CPSC have been roughly treated for moving too slowly, they were not generally thought to have become industry captives. See, e.g., R. Shep Melnick, *Regulation and the Courts, the Case of the Clean Air Act* (Washington, D.C.: Brookings Institution, 1983). ("None but the most unrelenting environmentalist would claim [at least until the Reagan administration made its appointments in 1981] that the polluters had 'captured' the EPA.")

164. See Stewart, "Reformation," 1789 ("agency solicitude for the interests of regulated or client firms is likely to persist").
165. See generally Stephen Breyer, "Analyzing Regulatory Failure: Mismatches, Less Restrictive Alternatives, and Reform," *Harv. L. Rev.* 92 (1979): 549, 552.
166. Ibid. See also *The Crisis of the Regulatory Commission*, ed. Paul W. MacAvoy (New York: Norton, 1970); *Promoting Competition in Regulated Markets*, ed. Almarin Phillips (Washington, D.C.: Brookings Institution, 1975).
167. E.g., "The Airline Deregulation Act of 1978," P.L. 95–504, 92 Stat. 1705 (codified at 49 U.S.C. sec. 1301 et seq. (Supp. V 1981); "The Motor Carrier Act of 1980," P.L. 96–296, 94 Stat. 793 (codified at 18 U.S. sec. 1114, 49 U.S.C. secs. 10101–11902a (Supp. V 1981); and the 1982 settlement between the Justice Department and AT&T partially deregulating the telephone industry. United States v. AT&T, 552 F. Supp. 131 (D.D.C. 1982).
168. Stephen Breyer, "Analyzing Regulatory Failure."
169. Krier, "The Pollution Problem"; George Eads and Michael Fix, *Relief or Reform? Reagan's Regulatory Dilemma* (Washington, D.C.: Urban Institute Press, 1984), 14.
170. Sagoff, "On Preserving the Natural Environment."
171. Susan J. Tolchin and Martin Tolchin, *Dismantling America* (Boston: Houghton-Mifflin, 1983), 5.
172. Amory B. Lovins, *Soft Energy Paths: Toward a Durable Peace* (San Francisco: Friends of the Earth International; distributed by Ballinger Pub. Co., Cambridge, Mass., 1977).
173. Ernst F. Schumacher, *Small Is Beautiful* (New York: Harper and Row, 1973); Ivan Illich, *Tools for Conviviality* (New York: Harper and Row, 1973).
174. Also within its political tradition. The Anti-Federalists were committed to the ideal of the small, self-governing community, instilled with the civic virtues. See, e.g., Jennifer Nedelsky, "Confining Democratic Politics: AntiFederalists, Federalists, and the Constitution," review of Herbert J. Storing, ed., *The Complete Anti-Federalists*; and George L. Haskins and Herbert A. Johnson, "Foundation of Power: John Marshall, 1801–15," *Harv. L. Rev.* 96 (1982): 340, 342.
175. Soft paths and appropriate technologies promote the virtues of "frugality . . . thrift, simplicity, diversity, neighborliness, humility, and craftsmanship," Amory B. Lovins, *Soft Energy Paths*, 57, and many in the ecological movement who count themselves supporters of such values are strongly critical of the free market economy so dear to deregulators. In fact, for them the free market is a cause of the malfunctioning of modern society that has brought us to the point of requiring drastic measures to gain, or regain, more communalistic life patterns.

    It bears noting, however, that both movements are interested in extolling resilience as an advantage of their program: "The soft path also minimizes the economic risks of capital in case of error, accident, or sabotage; the hard path effectively maximizes those risks by relying on vulnerable high technology devices. . . . Finally, the soft path appears generally more flexible—and thus more robust. Its technical diversity, adaptability, and geographic dispersion make it resilient and offer a good prospect of stability under a wide range of conditions, foreseen or not. The hard path, however, is brittle; it must fail, with widespread and serious disruption, if any of its exacting technical and social conditions is not satisfied continuously and indefinitely." A. Lovins, *Soft Energy Paths*, 51.

    With just a few word changes, this defense of soft paths could serve as an attack on anticipatory policies: centralization ensures that errors will be large; inability to foresee the future accurately guarantees that errors there will be. See, e.g., Wildavsky, "If Regulation is Right," xvi-xvii.
176. It is something of an exaggeration to think that any plaintiffs' organization would actually litigate in fifty different jurisdictions. Since growth in the common law operates partially through fertilization of trends across state lines, not all these cases would be extremely novel or difficult, assuming a discernible trend could be established by a series of consistent opinions. After a time, companies advised by

competent counsel would be able to anticipate a change in law applicable to places where they do business, in advance of a specific ruling of the relevant state court on that subject. On the other hand, if early litigation proved inconclusive or unsuccessful, plaintiffs' organizations would turn their attention to more profitable pursuits, rather than continuing to litigate in successive state tribunals. In either case, the organization would most likely stop well short of fifty different lawsuits on the same issue or theory. The process of either establishing a trend or pursuing litigation long enough to be convinced of its fruitlessness still threatens to consume more resources than a single, national-level effort.

177. E. I. duPont de Nemours & Co. v. Train, 430 U.S. 112 (1977).
178. See e.g., Wesley A. Magat, Allen J. Krupnick, and Winston Harrington, *Rules for Making Rules: The Revealed Preference Approach to Understanding Regulatory Agency Behavior* (Washington, D.C.: Resources for the Future, 1986).
179. R. H. Coase, "The Problem of Social Cost," *J. Law & Econ.* 3 (1960): 1.
180. E.g., Harold Demsetz, "The Exchange and Enforcement of Property Rights," *J. Law & Econ.* 7 (1964): 11; "Some Aspects of Property Rights," *J. Law & Econ.* 9 (1966): 61; "Toward a Theory of Property Rights," *Am. Econ. Ass'n Pap. & Proc.* 57 (1967): 347. Guido Calabresi, "Transactions Costs, Resource Allocation and Liability Rules—A Comment," *J. Law & Econ.* 11 (1968): 67. Shortly into the 1970s came Calabresi and Jon T. Hirschoff, "Toward a Test for Strict Liability in Torts," *Yale L. J.* 81 (1972): 1055; and Calabresi and Melamed, "Property Rules, Liability Rules, and Inalienability." John H. Dales, *Pollution, Property and Prices* (Toronto: University of Toronto Press, 1968); Garrett Hardin, "The Tragedy of the Commons," *Science* 162 (1968): 1243; Allen V. Kneese, *Water Pollution: Economic Aspects and Research Needs* (Washington, D.C.: Resources for the Future, 1962); Allen V. Kneese and Blair T. Bower, *Managing Water Quality: Economics, Technology, Institutions* (Baltimore: Published for Resources for the Future by Johns Hopkins University Press, 1968).
181. The Pigouvian analysis of pollution, which emphasized emission fees as a means of internalizing externalities, had been around much longer. Arthur C. Pigou, *The Economics of Welfare*, 4th ed. (London: Macmillan and Co., Ltd., 1932). The emission fees idea did receive some Congressional attention, e.g., U.S. Congress, *Economic Analysis and the Efficiency of Government, Part 6—Economic Incentives to Control Pollution: Hearings Before the Subcommittee on Priorities and Economy in Government of the Joint Economics Committee*, 92d Cong., 1st sess., 1971, but the JEC was not the legislative drafting committee for environmental legislation. In any event, toxics were generally excluded from the emissions fees argument.

    The pre–1970 economics literature on externalities is reviewed in E. J. Mishan, "The Postwar Literature on Externalities: An Interpretative Essay," *J. Econ. Lit.* 9 (1971): 1. Mishan notes that contributions to the literature picked up dramatically in the 1960s, but were still largely focused on interfirm, interindustry, and interperson externalities, despite the obvious relevance of the analysis to environmental policy. It seems that economists, like politicians, "respond to real world problems with a time lag. . . ." Ibid., 28. For a further review of the literature see Anthony C. Fisher and Frederick M. Peterson, "The Environment in Economics: A Survey," *J. Econ. Lit.* 14 (1976): 1. They identify A. Kneese, *Water Pollution: Economic Aspects and Research Needs*, as the first rigorous application of the externalities framework to any pollution issue. Ibid., 1, 12.
182. For an analysis of how political aspirations shaped the Clean Air Act Amendments of 1970, see Helen Ingram, "The Political Rationality of Innovation: The Clean Air Act Amendments of 1970," in *Approaches to Controlling Air Pollution*, ed. Ann F. Friedlaender (Cambridge, Mass.: MIT Press, 1978), 12.
183. Tolchin and Tolchin, *Dismantling America*, 11.
184. Alternatively, they can serve constituents by intervening in the regulatory process on their behalf. Study of the selection of regulatory institutions by the Congress is growing. For work that extends beyond the points made in the text, see Morris

Fiorina, "Legislative Uncertainty, Legislative Control and the Delegation of Legislative Power," *J. of Law, Econ. & Org.* 2 (1986): 33; Jerry Mashaw, "Prodelegation: Why Administrators Should Make Political Decisions," *J. of Law, Econ. & Org.* 1 (1985): 81; Matthew McCubbins, "The Legislative Design of Regulatory Structure," *Amer. J. of Pol. Sci.* 29 (1985): 721.

185. See Hays, "Three Decades," chapter 1 of this volume.
186. This is as true with respect to the theories of regulation they embody as it is to the politics of their enactment. On the lack of a consensus as to the values that environmental legislation should reflect, see Schroeder, "A Decade of Change."
187. John Murphy, "Legislative View of the Toxic Substances Act," in *Toxic Substances Law and Regulations 1977*, ed. Marshall L. Miller (Washington, D.C.: Government Institutes, 1977), 56, 59.
188. See Ray M. Druley and Girard L. Ordway, *The Toxic Substances Control Act* (Washington, D.C.: Bureau of National Affairs, 1977): 12.
189. Kornblau, "Evolution of Pesticide Legislation."
190. For air and water, see Advisory Commission on Intergovernmental Regulations, *Regulatory Federalism: Policy, Process, Impact and Reform* (Washington, D.C.: ACIR, 1984), 87 ("national standards . . . not strongly opposed even by industry lobbyists, for by that time many business executives had decided that a single national standard for each pollutant was probably preferable to 50 different state standards").
191. Kornblau, "Evolution of Pesticide Legislation."
192. Exxon Corp. v. Hunt, 106 S. Ct. 1103 (1986). Congress has since deleted the Superfund section upon which the preemption argument was based, thereby empowering states to enact Superfund taxing provisions. Superfund Amendments and Reauthorization Act of 1986, P.L. 99–499, 100 Stat. 1613 (1986), sec. 114(a). Some commentators are suggesting that the entire deregulation movement has spent its energies, for example, Brian J. Cook, *Bureaucratic Politics and Regulatory Reform* (Westport, Conn.: Greenwood Press, 1988), 1.
193. See Comment, "The Extent of OSHA Preemption of State Hazard Reporting Requirements," *Colum. L. Rev.* 88 (1988): 630.
194. See text at n. 38 to 51.
195. There is an extensive literature on the trading off of decisional accuracy, process, and regulatory action. For part of the debate, see Howard Latin, "Good Science, Bad Regulation, and Toxic Risk Assessment," *Yale J. of Reg.* 5 (1988): 89, 95–106; Bruce Ackerman and Richard Stewart, "Reforming Environmental Law," *Stan. L. Rev.* 37 (1985): 1333; Latin, "Ideal Versus Real Regulatory Efficiency: Implementation of Uniform Standards and 'Fine-Tuning' Regulatory Reform," *Stan. L. Rev.* 37 (1985): 1267.
196. For federal legislation regulating toxics since 1906 see Table 8.1 above and the following sources: Food and Drug Act, ch. 3915, 34 Stat. 768 (1906); Federal Insecticide Act, 7 U.S.C. secs. 121–134 (1910); Food, Drug and Cosmetic Act, 21 U.S.C. secs. 301–392 (1938); Federal Insecticide, Fungicide, and Rodenticide Act, 7 U.S.C. secs. 136–1364 (1947); Pesticide Chemicals Amendment, 21 U.S.C. secs. 321, 342, 346a, 346b (1954); Food Additive Amendments of 1958, P.L. 85–929, 72 Stat. 1784 (1958); Federal Hazardous Substances Labeling Act, 15 U.S.C. secs. 1261–1273 (1960); Clean Air Act, 42 U.S.C. secs. 1857–18571 (1963); Motor Vehicle Air Pollution Control Act, P.L. 89–272, 79 Stat. 992–995 (codified as amended in scattered sections of 42 U.S.C.) (1965); Solid Waste Disposal Act, 42 U.S.C. secs. 3251–3259 (1965); Clean Water Restoration Act, P.L. 89–753, 80 Stat. 1246 (codified as amended in scattered sections of 33 U.S.C.) (1966); Animal Drug Amendment, P.L. 90–399, 82 Stat. 342 (codified as amended in scattered sections of 21 U.S.C.) (1968); Child Protection and Toy Safety Act, 15 U.S.C. secs. 401 note, 1261, 1262, 1274 (1969); National Environmental Policy Act, 42 U.S.C. secs. 4321, 4331, 4341 (1970); Poison Prevention Packaging Act, 15 U.S.C. sec. 1471 (1970); Occupational Safety and Health Act, 42 U.S.C. sec. 3142–1 (1970);

Consumer Products Safety Act, 15 U.S.C. secs. 2051–2081 (1972); Federal Water Pollution Control Act, 33 U.S.C. sec. 1252 et seq. (1972); Federal Environmental Pesticide Control Act, P.L. 92–516, 86 Stat. 973 (1973) (amending FIFRA) (1972); Safe Drinking Water Act, 21 U.S.C. sec. 349, 42 U.S.C. secs. 201, 300f-q (1974); Hazardous Materials Transportation Act, 46 U.S.C. sec. 170, 49 U.S.C. sec. 1471 (1975); Toxic Substances Control Act, 15 U.S.C. secs. 2600–2629 (1976); Resource Conservation and Recovery Act, 90 Stat. 2795 (codified as amended in scattered sections of 42 U.S.C.) (1976); Clean Air Act Amendments, P.L. 95–95, 91 Stat. 685 (codified as amended in scattered sections of 15, 42 U.S.C.) (1977); Clean Water Act, P.L. 95–217, 91 Stat. 1566–1609 (codified as amended in scattered sections of 33 U.S.C.) (1977); Federal Pesticide Act of 1978, P.L. 95–396, 92 Stat. 819 (1978) (amending FIFRA); Comprehensive Environmental Response, Compensation and Liability Act ("Superfund"), P.L. 96–510, 94 Stat. 2767 (codified as amended in scattered sections of 26, 33, 42, and 49 U.S.C.) (1980); Consumer Product Safety Amendments of 1981, P.L. 97–35, 95 Stat. 703 (1981) (amending 15 U.S.C. sec. 2081) (1982); Safe Drinking Water Amendments of 1986, P.L. 99–339, 100 Stat. 666; Superfund Amendments and Reauthorization Act of 1986, P.L. 99–499, 100 Stat. 1613; Water Quality Act of 1987, P.L. 100–4, 101 Stat. 60.

ABOUT THE AUTHORS

Malcolm Forbes Baldwin is a consultant on environmental affairs with the United States Agency for International Development mission in Colombo, Sri Lanka. From 1974 to 1981 he served as a senior staff member of the Council on Environmental Quality, and, during 1981, as acting chairman of the Council. He served as chairman of the International Working Group on Wetlands, 1986–1988; chairman of the American Land Resource Association, 1984–1988; and chairman of the board of directors of Defenders of Wildlife, 1983–87. He is author, with Pamela L. Baldwin, of *Onshore Planning for Offshore Oil: Lessons from Scotland* (1977) and editor, with Jake Page, of *Law and Environment* (1970). Baldwin is a former Fellow of The Woodrow Wilson Center.

Thomas R. Dunlap is associate professor of history, Virginia Polytechnic Institute and State University, and author of *Saving America's Wildlife* (1988) and *DDT: Scientists, Citizens, and Public Policy* (1981).

Frank Gregg is professor at the School of Renewable Natural Resources and adjunct professor of political science, University of Arizona. A former chairman of the New England River Basins Commission (1967–78), he served as director of the U.S. Bureau of Land Management from 1978 to 1981.

Samuel P. Hays is professor of history, University of Pittsburgh, and author of *Beauty, Health, and Permanence: Environmental Politics in the United States, 1955–1985* (1987), *American Political History as Social Analysis* (1980), and *Conservation and the Gospel of Efficiency* (1958). He is a former Fellow, The Woodrow Wilson Center, and former chairman of the academic advisory council of the Center's Program on American Society and Politics.

Michael J. Lacey is director of the Program on American Society and Politics of The Woodrow Wilson Center and editor of *Religion and Twentieth-Century American Intellectual Life* (1989) and *The Truman Presidency* (1989).

Robert Cameron Mitchell is professor of geography, Clark University, and author, with Richard T. Carson, of *Using Surveys to Value Public Goods: The Contingent Valuation Method* (1988). A specialist on public opinion and environmental issues, Mitchell is editor of

"Whither Environmentalism?" a special issue of *Natural Resources Journal* (April 1980).

Joseph L. Sax is professor of law at the University of California, Berkeley, and author of *Legal Control of Water Resources* (1986), *Mountains without Handrails* (1980), and *Defending the Environment* (1971).

Christopher Schroeder is professor of law, Duke University, and editor, with Alan Miller and Robert Percival, of *Environmental Policy: Cases and Materials* (forthcoming in 1990).

Michael Smith is associate professor of history at the University of California, Davis, and author of *Pacific Visions: California Scientists and the Environment, 1850–1915* (1988).

INDEX